# International
# REVIEW OF
# Neurobiology
## Volume 55

# *International*

REVIEW OF

# *Neurobiology*

## *Volume 55*

# Viral Vectors FOR Treating Diseases OF THE Nervous System

EDITED BY

DAVID S. LATCHMAN
Institute of Child Health
London, UK

ACADEMIC PRESS

An imprint of Elsevier

Amsterdam Boston Heidelberg London New York Oxford
Paris San Diego San Francisco Singapore Sydney Tokyo

*Cover Figure*: Courtesy of Dr. Pedro R. Lowenstein. For mcre details see Chapters 1, figure 5.

This book is printed on acid-free paper. ♾



Academic Press
*An Elsevier Imprint.*
525 B Street, Suite 1900, San Diego, California 92101-4495, USA
http://www.academicpress.com

Academic Press
84 Theobald's Road, London WC1X 8RR, UK
http://www.academicpress.com

International Standard Book Number: 0-12-366856-5

PRINTED IN THE UNITED STATES OF AMERICA
03 04 05 06 07 08 9 8 7 6 5 4 3 2 1

# CONTENTS



## SECTION I
## VIRUS VECTORS FOR USE IN THE CENTRAL NERVOUS SYSTEM

### Nonneurotropic Adenovirus: A Vector for Gene Transfer to the Brain and Gene Therapy of Neurological Disorders

Pedro R. Lowenstein, Donata Suwelack, Jinwei Hu, Xianpeng Yuan, Maximiliano Jimenez-Dalmaroni, Shyam Goverdhana, and Maria G. Castro



### Adeno-Associated Viral Vectors

E. Lehtonen and L. Tenenbaum

## Problems in the Use of Herpes Simplex Virus as a Vector

LAWRENCE T. FELDMAN



## Lentiviral Vectors

JOHAN JAKOBSSON, CECILIA ERICSON, NINA ROSENQVIST, AND CECILIA LUNDBERG



## Retroviral Vectors for Gene Delivery to Neural Precursor Cells

RYOICHIRO KAGEYAMA, HIROMI HIRATA, AND JUN HATAKEYAMA



# SECTION II
# GENE THERAPY WITH VIRUS VECTORS FOR SPECIFIC DISEASES OF THE NERVOUS SYSTEM

## The Principles of Molecular Therapies for Glioblastoma

GEORGE KARPATI AND JOSEPHINE NALBANTOGLU

## Oncolytic Herpes Simplex Virus for Tumor Therapy

JENNIFER C. C. HU AND ROBERT S. COFFIN



## Recombinant Retrovirus Vectors for Treatment of Malignant Brain Tumors

NIKOLAI G. RAINOV AND CHRISTOF M. KRAMM



## Adeno-Associated Viral Vectors for Parkinson's Disease

SHIN-ICHI MURAMATSU, LIJUN WANG, KUNIHIKO IKEGUCHI, KEN-ICHI FUJIMOTO, IMAHARU NAKANO, TAKASHI OKADA, HIROAKI MIZUKAMI, YUTAKA HANAZONO, AKIHIRO KUME, IMAHARU NAKANO, AND KEIYA OZAWA



## Herpes Simplex Virus Vectors for Parkinson's Disease

DAVID S. LATCHMAN

## Gene Therapy for Stroke

K. Abe and W. R. Zhang



## Gene Therapy for Mucopolysaccharidosis

Assumpció Bosch and Jean Michel Heard

# CONTRIBUTORS

*Numbers in parentheses indicate the pages on which the authors' contributions begin.*

**K. Abe** (243), Department of Neurology, Graduate School of Medicine and Dentistry, Okayama University, Okayama 700-8558 Japan

**Assumpció Bosch** (271), Department de Bioquímica i Biologia Molecular, Facultat de Veterinàrio, Centre de Biotecnologia Animal i Teràpia Gènica (CBATEG), Universitat Autònoma de Barcelona, 08193 Bellaterra, Barcelona, Spain

**Maria G. Castro** (3), Cedars-Sinai Medical Center and Department of Medicine, Gene Therapeutics Research Institute, University of California, Los Angeles, California 90048

**Robert S. Coffin** (165), Bio Vex Ltd., Abingdon OX14 4RX, United Kingdom

**Cecilia Ericson** (111), Division of Neurobiology, Wallenberg Neuroscience Center, BMC, A11, 221 84 Lund, Sweden

**Lawrence T. Feldman** (99), Department of Microbiology, Immunology, and Molecular Genetics, University of California, Los Angeles Department of Medicine, Los Angeles, California, 90095

**Ken-ichi Fujimoto** (205), Division of Neurology, Jichi Medical School, Tochigi 329-0498, Japan

**Shyam Goverdhana** (3), Cedars-Sinai Medical Center and Department of Medicine, Gene Therapeutics Research Institute, University of California, Los Angeles, California 90048

**Yutaka Hanazono** (205), Division of Genetic Therapeutics, Jichi Medical School, Tochigi 329-0498, Japan

**Jun Hatakeyama** (123), Institute for Virus Research, Kyoto University, Kyoto 606-8507, Japan

**Jean Michel Heard** (271), Unité Rétrovirus et Transfert Génétique, Département de Neuroscience, Institut Pasteur 75725, Paris, France

**Hiromi Hirata** (123), Institute for Virus Research, Kyoto University, Kyoto 606-8507, Japan

**Jennifer C. C. Hu** (165), Cancer Cell Biology, Hammersmith Hospital Campus, Imperial College School of Medicine, London W12 0NN, United Kingdom

**Jinwei Hu** (3), Cedars-Sinai Medical Center and Department of Medicine, Gene Therapeutics Research Institute, University of California, Los Angeles, California 90048

**Kunihiko Ikeguchi** (205), Division of Neurology, Jichi Medical School, Tochigi 329-0498, Japan

**Johan Jacobsson** (111), Division of Neurobiology, Wallenberg Neuroscience Center, BMC, A11, 221 84 Lund, Sweden

**Maximilano Jiminez-Dalmaroni** (3), Cedars-Sinai Medical Center and Department of Medicine, Gene Therapeutics Research Institute, University of California, Los Angeles, California 90048

**Ryoichiro Kageyama** (123), Institute for Virus Research, Kyoto University, Kyoto 606-8507, Japan

**George Karpati** (151), Montreal Neurological Institute, Montreal QC H3A 2B4, Canada

**Christof M. Kramm** (185), Department of Pediatrics, Heinrich Heine University, Dusseldorf, Germany

**Akihiro Kume** (205), Division of Genetic Therapeutics, Jichi Medical School, Tochigi 329-0498, Japan

**David S. Latchman** (223), Institute of Child Health, University College London, London WC1N 1EH, United Kingdom

**E. Lehtonen** (65), Laboratory of Experimental Neurosurgery and Interdisciplinary Research Institute (I.R.I.B.H.M.), Free University of Brussels, B-1070 Brussels, Belgium

**Pedro R. Lowenstein** (3), Cedars-Sinai Medical Center and Department of Medicine, Gene Therapeutics Research Institute, University of California, Los Angeles, California 90048

**Cecilia Lundberg** (111), Division of Neurobiology, Wallenberg Neuroscience Center, BMC, A11, 221 84 Lund, Sweden

**Hiroaki Mizukami** (205), Division of Genetic Therapeutics, Jichi Medical School, Tochigi 329-0498, Japan

**Shin-ichi Muramatsu** (205), Division of Neurology, Jichi Medical School, Tochigi 329-0498, Japan

**Imaharu Nakano** (205), Division of Neurology, Jichi Medical School, Tochigi 329-0498, Japan

**Josephine Nalbantoglu** (151), Montreal Neurological Institute, Montreal QC H3A 2B4, Canada

**Takashi Okada** (205), Division of Genetic Therapeutics, Jichi Medical School, Tochigi 329-0498, Japan

**Keiya Ozawa** (205), Division of Genetic Therapeutics, Jichi Medical School, Tochigi 329-0498, Japan

**Nikolai G. Rainov** (185), Department of Neurological Science, University of Liverpool; The Walton Centre for Neurology and Neurosurgery NHS Trust, Liverpool L9 7IJ, United Kingdom

**Nina Rosenqvist** (111), Division of Neurobiology, Wallenberg Neuroscience Center, BMC, A11, 221 84 Lund, Sweden

**Donata Suwelack** (3), Cedars-Sinai Medical Center and Department of Medicine, Gene Therapeutics Research Institute, University of California, Los Angeles, California 90048

**L. Tenebaum** (65) Laboratory of Experimental Neurosurgery and Interdisciplinary Research Institute (I.R.I.B.H.M.), Free University of Brussels, B-1070 Brussels, Belgium

**Lijun Wang** (205), Division of Neurology and Division of Genetic Therapeutics, Jichi Medical School, Tochigi 329-0498, Japan

**Xianpeng Yuan** (3), Cedars-Sinai Medical Center and Department of Medicine, Gene Therapeutics Research Institute, University of California, Los Angeles, California 90048

**W. R. Zhang** (243), Department of Neurology, Graduate School of Medicine and Dentistry, Okayama University, Okayama 700-8558 Japan

# SECTION I

# VIRUS VECTORS FOR USE IN THE CENTRAL NERVOUS SYSTEM

# NONNEUROTROPIC ADENOVIRUS: A VECTOR FOR GENE TRANSFER TO THE BRAIN AND GENE THERAPY OF NEUROLOGICAL DISORDERS

Pedro R. Lowenstein, Donata Suwelack, Jinwei Hu, Xianpeng Yuan, Maximiliano Jimenez-Dalmaroni, Shyam Goverdhana, and Maria G. Castro

Gene Therapeutics Research Institute
Cedars-Sinai Medical Center and Department of Medicine
UCLA
Los Angeles, California 90048



## I. Introduction: Some History and Other Vectors

For many years it has been virtually impossible to transfer genes into brain cells either to study or manipulate molecular, cellular, or, *in vivo*, behavioral processes. In addition to the physical barriers that protect the brain (bone and three layers of meninges), the earliest gene delivery systems, the retroviral vectors, require cell division to integrate the transgenes into their genomes to express any transgenes. Thus they limited transduction to dividing cells of the nervous system, e.g., astrocytes and oligodendrocytes, neurons in the neonatal brain undergoing cell division, or nonneural cells such as fibroblasts that could then be transplanted to the brain. Because neurons in the adult brain do not divide, retroviruses were of limited use to neurobiologists wanting to manipulate the molecular makeup of neurons *in vivo* (Fisher and Gage, 1993; Suhr and Gage, 1999; Lowenstein and Castro, 2002; Hsich *et al.*, 2002; see Kageyama *et al.*, this volume, for further discussion).

Far greater transduction efficiencies in the nervous system can now be achieved with the recently developed lentiviral vectors. Lentiviruses are a subclass of retroviruses whose genomes contain additional viral proteins. Two of these virion proteins, matrix (MA) and virion protein R (VPR), are used to transport the cDNA/integrase complex (also called preintegration complex) across the




0074-7742/03 $35.00

nuclear membrane in the absence of mitosis. This allows lentiviruses not only to infect, but (at least in theory) also to express transgenes in both proliferating, e.g., astrocytes, and nonproliferating cells, such as neurons. *In vivo* experiments with lentiviral vectors, however, have demonstrated that only neurons express transgenes, presumably due to a not yet characterized selectivity of lentiviral vectors to infect neuronal cells (Kay *et al.*, 2001; Consiglio *et al.*, 2001; Brooks *et al.*, 2002).

The first lentiviral vectors were derived from human immunodeficiency virus-1 (HIV-1) but were pseudotyped by using the envelope glycoproteins from other viruses such as the vesicular stomatitis virus G protein (VSV-G), a fusion protein used to improve infection efficiency. The cell line 293 is used to generate these vectors using a three-plasmid cotransfection system. Three separate plasmids encode for the pseudotyped *env* gene, the transgene cassette, and a packaging construct respectively supplying the structural and regulatory genes *in trans* (Naldini *et al.*, 1996). Currently other lentivurses, e.g., feline immune deficiency virus and equine infectious anemia virus, have also been engineered as gene transfer vectors. An important attraction of these latter systems is the fact that they are derived from animal, rather than human lentiviruses (Brooks *et al.*, 2002; Curran and Nolan, 2002).

Lentivirus vectors have all the desirable properties of the early Moloney murine leukemia virus (Mo-MLV)-based vectors, but, in addition, can infect both dividing and quiescent cells and induce a limited inflammatory response even in the CNS (Blomer *et al.*, 1997; unpublished observations). Lentiviruses have been used very successfully infecting and transducing the central nervous system (CNS), with almost negligible inflammatory responses (Deglon and Aebischer 2002; Hsich *et al.*, 2002; Tate *et al.*, 2002). Lentiviral vectors are currently being evaluated for safety, with a view to removing all nonessential regulatory genes to facilitate and accelerate approval for clinical trials. If the issues of increased production and safety can be overcome, higher titers are achieved, and public fears of some lentiviral vectors being HIV-1 derived are addressed, these vectors hold great promise for clinical applications within the nervous system (Ailles and Naldini, 2002; De Palma and Naldini, 2002; Galimi and Verma, 2002; Huentelman *et al.*, 2002). Novel and/or improved vector systems are constantly being developed to provide a highly effective technology with which to explore the molecular basis of neuronal gene therapy (see Jakobsson *et al.*, for further discussion).

Having briefly reviewed the current status of vectors available for gene transfer into the brain, we will now explore the potential application of replication-deficient adenovirus vectors as vehicles for gene delivery into the CNS (Table I). A significant feature of adenovirus as a potential vector for DNA delivery in human gene therapy protocols has been that a live (replication-competent) vaccine has been safely administered to many million human beings (U.S. military recruits) over several decades to provide protection against natural adenovirus infections (Rubin and Rorke, 1994). More recently, similar replication-competent

TABLE I
A COMPARISON BETWEEN ADENOVIRUS VECTORS AND OTHER GENE TRANSFER METHODS[a]

| | Vectors | | | | | | | | |
|---|---|---|---|---|---|---|---|---|---|
| | Ads | HC-Ad | HSV1/r | HSV1/a | AAV | Retro/ lentivirus | Vaccinia | SFV | Micro injection |
| Size (kb) | 36 | 36 | 152 | 5–30 | 4.68 | 7–10 | 185 | 11.51 | 10–30 |
| Cloning capacity (kb) | 7.5 | ~30 | 10–30 | ~150 | 4.5 | 7–10 | 30 | 9–20 | 10–30 |
| Neuronal transduction | | | | | | | | | |
| *In vivo?* | Yes | Yes | Yes | Yes | Yes | Yes | Yes | Yes | No |
| *In vitro?* | Yes | Yes | Yes | Yes | Yes | Yes | Yes | Yes | Yes |
| Long-term gene expression (≥12 months) | Yes | Yes | Yes | Yes | Yes | Yes | No | No | N/A |
| Gene therapy? | Yes | Yes | Yes | Yes | Yes | Yes | No | No | No |
| Vaccination | Yes | Yes | Yes | Yes | Yes | No | Yes | Yes | No |

[a]HSV-1/r, recombinant HSV1-based vectors; HSV-1/a, amplicon HSV1-based vectors; Ad, adenovirus vectors; HC-Ad, high-capacity, helper-dependent adenovirus vectors; SFV, Semiliki Forest virus vectors.

adenovirus recombinants encoding defined immunogens have also been used in human vaccine trials, or as replication-competent vectors with or without additional anticancer genes to induce tumor cell killing. Although most adenovirus vaccine development has been based on exploiting replication-competent systems, replication-deficient adenovirus vectors have also proved to be highly effective agents with which to generate both humoral and cell-mediated immune responses to expressed transgenes. These vectors have definite attributes as agents for immunization. There is therefore considerable experience in using live adenovirus isolates, replication-competent adenovirus recombinants, and replication-deficient recombinant adenovirus as immunizing agents (Top *et al.*, 1971a, b; Schwartz *et al.*, 1974; Morin *et al.*, 1987; Prevec *et al.*, 1989; Eloit *et al.*, 1990; Jacobs *et al.*, 1992; Gallichan *et al.*, 1993; Rubin, 1994; Wilkinson, 1994a,b; Wilkinson and Borysiewicz, 1995; Amalfitano and Parks, 2002; Nicklin and Baker, 2002; Wickham, 2000).

Furthermore, human gene therapy programs have been launched by a number of groups in the United States and Europe, resulting in the application of replication-deficient adenovirus recombinants in the treatment of patients suffering cystic fibrosis, coronary artery restenosis postpercutaneous angioplasty, brain tumors, head and neck carcinomas, ovarian carcinomas, melanoma, and metabolic disorders among others (Rosenfeld *et al.*, 1992; Engelhardt *et al.*, 1993; Mastrangeli *et al.*, 1993; Simon *et al.*, 1993; Boucher *et al.*, 1994; Bout *et al.*, 1994a,b; Brody *et al.*, 1994a,b; Mittereder *et al.*, 1994; Welsh *et al.*, 1994; Wilson *et al.*, 1994; Yei *et al.*, 1994a,b; Zabner *et al.*, 1994; Kirn *et al.*, 2001; Rutanen *et al.*, 2002; Bauknecht and Meinhold-Heerlein, 2002; Raper *et al.*, 2002). Thus, a wealth of data has accumulated and continues to grow concerning the therapeutic administration of adenovirus and their effectiveness as well as their side effects in human patients.

Interest in exploiting the new capabilities offered by adenoviruses as gene therapy vectors has been very high. The key features that make adenovirus potentially a credible vector for neurological gene therapy specifically are that (1) sufficiently high titers can be easily produced as to allow their administration *in vivo*, (2) they can transduce many different differentiated cell types, including postmitotic neurons, (3) expression in the target cell can be restricted to the transgene only (reviewed in Legrand *et al.*, 2002; Nicklin and Baker, 2002), and (4) they can be scaled up to very high titres, i.e., $10^{12}$–$10^{13}$ IU/ml. So far adenovirus recombinants have been used to transduce cells of the lung (see above for CFTR transfer; Gilardi *et al.*, 1990; Rosenfeld *et al.*, 1991; Yei *et al.*, 1994a,b), liver (Herz and Gerard, 1993; Ishibashi *et al.*, 1993; Engelhardt *et al.*, 1994; Hayashi *et al.*, 1994; Kozarsky *et al.*, 1994), arteries and other blood vessels (Lemarchand *et al.*, 1992; Roessler *et al.*, 1993; Kingston *et al.*, 2001), joints, bone marrow cells, and various differentiated circulating cells of the immune system (Haddada *et al.*, 1993), heart (Stratford-Perricaudet *et al.*, 1992; Kass-Eisler *et al.*, 1993), skeletal muscle (Quantin *et al.*, 1992; Stratford-Perricaudet *et al.*, 1990, 1992; Acsadi *et al.*, 1994a,b), brain (Akli *et al.*, 1993; Bajocchi *et al.*, 1993; Davidson *et al.*, 1993; Le Gal La Salle *et al.*, 1993; Ambar *et al.*, 1999; Dewey *et al.*, 1999; Thomas *et al.*, 2000a,b, 2001a,b), and spinal cord (Lisovoski *et al.*, 1994), and to release factor IX into the bloodstream (Smith *et al.*, 1993). The widespread potential applications of adenovirus systems has promoted further developments of the adenovirus vectors, in particular, efficient strategies have been devised for the insertion of transgenes into recombinant adenovirus. Also, the further developments have concentrated on two main limitations of the adenoviral vectors, namely, the limited transgene capacity of first-generation adenoviral vectors and the inflammatory and immune responses caused by the adenoviral virions and its genome. Inserts of up to 8 kbp are feasible in replication-deficient recombinants, which can be grown without helper virus on *trans*-complementing 293 cells (Graham *et al.*, 1977; Berkner, 1988; Wilkinson, 1994a,b), gutless vectors, and the new helper-dependent high-capacity vectors have a theoretical capacity of up to 30–36 kbp (Lowenstein *et al.*, 2002; Hartigan-O'Connor *et al.*, 2002; Zhou *et al.*, 2002).

Recent reviews have provided detailed protocols and methodologies involved in constructing and utilizing adenovirus vectors for gene transfer to cells of the brain both *in vitro* and *in vivo* (Lowenstein, 1995; Southgate *et al.*, 2001; Thomas *et al.*, 2001a,b; see Phillips, 2002). In this chapter, we will consequently concentrate on progress that has been made in the application of such vectors within the context for their exploitation for neurobiology and discuss potential future developments in this field.

## II. The Vector System

Adenoviruses have the ability to infect a wide range of different tissues and cell types. It has been shown convincingly that adenovirus type 5 serotype initiates infection by binding to the coxsackievirus and adenovirus receptor (CAR) through the knob domain located at the distal extreme of the fiber protein (Bergelson *et al.*, 1997; Roelvink *et al.*, 1999). CAR is a 46-kDa integral membrane protein with a typical transmembrane region and an extracellular region composed of two immunoglobulin (Ig)-like domains (Tomko *et al.*, 1997). The C-terminal knob of the fiber protein confers the specificity of the cellular receptor recognition. Following its binding to CAR, adenovirus is subsequently internalized through an interaction with integrins on the cell surface of the target cells, especially $\alpha_v\beta_5$ (Wickham *et al.*, 1993). Entry into the cell is by receptor-mediated endocytosis (Wickham *et al.*, 1994), after which the viral nucleocapsid is released from endosomes into the cytoplasm, and transported to the nucleus (Greber *et al.*, 1993; Trotman *et al.*, 2001). The double-stranded DNA genome (~36 kb; Fig. 1) is then released into the nucleus of the infected cell (Trotman *et al.*, 2001). Upon infection of a permissive cell, the replication cycle can be divided into three phases of gene expression: immediate early (E1 genes), early (E1–E4 genes), and finally the late events, which follow the onset of viral DNA replication. The very first region of the genome to become active is the left-hand region of the left strand, E1, which encodes a series of transcriptional activators, which will promote progression into the early phase of gene expression. These phases of virus replication are associated with host cell cycle progression to S1, onset of viral DNA replication, as well as the production of gene products associated with evasion of host antivirus defenses. Five late transcriptionally active regions are found at the right of the left strand, and are produced from a single major late promoter by alternative splicing (L1, L2, L3, L4, and L5). Expression of late viral genes becomes activated after the start of viral DNA replication and will eventually lead to the assembly of progeny virions, in excess of $10^5$ new virions per infected cell (Shenk, 1995).

Most available recombinant adenoviruses for use as vectors for gene transfer are rendered replication defective through deletions in the E1, E2, E3, or E4 regions, or through combinations thereof. These vectors, have been termed "first (and sometimes second and third)-generation" adenovirus vectors, depending on the authors, and the particular deletions to increase the cloning capacity (Gilardi *et al.*, 1990; Davidson *et al.*, 1993; Brody *et al.*, 1994b; Wilkinson, 1994a,b; see

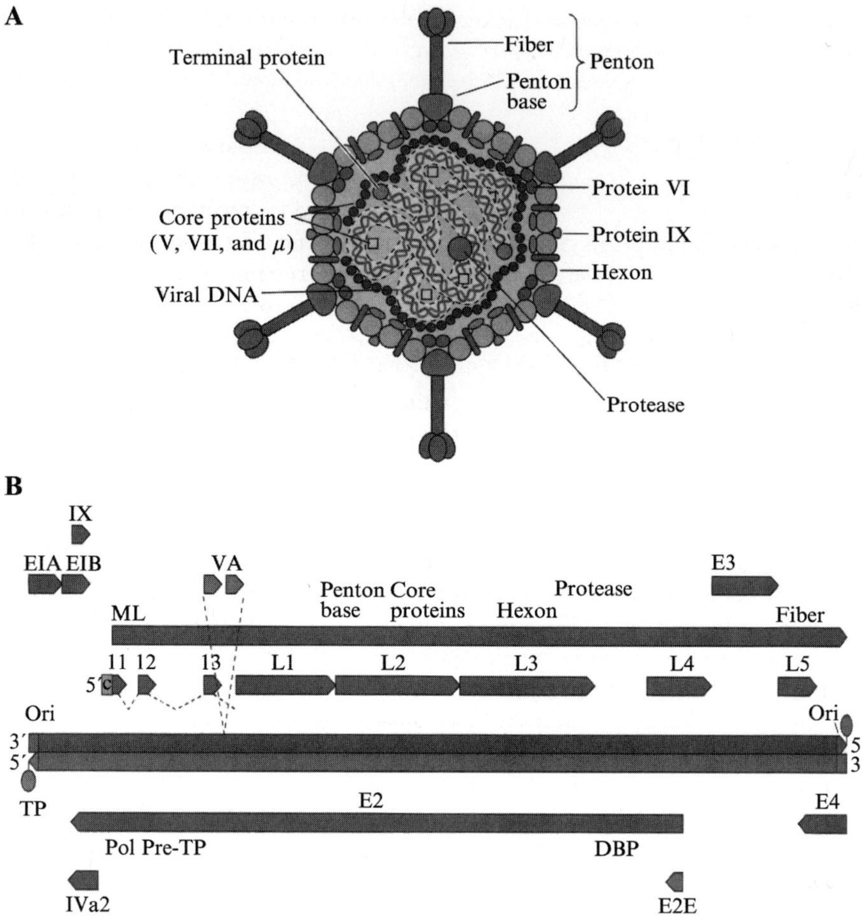


FIG. 1. (A) Schematic representation of an adenovirus particle based on the current understanding of its polypeptide constituents and genome. (B) Genome is 100 map units (mu) in length (1 mu = 360 bp). Early primary mRNA transcripts are designated in bold. Certain polypeptides are identified by conventional numbering system (roman numerals). Modified from Flint *et al.* (2000).

Philips, 2002). The regions deleted comprise some genes that are necessary for replication (e.g., those located in the E2 region), to others that do not appear to be necessary for replication (e.g., the E3 and E4 region) and that encode genes associated with modulating the host immune response and transcriptional control. No deletions are neutral, and they will have varying effects on the production, replication, expression, and toxicity of the particular viral vectors.

Replication-deficient adenovirus recombinants deleted in various genes or genomic regions are commonly grown on *trans*-complementing 293 cells that express an integrated copy of the Ad5 E1 gene, complemented by other regions in the case of deletions in the E2 region (Graham *et al.*, 1977). Transcomplementation with the E3 or E4 region has usually not been necessary.

Efficient systems to construct adenovirus recombinants are now available both from a number of different laboratories, and more recently also from various commercial sources (e.g., Microbix Biosystems Inc., Toronto, Canada; QBio Gene, Montreal, Canada). These allow the construction of both replication-competent and replication-deficient recombinant adenoviruses for gene transfer (Prevec *et al.*, 1989), using conventional and well-established DNA cloning and transfection procedures (Sambrook, 1989). Most published reports utilize adenovirus (Ad) recombinant derived from either serotypes 5 or 2 (see Table II), although there are some 47 human serotypes identified and many more animal isolates, some of which have also been used within the context of gene transfer (Cotten *et al.*, 1993). The exploration of other adenoviral serotypes has substantially increased since 1995 as some of their distinct properties concerning the receptors used for entry, the consequent cell types transduced, have been discovered to differ in important ways from adenovirus type 2 and 5 (Zabner *et al.*, 1999; Segerman *et al.*, 2000; Shayakhmetov *et al.*, 2000). In Ad E1-deleted vectors, transgenes are usually inserted into the E1 region under the control of exogenous viral promoters (e.g., RSV-LTR, IE-hCMV). Cell specific alone, or combined with inducible promoters, or other elements that increase expression levels can also be used effectively to target expression to specific cell types, to regulate its expression, and to increase expression levels (Babiss *et al.*, 1986; Bessereau *et al.*, 1994; Grubb *et al.*, 1994; Smith-Arica *et al.*, 2000, 2001; Southgate *et al.*, 2001; Gerdes *et al.*, 2000; Harding *et al.*, 1998; Ralph *et al.*, 2000; Glover *et al.*, 2002). Most recently the use of replication-competent adenoviruses for the treatment of turmors has also begun to be explored (Nanda *et al.*, 2001; Doronin *et al.*, 2000, 2001).

## III. Developments in the Use of Adenovirus Recombinants as Vectors

The deletion of the Ad E1 gene region results in a failure of the replication-deficient Ad vector to activate both early and late phase transcription from the viral genome. Consequently expression of only a transgene under the control of a constitutive or cell-type-specific or inducible promoter is achieved in the infected target cell. Although this genetic barrier to adenovirus gene expression is efficient, it is not absolute. Even limited breakthrough of Ad gene expression does occur in many if not most tissues, and can thus be problematic,

TABLE II

Review of Publications Using Recombinant Adenovirus Vectors for Gene Transfer into the Brain or Constituent Cells

| References[a] | Recombinant adenovirus E1 vector | Target *in vitro* and *in vivo* | Therapeutic efficacy/type of experiment |
|---|---|---|---|
| (1) "Early" Adenoviral Gene Transfer into the Brain (1993–1994) | | | |
| Akli *et al.* (1993) | Ad-RSVlacZ, Ad type 5 | Brain *in vivo*; rats | Gene transfer into the rodent brain *in vivo* |
| Bajocchi *et al.* (1993) | Ad-RSVlacZ, Ad-$\alpha$1AT, Ad type 5 | Intraventricular or intrastriatal administration; $5 \times 10^8$–$10^9$ pfu/animal | Gene transfer of marker enzymes and transgenes into the rodent ventricles and brain *in vivo* |
| Bennett *et al.* (1994) | Ad-CMVlacZ; Ad5 | Adult mouse mammalian retina; subretinal injection of $10^2$–$10^9$ pfu/retina in 1–10 $\mu$l | Gene transfer into retina *in vivo* |
| Caillaud *et al.* (1993) | Ad-RSVlacZ; Ad5 | Primary cultures of neurons and glial cells | Gene transfer into neural cells *in vitro* |
| Davidson *et al.* (1993) | Ad-CMVlacZ Ad type 5 E1; E3 | *In vivo*: intrastriatal, mice | Gene transfer into rodent brain *in vivo* |
| Jomary *et al.* (1994) | Ad2/CMV lacZ | Retinal cells | Gene transfer of marker cells into retina |
| Le Gal La Salle *et al.* (1993) | Ad-RSVlacZ, Ad type 5 | SCG and astrocytes in culture. Hippocampus and substantia nigra (SN) *in vivo*: 3–5×107 pfu/injection | Gene transfer into brain cells *in vivo* and *in vitro* |
| Li *et al.* (1994) | Ad-CMV $\beta$-Actin-nls lacZ; Ad5 | Subretinal injection in normal and mutant mice of $4\times10^4$–$4\times10^7$ pfu in 0.3–0.5 $\mu$l/retina | Gene transfer into retina *in vivo* |
| Ridoux *et al.* (1994b) | Ad-RSVlacZ | Astrocytes *in vitro*; $2\times10^7$ pfu/confluent dish | Gene transfer into astrocytes prior to intracranial transplantation |
| (2) *Ex Vivo* Gene Transfer and Transplantation of Transduced Cells | | | |
| Hughes *et al.* (2002) | Ad-lacZ or eGFP | *In vitro*: embryonic murine neural progenitor cells | FIV-transduced progenitors retain potential for differentiation; Ad induces differentiation into astrocytes |

| | | | |
|---|---|---|---|
| Corti *et al.* (1999) | Ad-tet-TH | *In vitro*: human neural progenitor cells | Ad encoding TH under negative control of the tetracycline regulatory system produces high levels of TH in neural progenitor cells |
| Boer *et al.* (1997) | Ad-lacZ | *In vivo*: third ventricle of rats | *Ex vivo* Ad delivered transgene expression of neurografts implanted in the third ventricle is compared to explant cultures |
| Yoon *et al.* (1996) | Ad-lacZ or p75 | *In vivo*: intraventricular injection in mice | Ad-mediated reporter genes are introduced into progenitor cells to monitor their migration and assess stability of transgene expression |
| | | (3) Recent Vector Developments | |
| Umana *et al.* (2001) | HD-Ad-lacZ | *In vivo*: rats | Use of FLPe recombinase to improve quality and quantitiy of HD Ad production |
| Xia *et al.* (2000) | Ad-GFP, Ad5GFPBxHI, where "x" is the epitope number | *In vitro*: engineered cell lines and human brain microcapillary endothelia | The Ad HI loop is modified to achieve binding to a specific receptor; this is used to achieve a disseminated pattern of transduction |
| Cregan *et al.* (2000) | HD-Ad-lacZ | *In vitro*: primary cerebellar neuronal cultures | Efficacy and cytotoxicity of an HD Ad is compared to a first-generation Ad |
| Zheng *et al.* (2000) | Chimaeric Ad-retroviral vector | *In vitro*: *in vivo*: submandibular gland, cortex, and caudate nucleus | Examination of the characteristics of an Ad construct containing LTR sequences of a retroviral vector |
| Beer *et al.* (1998) | Encapsulated RAds | *In vitro*; *in vivo*: mice | Administration of PLGA encapsulated Ad to diminish immunogenicity of the viral vectors |

(*Continued*)

TABLE II (*Continued*)

| References[a] | Recombinant adenovirus E1 vector | Target *in vitro* and *in vivo* | Therapeutic efficacy/type of experiment |
|---|---|---|---|
| Sato *et al.* (1998) | Ad-bc12; regulatable by *cre* | *In vitro*: PC12, a hybrid motoneuronal cell line and primary chicken motoneurons | A new strategy for delivery of Bcl-2 mediated by Cre/loxP recombination using Ad vectors |
| Peltekian *et al.* (2002) | CAV2 | *In vivo*: rat | Selective transduction of neuronal cells by CAV2 vectors introduced as a new tool for targeting key structures of neurodegeneration |
| Soudais *et al.* (2001) | CAV2 | Rodent CNS neurons *in vitro*, and *in vivo* | Neurotropism of CAV2 and its efficient retrograde transport are demonstrated |
| (4) Transcriptional Targeting | | | |
| Kugler *et al.* (2001) | Ad-NSE, tubulin $\alpha 1$, synapsin promoters-lacZ or Bcl-X(L) | *In vitro*: primary neuronal cultures; *in vivo*: rats | Different cellular promoters are investigated in an adenoviral context to determine their capability of neuron-restrictive transgene expression |
| Smith *et al.* (2000) | Ad-CMV, RSV, E1A promoters-lacZ or eGFP | *In vitro*: hippocampal brain slices; *in vivo*: rat hippocampus | Cytotoxicity and transgene expression from three viral promoters in the same Ad backbone are compared |
| Ralph *et al.* (2000) | Ad-synapsin 1, GFAP-tet-eGFP | *In vitro*: primary hippocampal cultures; *in vivo*: rat hippocampus | Ad system employing regulatable, cell-specific transgene expression |
| Navarro *et al.* (1999); Nillecamps (1999) | Ad-NSE, RSV promoters-LacZ | *In vivo*: rat | NSE and RSV promoters driving LacZ in an Ad are compared |
| | Ad-xNRSE-PK-luc, x denotes $\leq 12$ NRSE elements | *In vitro*: different nonneuronal and neuronal cell lines and primary SCG neurons; *in vivo*: mouse tongue injection, rats | More specific neuronal transgene expression is achieved with an Ad vector construct encoding luciferase with increasing numbers of NRSEs upstream of the phosphoglycerate promoter |

| | | | |
|---|---|---|---|
| Miyaguchi *et al.* (1999) | Ad-SCG10-NRSE-lacZ | *In vitro*: rat hippocampal slice cultures | NRSE in an Ad-encoding LacZ to achieve selective neuronal expression |
| Smith-Arica *et al.* (2000) | Ad-tet-GFAP, NSE, $\beta$-Actin-CMV promoters | *In vitro*: different cell lines and primary neocortical cultures; *in vitro*: rats | A dual Ad system encoding for cell-type-specific and regulatable transcription units is tested |
| Morelli *et al.* (1999) | Ad-NSE, GFAP, hCMV promoters—FasL | *In vitro*: primary, neuronal, glial, fibroblastic, and epithelial cell lines; *in vivo*: systemic administration in mice | Ads encoding FasL controlled by non-specific and cell-type specific promoters are tested in different cell types and *in vivo* |
| Shering *et al.* (1997) | Ad, HSV-1, HSV-1 amplicon vectors—hCMV | *In vitro*: rat nervous system cell cultures | Cell-specific expression from a short HCMV major IE promoter enhancer is assessed in Ad and HSV-1 viral vectors |
| | (5) Targeting through the Blood–Brain Barrier | | |
| Nilaver *et al.* (1995) | Ad-RSVlacZ, HSV-1-lacZ | *In vivo*: intracerebral xenografts of human LX-1 small cell lung carcinoma in nude rats | Distribution of LacZ delivered by Ad and HSV-1 vectors after intratumor or intracarotid administration +/− osmotic disruption of the BBB |
| Muldoon *et al.* (1995) | Ad, HSV | *In vivo*: rat and cat | Delivery of Ad, HSV, and MION into normal brain or intraarterially is compared in animals with intact or disrupted BBB |
| Doran *et al.* (1995) | Ad-lacZ | *In vivo*: rats | Short-term gene expression is evaluated after intracarotid administration of an Ad encoding lacZ after or without osmotic BBB disruption |
| | (6) Brain Immunity/Inflammation | | |
| Zermansky *et al.* (2001) | Ad-HSV1-TK | *In vivo*: rats | 12 months persisting high-level and widespread expression of Ad delivered HSV1-TK is transgene dependent |

(*Continued*)

TABLE II (*Continued*)

| References[a] | Recombinant adenovirus E1 vector | Target *in vitro* and *in vivo* | Therapeutic efficacy/type of experiment |
|---|---|---|---|
| Thomas *et al.* (2001a) | Ad-hCMV-lacZ or HPRT | *In vivo*: rats | Mechanisms contributing to the decline in transgene expression delivered by first generation Ad vectors are examined |
| Thomas *et al.* (2002) | Ad-hCMV-lacZ, luc with different capsid modifications | *In vivo*: rats | Engineered viral capsids ablated for CAR and/or integrin and their inflammatory potential are examined |
| Thomas *et al.* (2001b) | Ad-hCMV-lacZ or HPRT, HD-Ad-hCMV-lacZ | *In vivo*: rats | Effects of preexisting antiadenoviral immunity on transduction of brain cells with HD Ad and first-generation Ad are compared |
| Zou *et al.* (2001) | Ad-lacZ, HD-Ad-lacZ | *In vivo*: intraventricular and intrahippocampal administration in rats | Toxicity and immunogenicity of HD and first-generation Ad vectors and their effect on transgene expression are compared |
| Cua *et al.* (2001) | Ad-IL10 | *In vivo*: animal model of autoimmune encephalomyelitis | Therapeutic effects of intracranial and systemic administration of an Ad encoding an antiinflammatory cytokine are compared |
| Gerdes *et al.* (2000) | Ad-hCMV, mCMV promoters-lacZ | *In vivo*: rats | Ad-delivered MIEmCMV promoter performance is characterized |
| Driesse *et al.* (2000) | Ad-TK or lacZ | *In vivo*: rats and nonhuman primates | Analysis of immune response to CSF administration of Ad encoding TK with subsequent GCV or lacZ |
| Zou *et al.* (2000) | Ad and HD-Ad-$\beta$-Geo | *In vivo*: hippocampus of rats | Immunogenicity and its effects on transgene expression and stability are compared in HD and first-generation Ad |

| | | | |
|---|---|---|---|
| Kajiwara *et al.* (2000) | Ad-lacZ | *In vivo*: mice | Investigation of the humoral immune response to a first-generation Ad and its transgene delivered into the brain |
| Dewey *et al.* (1999) | Ad-TK | *In vivo*: rat glioma model | Characterization of long-term effects of Ad-mediated conditional cytotoxic gene therapy for glioma |
| Ohmoto *et al.* (1999) | Ad-lacZ | *In vivo*: different mouse strains | Influence of environmental conditions and animal priming with Ad on quality and duration of immune response to Ad gene delivery is examined |
| Cartmell *et al.* (1999) | Ad-hCMV-lacZ, Ad-0, Ad-TK | *In vivo*: intrastriatal and intraventricular administration in rats | Role of TNF-$\alpha$ and IL-$\beta$ in mediating an early inflammatory response to Ad administered to different CNS targets is delineated |
| Proescholdt *et al.* (1999) | Ad-CMV-VEGF, Ad-CMV-lacZ | *In vivo* | Inflammatory response to Ad-mediated VEGF versus pump-injected protein and its effects on the integrity of the BBB are studied |
| Shine *et al.* (1997) | Ad-RSV-TK | *In vivo*: cotton rat | Toxicity of Ad-mediated TK +/− GCV at different doses in naive and pre-immunized rats is tested |
| Kajiwara *et al.* (1997) | Ad-lacZ | *In vivo*: mice | Quantitative analysis of the immune response to an Ad-encoding lacZ and transgene longevity |
| Byrnes *et al.* (1996) | Ad-lacZ | *In vivo*: mice | Immune response to Ad-delivered lacZ is examined |
| Byrnes *et al.* (1996a) | Ad-RSV-lacZ, HSV-1 RSV-lacZ, HSV-1 tsK | *In vivo*: caudate nucleus of rats | Duration of Ad-mediated transgene expression and related inflammation in immune-naive animals and subsequent peripheral exposure to the vector are compared |

(*Continued*)

TABLE II (*Continued*)

| References[a] | Recombinant adenovirus E1 vector | Target *in vitro* and *in vivo* | Therapeutic efficacy/type of experiment |
|---|---|---|---|
| Byrnes *et al.* (1995) | AdR1, Ad0, Ad5 | Brain, *in vivo*; $3\times10^6$ pfu/0.5 $\mu$l, injection site | Study of the immune response to the direct administration of adenovirus recombinants into the brain. |
| Davidson *et al.* (1994) | Ad5-RSV-lacZ and RSV-rHPRT | Striatum and cortex *in vivo* | Gene therapy into primate brain; immune response to adenovirus vectors and encoded transgenes |
| | | (7) Neuroprotection | |
| Wang *et al.* (2001) | Ad-WldS | *In vitro*: rat DRG neurons | Ad-mediated WldS protein utilized against axonal degeneration |
| Tominaga-Yoshino *et al.* (2001) | Ad-APP, Ad-lacZ | *In vitro*: cultured hippocampal neurons | Adenoviral introduction of APP cDNA to enhance neurotoxic and neuroprotective effects of glutamate |
| Yuan *et al.* (1999) | Ad-CMV-APP or lacZ | *In vitro*: central neuron cultures | Ad is used as a vector for protein trafficking and processing studies to elucidate pathogenesis of AD |
| Horellou *et al.* (1994) | Ad-RSV-hTH, Ad-RSVlacZ | Brain *in vivo* | Gene therapy of Parkinson's disease; significant reduction in abnormal behavior of intrastriatal AdRSVhTH vs AdRSVlacZ |
| Apoptosis | | | |
| Zhu *et al.* (2002) | Ad-TGF-$\beta$1 | *In vivo*: mouse brains | Transfer of an antiapoptotic gene to the CNS for neuroprotection |
| | | *In vitro*: rat primary hippocampal cells | |
| Araki *et al.* (2001) | Ad-(wt or mutant) PS2, Ad-lac | *In vitro*: rat primary cortical neurons | The proapoptotic effect of Ad-mediated PS2 overexpression is examined *in vitro* |

| | | | |
|---|---|---|---|
| Xia *et al.* (2001) | Ad-EGFP, Ad-JBD | *In vitro*: human neuroblastoma cells; *in vivo*: mouse MPTP model of PD | Ad-induced inhibition of the JNK pathway protects MPTP lesioned DA neurons |
| Neurotrophic factors | | | |
| Dorsey *et al.* (2002) | Ad-trkB | *In vitro*: cultured primary hippocampal neurons | Correction of dysregulated receptor expression with an adenoviral vector relating to an animal model of Down's syndrome |
| Benraiss *et al.* (2001) | AdCMV-BDNF-IRES-hGFP | *In vivo*: intraventricular administration in rats | Ad delivered BDNF to induce neuronal recruitment from endogenous progenitor cells in the adult forebrain |
| Connor *et al.* (2001) | Ad-lacZ, Ad-GDNF | *In vivo*: intrastriatal and intra-SN administration in a 6-OHDA rat model of PD | Neuroprotective mechanisms of Ad-mediated GDNF are elucidated |
| Kozlowski *et al.* (2001) | Ad-hGDNF | *In vivo*: monkey caudate nucleus | Quantitative assessment of Ad-mediated hGDNF expression, using ELISA and (RT-)PCR techniques |
| Kozlowski *et al.* (2000) | Ad-lacZ, Ad-GDNF | *In vivo*: intrastriatal and SN injection in a 6-OHDA rat model | Behavioral testing of Ad-mediated GDNF therapy into the striatum or SN is compared |
| Connor *et al.* (1999) | Ad-LacZ, Ad-GDNF, Ad-mu GDNF | *In vivo*: 6-OHDA PD model in aged rats | Effects of Ad-mediated GDNF delivery to the striatum and SN are compared |
| Kitagawa *et al.* (1999) | Ad-LacZ, Ad-GDNF | *In vivo*: rats | Effects of Ad-mediated GDNF on ischemic brain injury are studied |
| Bemelmans *et al.* (1999) | Ad-LacZ, Ad-BDNF | *In vivo*: rat model of Huntington's disease | Therapeutic value of an intrastriatal injection of Ad-mediated BDNF in Huntington's disease is assessed |
| Choi-Lundberg *et al.* (1998) | Ad-lacZ, Ad-mGDNF, Ad-GDNF | *In vivo*: intrastriatal injection in 6-OHDA rat model of PD | Effects of intrastriatal administration of an Ad encoding GDNF on dopaminergic cell survival are examined |

(*Continued*)

TABLE II (*Continued*)

| References[a] | Recombinant adenovirus E1 vector | Target *in vitro* and *in vivo* | Therapeutic efficacy/type of experiment |
|---|---|---|---|
| Choi-Lundberg *et al.* (1997) | Ad-lacZ, Ad-mGDNF, Ad-GDNF | *In vivo*: 6-OHDA rat model of PD | Neuroprotective effect of Ad-mediated GDNF injected near the SN is examined |
| (8) Traumatic or Ischemic Brain Injury/Excitotoxicity | | | |
| Qiu *et al.* (2002) | Ad-FasL | *In vivo*: mouse parietal cortex and contused human cortical tissue samples<br>*In vitro*: embryonal cortical neurons | Role of Fas receptor in traumatic brain injury is addressed with an Ad encoding FasL |
| Kim *et al.* (2002) | Ad-Cu/Zn SOD | *In vivo*: rats | Intracisternal administration of an Ad encoding human Cu/Zn SOD to restore CBF autoregulation after fluid percussion injury |
| Matsuoka *et al.* (2002) | Ad-bFGF, Ad-Bcl-xL | *In vitro*: rat neuronal cells | Synergistic neuroprotective effect through Ad-mediated bFGF and Bcl-xl against EAA toxicity |
| Aoki *et al.* (2001) | Ad-GRP78 | *In vitro*: rat hippocampal neurons; *in vivo*: hippocampus of gerbils | Ad-mediated overexpression of GRP78 preserves rat neurons; hypothermic treatment is neuroprotective by restoring GRP78 expression in ischemic brain |
| Masada *et al.* (2001) | Ad.RSVlacZ, Ad-IL-1ra | *In vivo*: cerebral intraventric-ular injection in rats | Ad-mediated overexpression of IL-1ra reduces ischemic brain injury |
| Yang *et al.* (2001) | Ad-CMV-lacZ, Ad-CMV-mSAG, Ad-CMV-SAG | *In vivo*: mouse | Overexpression of SAG, an antioxidant protein, via an Ad vector to prevent ischemic injury of brain cells |

| | | | |
|---|---|---|---|
| Gupta *et al.* (2001) | Ad-RSV-LacZ, Ad-RSV-GT | *In vitro*: hippocampal cultures | Ad-induced overexpression of the Glut-1 glucose transporter *in vitro* enhances neuronal survival after exposure to the excitotoxin KA |
| Ralph *et al.* (2001) | Ad-TRE-pep2m, Ad-TRE-pep4c, Ad-TRE-EGFP, Ad-Syn-tTA | *In vitro*: primary hippocampal neurons | AMPA receptors convey postischemic neurotoxicity; Ad-mediated reduction of AMPA receptor expression provides neuroprotection |
| Ooboshi *et al.* (2001) | Ad-lacZ | *In vivo*: spontaneously hypertensive rats | Effects of brain ischemia on expression of an Ad-delivered marker gene are examined |
| Yang *et al.* (1999) | Ad-RSV-lacZ, AdRSV-IL-1ra | *In vivo*: CD-1 mouse | Ad-mediated overexpression of IL-1ra is used to assess the role of interleukin-1 in cerebral ischemia |
| Yang *et al.* (1999) | Ad-RSV-lacZ, AdRSV-IL-1ra | *In vivo*: intraventricular administration in CD-1 mice | Effects of Ad-delivered IL-1ra on ICAM-1 protein in cerebral ischemia |
| Yang *et al.* (1997) | Ad-RSV-lacZ, AdRSV-IL-1ra | *In vivo*: intraventricular administration in mice | Effect of Ad-mediated IL-1ra on cerebral ischemia is assessed |
| Hagan *et al.* (1996) | Ad-lacZ, Ad-IL-1ra | *In vivo*: perinatal rats | Impact of Ad-mediated overexpression of IL-1ra is assessed in a model of excitotoxicity |
| Betz *et al.* (1995) | Ad-RSV-lacZ, AdRSV-IL-1ra | *In vivo*: intraventricular injection in rats | Suitability of Ad-mediated IL-1ra in reducing ischemic brain damage is tested |
| | | (9) Brain Tumors | |
| Abe *et al.* (2002) | Ad-lacZ, Ad-p53 | *In vivo*: athymic rat brains | Intraarterial infusion of p53-containing Ad |
| Shinoura *et al.* (2002) | Ad-APAF1, Ad-Casp9, Ad-Bax, Ad-Fas, Ad-p53 | *In vivo*: glioma cells harboring a mutated p53 | Cotransduction of Apaf-1 and caspase-9 via Ad to increase p53-mediated apoptosis in glioma cells |

(*Continued*)

TABLE II (*Continued*)

| References[a] | Recombinant adenovirus E1 vector | Target *in vitro* and *in vivo* | Therapeutic efficacy/type of experiment |
|---|---|---|---|
| Miller *et al.* (2002) | Ad-CMV-CD | *In vitro*: different human tumor cell lines; *in vivo*: immunodeficient mouse model of glioma | Glioma cell lines can be as sensitive as GI tumor cells to the antineoplastic effects of 5-FU; efficacy and specificity of Ad-mediated CD and subsequent systemically delivered 5-FC are tested in a mouse model of human glioma |
| Adachi *et al.* (2002) | Ad-uPAR/p16 | *Ex vivo* intracerebral tumor model and *in vivo* subcutaneous tumor model | Ad-mediated delivery of a bicistronic construct containing antisense uPAR and sense p16 genes to suppress glioma invasion and growth |
| Nanda *et al.* (2001) | IG.Ad5E1(+); E3Luc, HSV1-TK gene [IG.Ad5E1(+).E3TK] | *In vitro*: human glioma cell lines, rat gliosarcoma cell line, other tumor cell lines; *in vivo*: a mouse s.c. glioma xenograft model | Combination of replication-competent Ad and the HSV1-TK/GCV suicide system to improve glioma treatment outcome |
| Maleniak *et al.* (2001) | AdhCMV-mFasL, Ad-CMV-TK | *In vitro*: primary human glioma-derived cell cultures | Ad delivered FasL or HSV1-TK/GCV therapy to induce cell death in some glioma cell lines resistant to the chemotherapeutic agent CCNU |
| Okada *et al.* (2001) | Ad-TK | *In vivo*: rat glioma model | s.c. tumor injection of Ad expressing HSV1-TK/GCV shows vaccination effect by inducing a CTL response |
| Galanis *et al.* (2001) | Ad-MV-F/H | *In vitro*: cultured glioma cell lines; *in vivo*: glioma xenografts in nude mice | Assessment of two fusogenic membrane glycoproteins as adenovirally delivered transgenes for glioma therapy |
| Lammering *et al.* (2001) | Ad-EGFR-CD533 | *In vitro*: human glioma cell lines; *in vivo*: glioma xenografts | Ad-mediated overexpression of a dominant-negative epidermal growth factor receptor to achieve radiosensitization of glioma cells |

| | | | |
|---|---|---|---|
| Glaser *et al.* (2001) | Ad-TK | Human malignant glioma cells | Molecular pathways mediating TK/GCV-induced cell death are elucidated |
| Eck *et al.* (2001) | Ad-CMV-hIFN-$\beta$ | Phase I trial in humans | Assessment of toxicity and efficacy of hIFN-$\beta$ intralesional gene therapy for patients with recurrent or progressive malignant glioma with tumor resection |
| Sandmair *et al.* (2000) | Ad-CMV-TK | Phase I trial in humans | Evaluation of safety and efficacy of TK gene therapy by using retrovirus packaging cells and Ad |
| Trask *et al.* (2000) | Ad-RSV-TK | Phase I clinical study | Patients with recurrent malignant brain tumors are treated with HSV TK/GCV |
| Lawinger *et al.* (2000) | Ad-REST-VP16 | *In vitro*: three types of human medulloblastoma cells | Effects of expression of REST-VP16 mediated by an Ad vector are studied in medulloblastoma cells |
| Brust *et al.* (2000) | Ad-TK | *In vitro*: RT2 rat glioma cells; *in vivo*: soft tissue RT2 cell tumors in rats | Combination of Ad-delivered TK/GCV therapy and subsequent radiosensitization with BrdC |
| Shinoura *et al.* (2000) | Ad-FasL, Ad-p53 | *In vitro*: A-172 and U251 glioma cell lines | Coinfection with Ad-mediated p53 and FasL to enhance apoptosis in glioma cells |
| Kurihara *et al.* (2000) | AxCALNLNZK, Ax2iNPNCre | *In vitro*: seven human glioma cell lines, HeLa and COS-7 cells | Glioma-cell specific Ad-mediated expression is compared in different glioma and nonglioma cell lines |
| Cowsill *et al.* (2000) | Ad-TK, AdΔTK | *In vitro*: primary neuronal and glial cultures; *in vitro*: rat striatum | Ad encoding TK or its truncated form +/− GCV is compared |
| Adachi *et al.* (2000) | Ad-lacZ, Ad-CAG-UPRT, Ad-CD | *In vivo*: rat brain tumor model | Ad-mediated combined CD/5-FC and UPRT therapy is evaluated |

(*Continued*)

TABLE II (*Continued*)

| References[a] | Recombinant adenovirus E1 vector | Target *in vitro* and *in vivo* | Therapeutic efficacy/type of experiment |
|---|---|---|---|
| Fathallah-Shaykh *et al.* (2000) | Ad-lacZ, AdIFN$\gamma$ | *In vivo*: mouse model of metastatic brain tumor | Therapeutic effect of Ad-mediated IFN-$\gamma$ is examined |
| Biroccio *et al.* (1999) | Ad-p53, Ad-0 | *In vitro*: different glioblastoma cell lines | wt-p53 gene transfer enhances BCNU sensitivity in glioblastoma cells depending on the administration sequence |
| Harada *et al.* (1999) | Ad-p16 | | Evaluation of the effect of restoring Ad-mediated p16 on tumor angiogenesis in p16-deleted glioma cells |
| Shinoura *et al.* (1999) | Adv-lacZ-F/K20 and Adv-lacZ-F/wt | *In vitro* U-373 MG glioma cells; *in vivo* | Therapeutic effect of replication-competent Ad with a fiber mutation is compared to its corresponding Ad with wt fiber |
| Wagenknech *et al.* (1999) | Ad-XIAP | *In vitro*: different human glioma cell lines | Expression and biological activity of XIAP in human malignant glioma are examined using Ad as a delivery tool |
| Driesse *et al.* (1998) | Ad-TK | *In vivo*: nonhuman primates | Toxicity study of Ad-mediated TK with GCV therapy |
| Puumalainen *et al.* (1998) | Ad-lacZ, BAG retrovirus-lacZ | *In vivo*: humans | Safety and efficiency of retrovirus and Ad-mediated $\beta$-galactosidase in human malignant glioma are examined |
| Gomez-Manzano *et al.* (1997) | Ad-p53/p21 | *In vitro*: human glioma cell lines | Ad-mediated p53 and p21 is examined to characterize their role and timing in p53-induced apoptosis |

| | | | |
|---|---|---|---|
| Smith *et al.* (1997) | Ad-GFP, Ad-GFP/5-HT3 | *In vivo*: rats and monkeys | Toxicity and dose profiles are presented for Ad-mediated TK and systemic GCV administration in healthy animals |
| Tanaka *et al.* (1997) | Ad-sPF4 | RT2 glioma cells *in vitro* and *in vivo*, injected into the subrenal capsule or the caudate nucleus of nude mice | Analysis of the activity of retroviral and Ad vectors expressing a secretable form of PF4, an antiangiogenic protein |
| Eck *et al.* (1996) | Ad-TK | *In vivo*: human phase I trial in patients with recurrent glioma | Evaluation of safety and efficiency of Ad-mediated TK followed by systemic GCV therapy |
| Vincent *et al.* (1996a) | Ad-MLP-luc, Ad-MLP-TK | *In vitro*: several human cell lines; *in vivo*: rats | Efficiency of Ad-mediated TK and systemic GCV administration is assessed in a rat model of leptomeningeal metastases |
| Goodman *et al.* (1996) | Ad-RSV-TK | *In vivo*: baboons | Toxicity studies of Ad-mediated TK and systemic GCV in nonhuman primates |
| Vincent *et al.* (1996b) | Ad-TK | *In vivo*: 9L tumors in rats | Ad-mediated TK transfer is compared to retroviral mediation followed by systemic GCV in a rat gliosarcoma model |
| Viola *et al.* (1995) | Ad-lacZ | *In vitro*: human glioma cell cultures; *in vivo*: rat models of solid brain tumors and meningeal cancer | Transduction profiles of Ad encoding $\beta$-galactosidase in solid and leptominengeal tumor models are assessed |
| Badie *et al.* (1994) | AdLacZ | Rat C6 glioma cells | Delivery of marker genes to the CNS and brain tumors |
| Boviatsis *et al.* (1994) | Ad5MLP lacZ | Rat 9L gliosarcoma cells | $\beta$-Galactosidase expression in tumors and neurons; tumor necrosis |
| Chen *et al.* (1994) | Ad5, Ad/RSV-TK | C6 glioma | Tumor treatment/significant tumor decrease/long-term tumor regrowth |
| Perez-Cruet *et al.* (1994) | Adv-tl | *In vivo*: 9L in Fischer rats | Tumor treatment effective for >120 days |

(*Continued*)

TABLE II (*Continued*)

| References[a] | Recombinant adenovirus E1 vector | Target *in vitro* and *in vivo* | Therapeutic efficacy/type of experiment |
|---|---|---|---|
| | | (10) Metabolic Brain Disease | |
| Kosuga *et al.*, 2001a | Ad-lacZ, Ad-GUSB | *In vivo*: monkey | *In vitro* enzyme deficiency of MPSVII reduced by transduction with an Ad-expressing GUSB; adenovirally transduced mAEC transplanted into monkey brain to determine their migration and distribution |
| Shen *et al.*, 2001 | Ad-GALC | *In vivo*: intraventricular administration in a murine model of human globoid cell leukodystrophy | Effects of Ad-mediated replacement of GALC in Krabbe disease |
| Kosuga *et al.*, 2001b | Ad-GUSB | *In vivo*: mice | Adenovirally transduced mAEC overexpressing GUSB transplanted into mice with MPSVII |
| Ghodsi *et al.*, 1998 | Ad-GUSB | *In vivo*: intraventricular or intrastriatal injection in MPSVII or wt mice | Activity of Ad-delivered GUSB is examined up to 84 days postadministration |
| Peltola *et al.*, 1998 | Ad-AGA, Ad-AGU(Fin) | *In vitro* and *in vivo*: primary neuronal cultures from/and mouse model of AGU | Effects of systemic and intraventricular administration of AD encoding for the deficient enzyme AGA |
| Plumb *et al.*, 1994 | Ad-RSV-rHPRT | *In vivo*: mice | Utility of Ad encoding HPRT in HPRT-deficient mice is examined |

| | | | |
|---|---|---|---|
| (11) Other Applications: Brain Imaging, Pathway Tracing, Substance Abuse | | | |
| Umegaki *et al.* (2002) | Ad-CMV-DopD(2)R | *In vivo*: rats | PET is shown to be useful for longitudinal *in vivo* assessment of D(2)R expression mediated by an Ad vector in rat brain |
| Ross *et al.* (1995) | Ad-TK | *In vivo*: 9L gliosarcoma rat model | Ad-mediated TK and GCV therapy is evaluated with MRI and localized H magnetic resonance spectroscopy |
| Mackler *et al.* (2000) | Ad-NAC-1 | *In vivo*: rats | Ad-mediated overexpression of NAC-1 protein affects long-term behavior in psychostimulant abuse |
| Lisovoski *et al.* (1994) | Ad-RSV-LacZ, Ad type 5 | Spinal cord | Neuroanatomy: cell labeling for morphological studies |
| Kuo *et al.* (1995) | Ad-RSVlacZ | *In vivo*; injection into the laterodorsal striatum | Suitability of an Ad vector encoding $\beta$-galactosidase for neuronal pathways tracing is tested |
| Ridoux *et al.* (1994a) | Ad-RSVLacZ | CNS, basal ganglia | Pathway tracing in the CNS |

| | Abbreviations |
|---|---|
| Ad | Adenovirus |
| AD | Alzheimer's disease |
| AGA | Aspartylglucosaminidase |
| AGU | Aspartylglucosaminuria |
| APP | Amyloid precursor protein |
| $\alpha_1$-AT | $\alpha_1$-Antitrypsin |
| BBB | Blood–brain barrier |
| BCNU | Carmustine |
| BDNF | Brain-derived neurotrophic factor |
| bFGF | Basic fibroblastic growth factor |
| BrdC | 5-Bromo-2′-deoxycytidine |

(*Continued*)

TABLE II (*Abbreviations Continued*)

| | |
|---|---|
| CAR | Coxsackie and adenovirus receptor |
| CAV2 | Canine adenovirus type 2 |
| CBF | Cerebral blood flow |
| CCNU | 1-(2-Chloroethyl)-3-cyclohexyl-1-nitrosourea |
| CD | Cytosine deaminase |
| CMV | Human cytomegalovirus |
| CTL | Cytotoxic T-lymphocyte |
| DA | Dopaminergic |
| DRG | Dorsal root ganglion |
| D(2)R | Dopamine D(2) receptors |
| EAA | Excitatory amino acid |
| FasL | Fas lingand |
| 5-FC | 5-Fluorocytosine |
| FIV | Feline immunodeficiency virus |
| FPI | Fluid percussion injury |
| 5-FU | 5-Fluorouracil |
| GALC | Galactocerebrosidase |
| GI | Gastrointestinal |
| GRP78 | Glucose-regulated protein 78 |
| hGDNF | Human glial cell-line-derived neurotrophic factor |
| GUSB | $\beta$-Glucuronidase |
| HCMV major IE | Human cytomegalovirus major immediate early |
| HD | Helper dependent |
| HPRT | Hypoxanthine-guanine-phosphoribosyltransferase |
| HSV1 | Herpes simplex virus type 1 |
| ICAM-1 | Intercellular adhesion molecule-1 |
| IL-1$\beta$ | Interleukin-1$\beta$ |
| IL-1ra | Interleukin-1 receptor antagonist |
| IRES | Internal ribosomal entry site |
| JBD | JNK-binding domain |

| | |
|---|---|
| JNK | c-Jun N-terminal kinase |
| KA | Kainic acid |
| LTR | Long terminal repeat |
| Luc | Luciferase |
| MIEmCMV | Major immediate early murine CMV |
| MION | Monocrystalline iron oxide nanoparticles |
| MLP | Major late promoter |
| MPSVII | Mucopolysaccharidosis type VII |
| MPTP | 1-Methyl-4-phenyl-1,2,3,6-tetrahydropyridine/1-methyl-4-phenylpyridinium |
| MRI | Magnetic resonance imaging |
| Nls | Nuclear localization signal |
| NSE | Neuron-specific enolase |
| NRSE | Neuron-restrictive silencer element |
| 6-OHDA | 6-Hydroxydopamine |
| PD | Parkinson's disease |
| PET | Positron emission tomography |
| PF4 | Platelet factor 4 |
| PK | Phosphoglycerate kinase |
| PLGA | Polylactic glycolic acid |
| PS2 | Presenilin 2 |
| RAd | Recombinant adenovirus |
| REST | Repressor element 1 silencing transcription factor |
| RPE | Retinal pigment epithelium |
| RSV-LTR | Rous sarcoma virus-long terminal repeat |
| SCG | Superior cervical ganglia |
| SN | Substantia nigra |
| TGF-$\beta$1 | Transforming growth factor-$\beta$1 |
| TH | Tyrosine hydroxylase |
| TK | Thymidine kinase |
| TNF-$\alpha$ | Tumor necrosis factor-$\alpha$ |
| trkB | Tyrosine receptor kinase B |

(*Continued*)

TABLE II (*Abbreviations Continued*)

| | |
|---|---|
| UPAR | Urokinase-type plasminogen activator receptor |
| UPRT | 5-Fluorouridine 5′-monophosphate |
| VEGF | Vascular endothelial growth factor |
| VP16 | Activation domain of a viral protein |
| WldS protein | Unique protein derived from a spontaneous mutant mouse, characterized by slow Wallerian degeneration |
| Wt | Wild type |

[a]References to Table II:


Abe, T., Wakimoto, H., *et al.* (2002). Intra-arterial delivery of p53-containing adenoviral vector into experimental brain tumors. *Cancer Gene Ther.* **9**(3), 228–235.

Adachi, Y., Tamiya, T., *et al.* (2000). Experimental gene therapy for brain tumors using adenovirus-mediated transfer of cytosine deaminase gene and uracil phosphoribosyltransferase gene with 5-fluorocytosine. *Hum. Gene Ther.* **11**(1), 77–89.

Adachi, Y., Chandrasekar, N., *et al.* (2002). Suppression of glioma invasion and growth by adenovirus-mediated delivery of a bicistronic construct containing antisense uPAR and sense p16 gene sequences. *Oncogene* **21**(1), 87–95.

Akli, S., Caillaud, C., *et al.* (1993). Transfer of a foreign gene into the brain using adenovirus vectors. *Nat. Genet.* **3**(3), 224–228.

Aoki, M., Tamatani, M., *et al.* (2001). Hypothermic treatment restores glucose regulated protein 78 (GRP78) expression in ischemic brain. *Brain Res. Mol. Brain Res.* 95(1–2), 117–128.

Araki, W., Yuasa, K., *et al.* (2001). Pro-apoptotic effect of presenilin 2 (PS2) overexpression is associated with down-regulation of Bcl-2 in cultured neurons. *J. Neurochem.* **79**(6), 1161–1168.

Badie, B., Hunt, K., *et al.* (1994). Stereotactic delivery of a recombinant adenovirus into a C6 glioma cell line in a rat brain tumor model. *Neurosurgery* **35**(5), 910–915; discussion 915–916.

Bajocchi, G., Feldman, S. H., *et al.* (1993). Direct in vivo gene transfer to ependymal cells in the central nervous system using recombinant adenovirus vectors. *Nat. Genet.* **3**(3), 229–234.

Beer, S. J., Matthews, C. B., *et al.* (1998). Poly (lactic-glycolic) acid copolymer encapsulation of recombinant adenovirus reduces immunogenicity in vivo. *Gene Ther.* **5**(6), 740–746.

Bemelmans, A. P., Horellou, P., *et al.* (1999). Brain-derived neurotrophic factor-mediated protection of striatal neurons in an excitotoxic rat model of Huntington's disease, as demonstrated by adenoviral gene transfer. *Hum. Gene Ther.* **10**(18), 2987–2997.

Bennett, J., Wilson, J., *et al.* (1994). Adenovirus vector-mediated in vivo gene transfer into adult murine retina. *Invest. Ophthalmol. Vis. Sci.* **35**(5), 2535–2542.

Benraiss, A., Chmielnicki, E., *et al.* (2001). Adnoviral brain-derived neurotrophic factor induces both neostriatal and olfactory neuronal recruitment from endogenous progenitor cells in the adult forebrain. *J. Neurosci.* **21**(17), 6718–6731.

Betz, A. L., Yang, G. Y., *et al.* (1995). Attenuation of stroke size in rats using an adenoviral vector to induce overexpression of interleukin-1 receptor antagonist in brain. *J. Cereb. Blood Flow Metab.* **15**(4), 547–551.

Biroccio, A., Bufalo, D. D., *et al.* (1999). Increase of BCNU sensitivity by wt-p53 gene therapy in glioblastoma lines depends on the administration schedule. *Gene Ther.* **6**(6), 1064–1072.

Boer, G. J., van Esseveldt, K. E., *et al.* (1997). Long-term transgene expression in fetal rat suprachiasmatic nucleus neurografts following ex vivo adenoviral vector-mediated gene transfer. *Exp. Neurol.* **145**(2, pt. 1), 536–545.

Boviatsis, E. J., Chase, M., *et al.* (1994). Gene transfer into experimental brain tumors mediated by adenovirus, herpes simplex virus, and retrovirus vectors. *Hum. Gene. Ther.* **5**(2), 183–191.

Brust, D., Feden, J., *et al.* (2000). Radiosensitization of rat glioma with bromodeoxycytidine and adenovirus expressing herpes simplex virus-thymidine kinase delivered by slow, rate-controlled positive pressure infusion. *Cancer Gene Ther.* **7**(5), 778–788.

Byrnes, A. P., Rusby, J. E., *et al.* (1995). Adenovirus gene transfer causes inflammation in the brain. *Neuroscience* **66**(4), 1015–1024.

Byrnes, A. P., MacLaren, R. E., *et al.* (1996a). Immunological instability of persistent adenovirus vectors in the brain: Peripheral exposure to vector leads to renewed inflammation, reduced gene expression, and demyelination. *J. Neurosci.* **16**(9), 3045–3055.

Byrnes, A. P., Wood, M. J., *et al.* (1996b). Role of T cells in inflammation caused by adenovirus vectors in the brain. *Gene Ther.* **3**(7), 644–651.

Caillaud, C., Akli, S., *et al.* (1993). Adenoviral vector as a gene delivery system into cultured rat neuronal and glial cells. *Eur. J. Neurosci.* **5**(10), 1287–1291.

Cartmell, T., Southgate, T., *et al.* (1999). Interleukin-1 mediates a rapid inflammatory response after injection of adenoviral vectors into the brain. *J. Neurosci.* **19**(4), 1517–1523.

Chen, S. H., Shine, H. D., *et al.* (1994). Gene therapy for brain tumors: Regression of experimental gliomas by adenovirus-mediated gene transfer in vivo. *Proc. Natl. Acad. Sci. USA* **91**(8), 3054–3057.

Choi-Lundberg, D. L., Lin, Q., *et al.* (1997). Dopaminergic neurons protected from degeneration by GDNF gene therapy. *Science* **275**(5301), 838–841.

Choi-Lundberg, D. L., Lin, Q., *et al.* (1998). Behavioral and cellular protection of rat dopaminergic neurons by an adenoviral vector encoding glial cell line-derived neurotrophic factor. *Exp. Neurol.* **154**(2), 261–275.

Connor, B., Kozlowski, D. A., *et al.* (1999). Differential effects of glial cell line-derived neurotrophic factor (GDNF) in the striatum and substantia nigra of the aged Parkinsonian rat. *Gene Ther.* **6**(12), 1936–1951.

Connor, B., Kozlowski, D. A., *et al.* (2001). Glial cell line-derived neurotrophic factor (GDNF) gene delivery protects dopaminergic terminals from degeneration. *Exp. Neurol.* **169**(1), 83–95.

Corti, O., Sabate, O., *et al.* (1999). A single adenovirus vector mediates doxycycline-controlled expression of tyrosine hydroxylase in brain grafts of human neural progenitors. *Nat. Biotechnol.* **17**(4), 349–354.

Cowsill, C., Southgate, T. D., *et al.* (2000). Central nervous system toxicity of two adenoviral vectors encoding variants of the herpes simplex virus type 1 thymidine kinase: Reduced cytotoxicity of a truncated HSV1-TK. *Gene Ther.* **7**(8), 679–685.

Cregan, S. P., MacLaurin, J., *et al.* (2000). Helper-dependent adenovirus vectors: Their use as a gene delivery system to neurons. *Gene Ther.* **7**(14), 1200–1209.

(*Continued*)

TABLE II ([a]*References Continued*)


Cua, D. J., Hutchins, B., *et al.* (2001). Central nervous system expression of IL-10 inhibits autoimmune encephalomyelitis. *J. Immunol.* **166**(1), 602–608.

Davidson, B. L., Allen, E. D., *et al.* (1993). A model system for in vivo gene transfer into the central nervous system using an adenoviral vector. *Nat. Genet.* **3**(3), 219–223.

Davidson, B. L., Doran, S. E., *et al.* (1994). Expression of Escherichia coli beta- galactosidase and rat HPRT in the CNS of Macaca mulatta following adenoviral mediated gene transfer. *Exp. Neurol.* **125**(2), 258–267.

Dewey, R. A., Morrissey, G., *et al.* (1999). Chronic brain inflammation and persistent herpes simplex virus 1 thymidine kinase expression in survivors of syngeneic glioma treated by adenovirus-mediated gene therapy: Implications for clinical trials. *Nat. Med.* **5**(11), 1256–1263.

Doran, S. E., Ren, X. D., *et al.* (1995). Gene expression from recombinant viral vectors in the central nervous system after blood-brain barrier disruption. *Neurosurgery* **36**(5), 965–970.

Dorsey, S. G., Bambrick, L. L., *et al.* (2002). Failure of brain-derived neurotrophic factor-dependent neuron survival in mouse trisomy 16. *J. Neurosci.* **22**(7), 2571–2578.

Driesse, M. J., Vincent, A. J., *et al.* (1998). Intracerebral injection of adenovirus harboring the HSVtk gene combined with ganciclovir administration: Toxicity study in nonhuman primates. *Gene Ther.* **5**(8), 1122–1129.

Driesse, M. J., Esandi, M. C., *et al.* (2000). Intra-CSF administered recombinant adenovirus causes an immune response-mediated toxicity. *Gene Ther.* **7**(16), 1401–1409.

Eck, S. L., Alavi, J. B., *et al.* (1996). Treatment of advanced CNS malignancies with the recombinant adenovirus H5.010RSVTK: A phase I trial. *Hum. Gene Ther.* **7**(12), 1465–1482.

Eck, S. L., Alavi, J. B., *et al.* (2001). Treatment of recurrent or progressive malignant glioma with a recombinant adenovirus expressing human interferon-beta (H5.010CMVhIFN-beta): A phase I trial. *Hum. Gene Ther.* **12**(1), 97–113.

Fathallah-Shaykh, H. M., Zhao, L. J., *et al.* (2000). Gene transfer of IFN-gamma into established brain tumors represses growth by antiangiogenesis. *J. Immunol.* **164**(1), 217–222.

Galanis, E., Bateman, A., *et al.* (2001). Use of viral fusogenic membrane glycoproteins as novel therapeutic transgenes in gliomas. *Hum. Gene Ther.* **12**(7), 811–821.

Gerdes, C. A., Castro, M. G., *et al.* (2000). Strong promoters are the key to highly efficient, noninflammatory and noncytotoxic adenoviral-mediated transgene delivery into the brain in vivo. *Mol. Ther.* **2**(4), 330–338.

Ghodsi, A., Stein, C., *et al.* (1998). Extensive beta-glucuronidase activity in murine central nervous system after adenovirus-mediated gene transfer to brain. *Hum. Gene Ther.* **9**(16), 2331–2340.

Glaser, T., Castro, M. G., *et al.* (2001). Death receptor-independent cytochrome c release and caspase activation mediate thymidine kinase plus ganciclovir-mediated cytotoxicity in LN-18 and LN-229 human malignant glioma cells. *Gene Ther.* **8**(6), 469–476.

Gomez-Manzano, C., Fueyo, J., *et al.* (1997). Characterization of p53 and p21 functional interactions in glioma cells en route to apoptosis. *J. Natl. Cancer Inst.* **89**(14), 1036–1044.

Goodman, J. C., Trask, T. W., *et al.* (1996). Adenoviral-mediated thymidine kinase gene transfer into the primate brain followed by systemic ganciclovir: Pathologic, radiologic, and molecular studies. *Hum. Gene Ther.* **7**(10), 1241–1250.

Gupta, A., Ho, D. Y., *et al.* (2001). Neuroprotective effects of an adenoviral vector expressing the glucose transporter: A detailed description of the mediating cellular events. *Brain Res.* **908**(1), 49–57.

Hagan, P., Barks, J. D., *et al.* (1996). Adenovirus-mediated over-expression of interleukin-1 receptor antagonist reduces susceptibility to excitotoxic brain injury in perinatal rats. *Neuroscience* **75**(4), 1033–1045.

Harada, H., Nakagawa, K., *et al.* (1999). Restoration of wild-type p16 down-regulates vascular endothelial growth factor expression and inhibits angiogenesis in human gliomas. *Cancer Res.* **59**(15), 3783–3789.

Horellou, P., Vigne, E., *et al.* (1994). Direct intracerebral gene transfer of an adenoviral vector expressing tyrosine hydroxylase in a rat model of Parkinson's disease. *Neuroreport* **6**(1), 49–53.

Hughes, S. M., Moussavi-Harami, F., *et al.* (2002). Viral-mediated gene transfer to mouse primary neural progenitor cells. *Mol. Ther.* **5**(1), 16–24.

Jomary, C., Piper, T. A., *et al.* (1994). Adenovirus-mediated gene transfer to murine retinal cells in vitro and in vivo. *FEBS Lett.* **347**(2–3), 117–122.

Kajiwara, K., Byrnes, A. P., *et al.* (1997). Immune responses to adenoviral vectors during gene transfer in the brain. *Hum. Gene Ther.* **8**(3), 253–265.

Kajiwara, K., Byrnes, A. P., *et al.* (2000). Humoral immune responses to adenovirus vectors in the brain. *J. Neuroimmunol.* **103**(1), 8–15.

Kim, C. D., Shin, H. K., *et al.* (2002). Gene transfer of Cu/Zn SOD to cerebral vessels prevents FPI-induced CBF autoregulatory dysfunction. *Am. J. Physiol. Heart Circ. Physiol.* **282**(5), H1836–1842.

Kitagawa, H., Sasaki, C., *et al.* (1999). Adenovirus-mediated gene transfer of glial cell line-derived neurotrophic factor prevents ischemic brain injury after transient middle cerebral artery occlusion in rats. *J. Cereb. Blood Flow Metab.* **19**(12), 1336–1344.

Kosuga, M., Sasaki, K., *et al.* (2001b). Engraftment of genetically engineered amniotic epithelial cells corrects lysosomal storage in multiple areas of the brain in mucopolysaccharidosis type VII mice. *Mol. Ther.* **3**(2), 139–148.

Kosuga, M., Takahashi, S., *et al.* (2001a). Widespread distribution of adenovirus-transduced monkey amniotic epithelial cells after local intracerebral injection: Implication for cell-mediated therapy for lysosome storage disorders. *Cell Transplant* 10(4–5), 435–439.

Kozlowski, D. A., Connor, B., *et al.* (2000). Delivery of a GDNF gene into the substantia nigra after a progressive 6-OHDA lesion maintains functional nigrostriatal connections. *Exp. Neurol.* **166**(1), 1–15.

Kozlowski, D. A., Bremer, E., *et al.* (2001). Quantitative analysis of transgene protein, mRNA, and vector DNA following injection of an adenoviral vector harboring glial cell line-derived neurotrophic factor into the primate caudate nucleus. *Mol. Ther.* **3**(2), 256–261.

Kugler, S., Meyn, L., *et al.* (2001). Neuron-specific expression of therapeutic proteins: Evaluation of different cellular promoters in recombinant adenoviral vectors. *Mol. Cell. Neurosci.* **17**(1), 78–96.

Kuo, H., Ingram, D. K., *et al.* (1995). Retrograde transfer of replication deficient recombinant adenovirus vector in the central nervous system for tracing studies. *Brain Res.* **705**(1–2), 31–38.

Kurihara, H., Zama, A., *et al.* (2000). Glioma/glioblastoma-specific adenoviral gene expression using the nestin gene regulator. *Gene Ther.* **7**(8), 686–693.

(*Continued*)

TABLE II ([a]*References Continued*)


Lammering, G., Valerie, K., *et al.* (2001). Radiosensitization of malignant glioma cells through overexpression of dominant-negative epidermal growth factor receptor. *Clin. Cancer Res.* **7**(3), 632–690.

Lawinger, P., Venugopal, R., *et al.* (2000). The neuronal repressor REST/NRSF is an essential regulator in medulloblastoma cells. *Nat. Med.* **6**(7), 826–831.

Le Gal La Salle, G., Robert, J. J., *et al.* (1993). An adenovirus vector for gene transfer into neurons and glia in the brain. *Science* **259**(5097), 988–990.

Li, T., Adamian, M., *et al.* (1994). In vivo transfer of a reporter gene to the retina mediated by an adenoviral vector. *Invest. Ophthalmol. Vis. Sci.* **35**(5), 2543–2549.

Lisovoski, F., Cadusseau, J., *et al.* (1994). In vivo transfer of a marker gene to study motoneuronal development. *Neuroreport* **5**(9), 1069–1072.

Mackler, S. A., Korutla, L., *et al.* (2000). NAC-1 is a brain POZ/BTB protein that can prevent cocaine-induced sensitization in the rat. *J. Neurosci.* **20**(16), 6210–6217.

Maleniak, T. C., Darling, J. L., *et al.* (2001). Adenovirus-mediated expression of HSV1-TK or Fas ligand induces cell death in primary human glioma- derived cell cultures that are resistant to the chemotherapeutic agent CCNU. *Cancer Gene Ther.* **8**(8), 589–598.

Masada, T., Hua, Y., *et al.* (2001). Attenuation of intracerebral hemorrhage and thrombin-induced brain edema by overexpression of interleukin-1 receptor antagonist. *J. Neurosurg.* **95**(4), 680–686.

Matsuoka, N., Ishii, K., *et al.* (2002). Overexpression of basic fibroblast growth factor and Bcl-xL with adenoviral vectors protects primarily cultured neurons against glutamate insult. *Neurosurgery* **50**(4), 857–862; discussion 862–863.

Millecamps, S., Kiefer, H., *et al.* (1999). Neuron-restrictive silencer elements mediate neuron specificity of adenoviral gene expression. *Nat. Biotechnol.* **17**(9), 865–869.

Miller, C. R., Williams, C. R., *et al.* (2002). Intratumoral 5-fluorouracil produced by cytosine deaminase/5-fluorocytosine gene therapy is effective for experimental human glioblastomas. *Cancer Res.* **62**(3), 773–780.

Miyaguchi, K., Maeda, Y., *et al.* (1999). Neuron-targeted gene transfer by adenovirus carrying neural-restrictive silencer element. *Neuroreport* **10**(11), 2349–2353.

Morelli, A. E., Larregina, A. T., *et al.* (1999). Neuronal and glial cell type- specific promoters within adenovirus recombinants restrict the expression of the apoptosis-inducing molecule Fas ligand to predetermined brain cell types, and abolish peripheral liver toxicity. *J. Gen. Virol.* 80(pt. 3), 571–583.

Muldoon, L. L., Nilaver, G., *et al.* (1995). Comparison of intracerebral inoculation and osmotic blood-brain barrier disruption for delivery of adenovirus, herpesvirus, and iron oxide particles to normal rat brain. *Am. J. Pathol.* **147**(6), 1840–1851.

Nanda, D., Vogels, R., *et al.* (2001). Treatment of malignant gliomas with a replicating adenoviral vector expressing herpes simplex virus-thymidine kinase. *Cancer Res.* **61**(24), 8743–8750.

Navarro, V., Millecamps, S., *et al.* (1999). Efficient gene transfer and long-term expression in neurons using a recombinant adenovirus with a neuron-specific promoter. *Gene Ther.* **6**(11), 1884–1892.

Nilaver, G., Muldoon, L. L., *et al.* (1995). Delivery of herpesvirus and adenovirus to nude rat intracerebral tumors after osmotic blood-brain barrier disruption. *Proc. Natl. Acad. Sci. USA* **92**(21), 9829–9833.

Ohmoto, Y., Wood, M. J., *et al.* (1999). Variation in the immune response to adenoviral vectors in the brain: Influence of mouse strain, environmental conditions and priming. *Gene Ther.* **6**(4), 471–481.
Okada, T., Shah, M., *et al.* (2001). AV.TK-mediated killing of subcutaneous tumors in situ results in effective immunization against established secondary intracranial tumor deposits. *Gene Ther.* **8**(17), 1315–1322.
Ooboshi, H., Ibayashi, S., *et al.* (2001). Adenovirus-mediated gene transfer to ischemic brain: Ischemic flow threshold for transgene expression. *Stroke* **32**(4), 1043–1047.
Peltekian, E., Garcia, L., *et al.* (2002). Neurotropism and retrograde axonal transport of a canine adenoviral vector: A tool for targeting key structures undergoing neurodegenerative processes. *Mol. Ther.* **5**(1), 25–32.
Peltola, M., Kyttala, A., *et al.* (1998). Adenovirus-mediated gene transfer results in decreased lysosomal storage in brain and total correction in liver of aspartylglucosaminuria (AGU) mouse. *Gene Ther.* **5**(10), 1314–1321.
Perez-Cruet, M. J., Trask, T. W., *et al.* (1994). Adenovirus-mediated gene therapy of experimental gliomas. *J. Neurosci. Res.* **39**(4), 506–511.
Plumb, T. J., Bosch, A., *et al.* (1996). Hypoxanthine-guanine phosphoribosyltransferase (HPRT) expression in the central nervous system of HPRT-deficient mice following adenoviral-mediated gene transfer. *Neurosci. Lett.* **214**(2–3), 159–162.
Proescholdt, M. A., Heiss, J. D., *et al.* (1999). Vascular endothelial growth factor (VEGF) modulates vascular permeability and inflammation in rat brain. *J. Neuropathol. Exp. Neurol.* **58**(6), 613–627.
Puumalainen, A. M., Vapalahti, M., *et al.* (1998). Beta-galactosidase gene transfer to human malignant glioma in vivo using replication-deficient retroviruses and adenoviruses. *Hum. Gene Ther.* **9**(12), 1769–1774.
Qiu, J., Whalen, M. J., *et al.* (2002). Upregulation of the Fas receptor death- inducing signaling complex after traumatic brain injury in mice and humans. *J. Neurosci.* **22**(9), 3504–3511.
Ralph, G. S., Bienemann, A., *et al.* (2000). Targeting of tetracycline-regulatable transgene expression specifically to neuronal and glial cell populations using adenoviral vectors. *Neuroreport* **11**(9), 2051–2055.
Ralph, G. S., Bienemann, A., *et al.* (2001). Disruption of the GluR2-NSF interaction protects primary hippocampal neurons from ischemic stress. *Mol. Cell. Neurosci.* **17**(4), 662–670.
Ridoux, V., Robert, J. J., *et al.* (1994a). Adenoviral vectors as functional retrograde neuronal tracers. *Brain Res.* **648**(1), 171–175.
Ridoux, V., Robert, J. J., *et al.* (1994b). The use of adenovirus vectors for intracerebral grafting of transfected nervous cells. *Neuroreport* **5**(7), 801–804.
Ross, B. D., Kim, B., *et al.* (1995). Assessment of ganciclovir toxicity to experimental intracranial gliomas following recombinant adenoviral-mediated transfer of the herpes simplex virus thymidine kinase gene by magnetic resonance imaging and proton magnetic resonance spectroscopy. *Clin. Cancer Res.* **1**(6), 651–657.
Sandmair, A. M., Loimas, S., *et al.* (2000). Thymidine kinase gene therapy for human malignant glioma, using replication-deficient retroviruses or adenoviruses. *Hum. Gene Ther.* **11**(16), 2197–2205.
Sato, N., Wang, S., *et al.* (1998). A novel strategy for introducing exogenous bcl-2 into neuronal cells: The Cre/loxP system-mediated activation of bcl-2 for preventing programmed cell death using recombinant adenoviruses. *Mol. Cell. Neurosci.* 12(1–2), 65–78.

(*Continued*)

TABLE II ([a]*References Continued*)


Shen, J. S., Watabe, K., *et al.* (2001). Intraventricular administration of recombinant adenovirus to neonatal twitcher mouse leads to clinicopathological improvements. *Gene Ther.* **8**(14), 1081–1087.

Shering, A. F., Bain, D., *et al.* (1997). Cell type-specific expression in brain cell cultures from a short human cytomegalovirus major immediate early promoter depends on whether it is inserted into herpesvirus or adenovirus vectors. *J. Gen. Virol.* **78**(pt. 2), 445–459.

Shine, H. D., Wyde, P. R., *et al.* (1997). Neurotoxicity of intracerebral injection of a replication-defective adenoviral vector in a semipermissive species (cotton rat). *Gene Ther.* **4**(4), 275–279.

Shinoura, N., Yoshida, Y., *et al.* (1999). Highly augmented cytopathic effect of a fiber-mutant E1B-defective adenovirus for gene therapy of gliomas. *Cancer Res.* **59**(14), 3411–3416.

Shinoura, N., Yamamoto, N., *et al.* (2000). Adenovirus-mediated transfer of Fas ligand gene augments radiation-induced apoptosis in U-373MG glioma cells. *Jpn. J. Cancer Res.* **91**(10), 1044–1050.

Shinoura, N., Sakurai, S., *et al.* (2002). Co-transduction of Apaf-1 and caspase-9 highly enhances p53-mediated apoptosis in gliomas. *Br. J. Cancer* **86**(4), 587–595.

Smith, J. G., Raper, S. E., *et al.* (1997). Intracranial administration of adenovirus expressing HSV-TK in combination with ganciclovir produces a dose-dependent, self-limiting inflammatory response. *Hum. Gene Ther.* **8**(8), 943–954.

Smith, R. L., Traul, D. L., *et al.* (2000). Characterization of promoter function and cell-type-specific expression from viral vectors in the nervous system. *J. Virol.* **74**(23), 11254–11261.

Smith-Arica, J. R., Morelli, A. E., *et al.* (2000). Cell-type-specific and regulatable transgenesis in the adult brain: Adenovirus-encoded combined transcriptional targeting and inducible transgene expression. *Mol. Ther.* **2**(6), 579–587.

Soudais, C., Laplace-Builhe, C., *et al.* (2001). Preferential transduction of neurons by canine adenovirus vectors and their efficient retrograde transport in vivo. *FASEB J.* **15**(12), 2283–2285.

Tanaka, T., Manome, Y., *et al.* (1997). Viral vector-mediated transduction of a modified platelet factor 4 cDNA inhibits angiogenesis and tumor growth. *Nat. Med.* **3**(4), 437–442.

Thomas, C. E., Birkett, D., *et al.* (2001a). Acute direct adenoviral vector cytotoxicity and chronic, but not acute, inflammatory responses correlate with decreased vector-mediated transgene expression in the brain. *Mol. Ther.* **3**(1), 36–46.

Thomas, C. E., Schiedner, G., *et al.* (2001b). Preexisting antiadenoviral immunity is not a barrier to efficient and stable transduction of the brain, mediated by novel high-capacity adenovirus vectors. *Hum. Gene Ther.* **12**(7), 839–846.

Thomas, C. E., Edwards, P., *et al.* (2002). Adenovirus binding to the coxsackievirus and adenovirus receptor or integrins is not required to elicit brain inflammation but is necessary to transduce specific neural cell types. *J. Virol.* **76**(7), 3452–3460.

Trask, T. W., Trask, R. P., *et al.* (2000). Phase I study of adenoviral delivery of the HSV-tk gene and ganciclovir administration in patients with current malignant brain tumors. *Mol. Ther.* **1**(2), 195–203.

Umana, P., Gerdes, C. A., *et al.* (2001). Efficient FLPe recombinase enables scalable production of helper-dependent adenoviral vectors with negligible helper-virus contamination. *Nat. Biotechnol.* **19**(6), 582–585.

Umegaki, H., Ishiwata, K., *et al.* (2002). In vivo assessment of adenoviral vector-mediated gene expression of dopamine D(2) receptors in the rat striatum by positron emission tomography. *Synapse* **43**(3), 195–200.

Vincent, A. J., Esandi, M. D., *et al.* (1996a). Treatment of leptomeningeal metastases in a rat model using a recombinant adenovirus containing the HSV-tk gene. *J. Neurosurg.* **85**(4), 648–654.

Vincent, A. J., Vogels, R., *et al.* (1996b). Herpes simplex virus thymidine kinase gene therapy for rat malignant brain tumors. *Hum. Gene Ther.* **7**(2), 197–205.

Viola, J. J., Ram, Z., *et al.* (1995). Adenovirally mediated gene transfer into experimental solid brain tumors and leptomeningeal cancer cells. *J. Neurosurg.* **82**(1), 70–76.

Wagenknecht, B., Glaser, T., *et al.* (1999). Expression and biological activity of X-linked inhibitor of apoptosis (XIAP) in human malignant glioma. *Cell Death Differ.* **6**(4), 370–376.

Wang, M. S., Fang, G., *et al.* (2001). The WldS protein protects against axonal degeneration: A model of gene therapy for peripheral neuropathy. *Ann. Neurol.* **50**(6), 773–779.

Xia, H., Anderson, B., *et al.* (2000). Recombinant human adenovirus: Targeting to the human transferrin receptor improves gene transfer to brain microcapillary endothelium. *J. Virol.* **74**(23), 11359–11366.

Xia, X. G., Harding, T., *et al.* (2001). Gene transfer of the JNK interacting protein-1 protects dopaminergic neurons in the MPTP model of Parkinson's disease. *Proc. Natl. Acad. Sci. USA* **98**(18), 10433–10438.

Yang, G. Y., Zhao, Y. J., *et al.* (1997). Overexpression of interleukin-1 receptor antagonist in the mouse brain reduces ischemic brain injury. *Brain Res.* **751**(2), 181–188.

Yang, G. Y., Mao, Y., *et al.* (1999). Expression of intercellular adhesion molecule 1 (ICAM-1) is reduced in permanent focal cerebral ischemic mouse brain using an adenoviral vector to induce overexpression of interleukin-1 receptor antagonist. *Brain Res. Mol. Brain Res.* **65**(2), 143–150.

Yang, G. Y., Pang, L., *et al.* (2001). Attenuation of ischemia-induced mouse brain injury by SAG, a redox-inducible antioxidant protein. *J. Cereb. Blood Flow Metab.* **21**(6), 722–733.

Yoon, S. O., Lois, C., *et al.* (1996). Adenovirus-mediated gene delivery into neuronal precursors of the adult mouse brain. *Proc. Natl. Acad. Sci. USA* **93**(21), 11974–11979.

Yuan, H., Zhai, P., *et al.* (1999). Recombinant adenovirus is an appropriate vector for endocytotic protein trafficking studies in cultured neurons. *J. Neurosci. Methods* **88**(1), 45–54.

Zermansky, A. J., Bolognani, F., *et al.* (2001). Towards global and long-term neurological gene therapy: Unexpected transgene dependent, high-level, and widespread distribution of HSV-1 thymidine kinase throughout the CNS. *Mol. Ther.* **4**(5), 490–498.

Zheng, C., Baum, B. J., *et al.* (2000). Genomic integration and gene expression by a modified adenoviral vector. *Nat. Biotechnol.* **18**(2), 176–180.

Zhu, Y., Yang, G. Y., *et al.* (2002). Transforming growth factor-beta 1 increases bad phosphorylation and protects neurons against damage. *J. Neurosci.* **22**(10), 3898–3909.

Zou, L., Zhou, H., *et al.* (2000). Prolonged transgene expression mediated by a helper-dependent adenoviral vector (hdAd) in the central nervous system. *Mol. Ther.* **2**(2), 105–113.

Zou, L., Yuan, X., *et al.* (2001). Helper-dependent adenoviral vector-mediated gene transfer in aged rat brain. *Hum. Gene. Ther.* **12**(2), 181–191.

it may either by itself affect the physiology of the target cell, induce cytotoxic responses, a cellular immune response, as well as provide antigenic epitopes recognized by the adaptive arm of the immune response (Engelhardt *et al.*, 1994; Yang *et al.*, 1994a,b; Thomas *et al.*, 2000b, 2001b; Lowenstein and Castro, 2002; Mahr and Gooding, 1999; Wold *et al.*, 1994, 1999; Hackett *et al.*, 2000; Wood *et al.*, 1996a). Although infections at low multiplicity with E1-deleted vectors are unlikely to lead to high level adenoviral gene expression, the same is not true at high multiplicity of infections such as used in gene therapy applications. This is likely to remain the case, even if deletion of Ela, Elb, and other regions of the adenovirus genomes provides a better block than simply the Ela deletion (Imperiale *et al.*, 1984; Spergel and Chen-Kiang, 1991). Nevertheless extremely high input multiplicities of infection (in excess of 100 or even higher infectious units per cell regularly achieved in *in vivo* experimental designs) can result in breakthrough of low level expression of the Ad genome (Engelhardt *et al.*, 1994; Kass-Eisler *et al.*, 1994; Yang *et al.*, 1994a,b). A further modification to current vectors is therefore of interest, i.e., the supplementation of deletions in E1, E3, with an additional conditional mutation in E2a, the so-called "second-generation vectors" (Engelhardt *et al.*, 1994; Yang *et al.*, 1994a,b). The E2a region plays a role in activating late phase gene expression. Recombinants carrying this additional mutation have proved to be less reactogenic and permit prolonged expression of the transgene (Engelhardt *et al.*, 1994; Yang *et al.*, 1994a,b). Expression of late genes from the adenoviral genome that are normally expressed only following adenoviral DNA replication suggests that limited replication of the adenoviral genome may occur in some cells. This remains to be proven. In view of the fact that the liver contains proteins that can substitute for E1a function, this topic ought to be investigated carefully.

Further, forskolin can increase the breakthrough to late gene expression. Forskolin results in enhanced levels of intracellular calcium and activation of the transcription factor cAMP responsive element binding protein (CREB or ATF), a factor that is known to bind and stimulate transcription from the Ad early promoter (Muchardt *et al.*, 1990). It is therefore not surprising that forskolin treatment of cells infected with a replication-deficient Ad recombinant can enhance breakthrough to late phase gene expression. Both a high input multiplicity of infection (around the site of virus inoculation or instillation) and elevated calcium levels are likely to occur in transduced cells following *in vivo* gene transfer. Both these phenomena are very likely to coincide in neurons surrounding the site of adenovirus direct injection into the brain. Physiological conditions at the site of gene transfer could thus further regulate transgene expression from viral vectors.

The induction of an immune response to elements of the Ad vector is now perceived as a significant limitation of the vector in long-term gene therapy protocols, especially when systemic administration is needed. Several groups have characterized the effects of the virions on brain inflammation, and the effects of peripheral injections on the stimulation of adaptive immune responses (Byrnes *et al.*, 1995, 1996a,b; Matyszak and Perry, 1996; Wood *et al.*, 1996a; Wood, 1996; Morral *et al.*, 1997; Matyszak 1998; Perry, 1998; Gerdes *et al.*, 2000; Thomas *et al.*, 2000b, 2001a,b, 2002). The effects of each of these on transgene expression are reviewed in detail elsewhere (Lowenstein and Castro, 2002). Although gene transduction remains possible in seropositive animal models and thus possibly in human clinical trials, repeated exposure to viral antigens will be expected to reduce the efficiency of gene transfer. The lack of need to utilize adjuvant when immunizing against adenovirus indicates that even small amounts of vectors leaking into the peripheral immune compartments could activate such immune responses. In the nervous system elements of the adaptive immune response, but not innate inflammatory processes and cells, are able to completely shut-off transgene expression, apparently in the absence of cytotoxicity (unpublished observations). Whether similar phenomena occur in other organs remains to be determined.

Utilizing different Ad serotypes as vectors in sequential treatment administrations could to some extent circumvent this problem. Breakthrough to early phase transcription was demonstrated by the identification of a humoral immune response to early phase proteins. In a mouse model breakthrough has been strongly correlated with the induction of $CD8^+$ cytotoxic T-lymphocyte (CTL) response to adenovirus proteins, which is responsible both for an inflammatory response and an elimination of target cells expressing a transgene (Yang *et al.*, 1994a,b). It has, however, been difficult to conclusively demonstrate that such CTLs are effectively cytotoxic *in vivo*, and thus some of these results and their implications for immune-mediated killing of target cells have been challenged (Wadsworth *et al.*, 1997). Many experiments have been performed using extremely high doses of input virus. If lower doses of adenovirus recombinant are inoculated into a mouse, major histocompatibility complex-I (MHC-I) responses specific for a viral transgene in the absence of a detectable response to the vector can be induced (Schadeck *et al.*, 1999). However, high input doses of recombinant may be essential for efficient *in vivo* gene transfer.

It is also important to consider that many experiments are performed in nonpermissive or semipermissive species (Ginsberg *et al.*, 1991; Ross and Ziff, 1992), e.g., mice, rats, and nonhuman primates. In such systems, Ad E1 recombinants are unable to replicate due to a "species block." It would thus be useful to test recombinants in a more relevant species, such as "permissive" cotton rat (*Sigmodon hispidus*), to determine whether cell-specific factors in addition to species-specific factors regulate the capacity of adenovirus to replicate in particular target organs (Oualikene *et al.*, 1994).

Many researchers see the current cloning capacity of ~7–8 kb as a limitation toward the more general applicability of adenoviral vectors. There is a stringent limitation exerted by the adenovirus particle, which will package only an additional 3% (~1 kb) in addition to the full-length viral genome (Shenk, 1995).

Thus, a new generation of high-capacity helper-dependent adenoviral vectors (also known as "gutless" or "gutted" vectors; HD-Ad) has been developed that is devoid of all viral coding sequences (Fig. 2) (Kochanek *et al.*, 1996; Mitani *et al.*, 1995; Chen *et al.*, 1997, 1999; Morral *et al.*, 1998; Morsy *et al.*, 1998; Schiedner *et al.*, 1998; Burcin *et al.*, 1999; Sandig *et al.*, 2000; Akagi *et al.*, 1997; Maione *et al.*, 2001; Parks *et al.*, 1996, 1999a,b; Cregan *et al.*, 2000; Ng *et al.*, 2001, 2002; Umana *et al.*, 2001; Lowenstein *et al.*, 2002). These vectors have a minimum requirement for the extreme termini of the linear adenovirus genome, containing only those *cis*-acting elements for viral DNA replication and packaging, mainly the inverted terminal repeat (ITR) sequences and packaging signal. Because these elements are contained ~500 bp from the ends of the genome

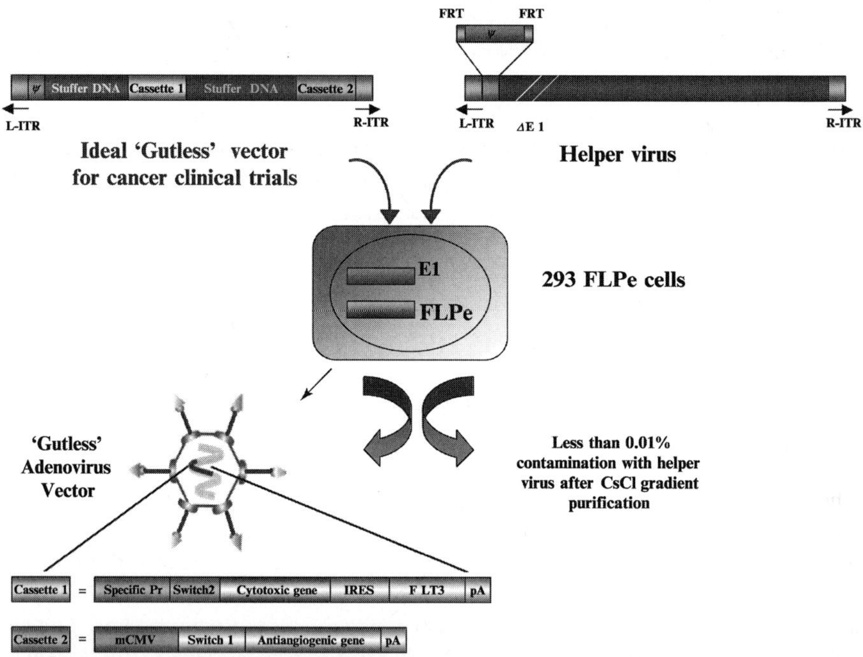


FIG. 2. An E1-, E3-deleted helper adenovirus (FL helper) with a packaging signal sensitive to FLP3-mediated excision. (A) FL helper was generated by homologous recombination in 293 cells after cotransfection with plasmid pΔE1sp1A-2xFRT and pBHG10-CMVluc-I. The bottom half of the figure indicates the left-end region of the helper virus genome, before and after FLP3-mediated recombination; the figure also indicates the location of the PCR primers used to determine excision of the packaging signal, $\phi$ (flanked by black arrows). (B) Production of gutless adenoviral vectors. Diagram of an ideal gene delivery vector for neurological gene therapy using the "gutless" adenoviral vector. The packaging signal of the helper virus, flanked by FRT sites, is excised after growth in 293 cells expressing FLPe recombinase. The cloning capacity of this system (N30 kbp) will enable the simultaneous use of two or more therapeutic genes and also includes inducible promoter systems. (See Color Insert.)

(Grable and Hearing, 1992), helper-dependent vectors have the potential encode range of few hundred base pairs to up to ~30 kb of foreign DNA, which is close to the size of the native genome.

HD-Ad are copropagated with an E1-deleted helper virus, which provides in *trans* all of the proteins required for packaging the vector. Up to now, several systems have been developed to prevent packaging of the helper virus. The Cre/loxP-based system for the generation of HD-Ad involves the use of a first-generation helper virus, where the packaging signal is flanked by loxP recognition sites (Hardy *et al.*, 1997). Infection of Cre-expressing 293 cells with the helper virus results in excision of the viral packaging signal, rendering the helper virus DNA unpackagable but still able to replicate and provide helper functions for HD-Ad vector propagation (Chen *et al.*, 1996). Purification by cesium chloride centrifugation is necessary to reduce the titer of the helper virus to negligible levels, typically ranging from 0.1 to 0.01% of the HD-Ad titer (Morsy *et al.*, 1998). Recently, another Flp/frt-base system has been developed. The Flp recombinase was used in place of Cre, and shown to excise the frt-flanked packaging signal in helper virus efficiently (Lowenstein and Castro, 2001; Ng *et al.*, 2001; Umana *et al.*, 2001). The most recent improvement to this system is the development a new Cre-expressing cell line based on E2T, an E1 and E2a complementary cell line (Zhou *et al.*, 2001). Thus an E1 and E2a double-deleted helper virus can be used with the new cell line to produce HD-Ad vector with low helper contamination, further improving HD vector safety (Zhou *et al.*, 2001). Compared with first-generation adenovirus vectors, the HD-Ad vector can efficiently transduce a wide variety of cell types from numerous species in a cell cycle-independent manner as first generation, but HD-Ad vector has the added advantage of increased cloning capacity, reduced toxicity and immune responses, and prolonged stable transgene expression *in vivo* (Schiedner *et al.*, 1998; Thomas *et al.*, 2000a,b; Thomas *et al.*, 2001a,b). This system is essentially analogous to herpes simplex virus (HSV1) amplicons and shares some of their limitations (Lowenstein, 1994, 1995), e.g., low titers and the presence of variable amounts of helper virus in viral stocks, although novel systems that produce vectors in a helper-free fashion are now available (Wade-Martins *et al.*, 2001; Logvinoff and Epstein, 2001; Wang *et al.*, 2002). Eventually, it should be possible to develop this system so that the recombinants can be produced in the absence of helper virus to generate a safer vector with markedly reduced potential to be reactogenic.

## IV. Adenoviral Recombinant Vectors: Applications to Basic Neuroscience

The technology for gene transfer to the brain has progressed substantially during the past 5 years, and new improved systems are being rapidly developed.

Table I compares the major systems for gene transfer into neurons *in vitro* and into the brain *in vivo*. Vectors vary dramatically in their efficiency of gene transfer, longevity of expression, associate toxicity, and size of the transgene they can harbor. Some vectors mainly direct short-term expression of transgenes, either because of their intrinsic toxicity [e.g., Semliki Forest virus (SFV), vaccinia], or promoter shut-off, a poorly understood phenomenon by which promoters within viral vectors eventually stop being active in spite of the vector genomes being present. Adenoviruses mediate long-term transgene expression in the central nervous system (up to over 1 year) in naive (nonimmunized) animals. Immunization completely eliminates expression from first-generation vectors but not high-capacity helper-dependent adenoviral vectors, whereas in preimmunized animals, transgene expression mediated by first-generation adenoviruses declines within 2–4 weeks postinjection, but expression from the high-capacity helper-dependent adenoviral vectors is reduced only to approximately 50% and stays stable thereafter (Thomas *et al.*, 2000a,b; Thomas *et al.*, 2001b).

In 1993 four groups independently reported for the first time *in vivo* gene transfer into several types of brain cells using adenovirus recombinant encoding $\beta$-galactosidase (Akli *et al.*, 1993; Bajocchi *et al.*, 1993; Davidson *et al.*, 1993; Le Gal La Salle *et al.*, 1993). Since these initial breakthroughs, there has been a logarithmic increase of original papers using adenovirus recombinants to transduce brain cells (see Table II). Although there is some specificity in the capacity of different adenovirus serotypes to infect/express in different target cells (Acsadi *et al.*, 1994a,b; Bessereau *et al.*, 1994; Grubb *et al.*, 1994; Millecamps *et al.*, 1999) a recombinant adenovirus-derived vector can indeed express transgenes in all brain cells, e.g., neurons, astrocytes, oligodendrocytes, ependymal cells, fibroblasts, macrophages, endothelial blood vessel cells, retinal pigment epithelium, photoreceptors, as well as retinal neurons proper, and peripheral nerve Schwann cells either *in vivo* or *in vitro* (Table II; Akli *et al.*, 1993; Bajocchi *et al.*, 1993; Caillaud *et al.*, 1993; Davidson *et al.*, 1993; Le Gal La Salle *et al.*, 1993; Bain *et al.*, 1994; Bennett *et al.*, 1994; Jomary *et al.*, 1994; Li *et al.*, 1994; Byrnes *et al.*, 1995; Lowenstein, 1995; Shering *et al.*, 1997; Morelli *et al.*, 1999; Smith-Arica *et al.*, 2000; Thomas *et al.*, 2000a,b; Umana *et al.*, 2001; Thomas *et al.*, 2000b, 2001a,b, 2002).

While initial experiments were performed using recombinant adenovirus to transfer marker proteins into the brains of both rodents and primates, recombinant vectors encoding therapeutic genes for treatment of neurological diseases have now been generated (Table III) (Chen *et al.*, 1994; Davidson *et al.*, 1994; Perez-Cruet *et al.*, 1994; Shewach *et al.*, 1994; Lowenstein, 1995; Geddes *et al.*, 1997; Ambar *et al.*, 1999; Dewey *et al.*, 1999; Gupta *et al.*, 2001; Bohn *et al.*, 1999; Lawrence *et al.*, 1999; Bohn, 2000; Connor *et al.*, 1999, 2001; Kozlowski *et al.*, 2000, 2001; Amalfitano and Parks, 2002). Expression of transgenes after *in vivo* administration of vectors to the brain was detected up to 5–6 months

TABLE III

NEUROLOGICAL GENE THERAPY: WHAT CAN BE ACHIEVED WITH ADENOVIRUS VECTORS?[a]

| | Noninherited | | Inherited | | | | |
|---|---|---|---|---|---|---|---|
| Disease | Tumors | PD[c] | HD | CMT | Leukodystrophies | Lesch–Nyhan | Frax-A |
| Outcome | Fatal | Fatal | Fatal | Nonfatal | Fatal | Fatal | Nonfatal |
| Distribution | Focal | Focal | Diffuse | Diffuse | Diffuse | Diffuse | Diffuse |
| Age at manifestation | Variable | Late | Variable | Variable | Early | Early | Prenatal |
| Presymptomatic detection[b] | No | No | Yes | Yes[d] | Yes[d] | Yes[d] | Yes[d] |
| Genotype predicts disease severity | ND | ND | Yes | Yes | Yes/no[e] | Yes | No |
| Gene cloned | Yes[f] | Yes[g] | Yes | Yes | Yes | Yes | Yes |
| Expression needed | Short term | Long term | Long term | Long term | Long term | Long term | Long term |
| Experimental | Tumor | Dopamine | Reduce | Normalize | BMT | Express | Express |
| Gene therapy | Cell killing | Replacement, GDNF, AADC | Expression of expanded allele | PMP22 expression | Expressing missing genes | Brain | Brain |
| Clinically relevant available therapeutic protocols | Yes | Yes | No | No | Yes | No | No |
| Available adenoviral vectors | Yes | Yes | No | No | No | Yes | No |

[a]HD, Huntington's disease; CMT, Charcot–Marie–Tooth disease (the commonest genetic alteration is duplication of the region of chromosome 17p11.2 containing the PMP 22 gene); Frax-A, fragile X syndrome type A; PD, Parkinson's disease (idiopathic); BMT, bone marrow transplant. This table is modified from MacMillan and Lowenstein (1996).

[b]Even if presymptomatic detection is available, it is not done routinely during pregnancies; it is performed only if there is any suspicion due to a positive family history.

[c]The primary pathology is loss of dopaminergic neurons from the pars compacta of the substantia nigra, but other lesions within PD are distributed throughout the brain.

[d]Molecular genetic diagnosis is possible in the fetus, but it is arguable whether this would be prior to the occurrence of pathological changes.

[e]Genotype predicts disease severity in metachromatic and Krabbe's leukodystrophy, but not yet in adrenoleukodystrophy.

[f]The genes mutated in inherited brain tumors have been cloned. However, therapeutic gene therapy is more likely to utilize the transduction of cytotoxic genes to eliminate tumor cells.

[g]The genes for tyrosine hyroxylase and dopa decarboxylase have been cloned. These are not, however, mutated in either familial or sporadic PD.

and even after 18 months postinoculation (Geddes *et al.*, 1997; Thomas *et al.*, 2000a,b; Thomas *et al.*, 2000b, 2001a,b; Zermansky *et al.*, 2001).

Other examples of adenoviruses encoding therapeutic transgenes for preclinical applications include recombinant adenoviral vectors encoding hypoxanthine phosphoribosyltransferase (HPRT), which could be utilized in the treatment of the Lesch–Nyhan syndrome. Although these vectors have mainly been utilized to transduce HPRT in primate (Davidson *et al.*, 1994) and rodent brains (Davidson *et al.*, 1994; Lowenstein, 1995; Southgate *et al.*, 2001), there is yet no published account of their capacity to complement an HPRT deficiency in animal models for the Lesch–Nyhan syndrome. Adenovirus vectors expressing tyrosine hyroxylase, and their therapeutic efficacy in a rat model of Parkinson's disease, have been reported (Horellou *et al.*, 1994; Corti *et al.*, 1999a,b; Hida *et al.*, 1999). Recombinant adenoviruses encoding neuroprotective factors such as glial cell line-derived neurotrophic factor (GDNF), aimed to prevent dopaminergic neuron degeneration in a rat model of Parkinson disease, have also been successfully developed (Bilang-Bleuel *et al.*, 1997; Choi-Lundberg *et al.*, 1997, 1998; Bohn *et al.*, 1999; Lawrence *et al.*, 1999; Bohn, 2000; Connor *et al.*, 1999, 2001; Kozlowski *et al.*, 2000, 2001).

Because adenovirus recombinants can be used to transduce essentially all brain cells, they have also been postulated to be applicable to gene therapy protocols for the treatment of brain tumors, by utilizing them to deliver cytotoxic genes directly into the tumors. Thus, a number of groups have utilized several adenovirus recombinants, e.g., expressing HSV1 thymidine kinase or Fas ligand, Fas receptor, or p53, under the control of viral promoters. In experimental paradigms the use of these viruses appears to be promising (Badie *et al.*, 1994; Boviatsis *et al.*, 1994; Brody *et al.*, 1994a; Chen *et al.*, 1994; Perez-Cruet *et al.*, 1994; Shewach *et al.*, 1994; Ambar *et al.*, 1999; Dewey *et al.*, 1999; Morelli *et al.*, 1999; Shinoura *et al.*, 2000a,b; Shono *et al.*, 2002; see George *et al.*, this volume).

Recombinant retroviruses have been used in the treatment of brain tumors (Ram *et al.*, 1993; Palu *et al.*, 1999; Tamura *et al.*, 2001), and replication-competent and conditional HSV1 vectors as well as HSV1 amplicon vectors have also been tried (Martuza *et al.*, 1991; Markert *et al.*, 1993; Kramm *et al.*, 1997; Pechan *et al.*, 1999; Herrlinger *et al.*, 2000; Papanastassiou *et al.*, 2002; see Hu and Coffin, this volume). In a recent phase I clinical trial the efficacy and safety profiles of retrovirus and adenovirus expressing HSVI TK have been compared (Sandmair *et al.*, 2000). Adenoviruses showed very promising results, both in terms of their safety and also their efficacy in glioma growth control (Sandmair *et al.*, 2000). However, the only vector system that has progressed to a phase III clinical trial has been a retroviral vector expressing HSV-1 thymidine kinase. The trial was halted because of the absence of positive effects in the gene therapy arm of patients treated (Rainov, 2000; Rainov and Kramm, 2001).

With Phase I and II being essentially toxicity trials, Phase III efficacy clinical trials are crucial in order to determine whether any of the new therapies are indeed effective, because such results can be obtained only rarely from the much smaller Phase I and II trials.

Two groups have also made interesting observations on the use of adenovirus recombinants that should be of interest to neuroanatomists, namely the use of adenovirus for pathway tracing and cell labeling. Ridoux and colleagues (1994a) observed retrograde labeling of substantia nigra neurons after injections of recombinants expressing $\beta$-galactosidase into the striatum; similar results were also observed by Byrnes *et al.* (1995). Thus, replication-defective adenovirus could be used for retrograde labeling of neuronal pathways. Interestingly, HSV1-derived recombinants can also be used to trace neuronal pathways (Ugolini *et al.*, 1989), with some recombinants being specific for anterograde transport and others for retrograde transport (Zemanick *et al.*, 1991), while certain pseudorabies recombinants have been shown to be specific markers for individual neuronal circuits and capable of retrograde axonal transport (Card *et al.*, 1991, 1992; Mazarakis *et al.*, 2001; Enquist *et al.*, 1998; DeFalco *et al.*, 2001). Lisovoski *et al.* (1994) utilized adenovirus recombinants to label neurons of the spinal cord at different stages of development to study their morphological development. The staining they obtained after *in vivo* administration filled the dendritic arbors of neurons, thus allowing for their morphological and morphometric examination during spinal cord development.

An effective way of delivering gene products to the brain, rather than by introducing it into constituent cells, is by transplanting genetically engineered cells directly into the CNS. So far, most genetically engineered cells expressing trangenes for transplantation have been transduced with recombinant retroviral vectors. Ridoux *et al.*, (1994b) have used adenovirus to tranduce primary cultures of rat astrocytes *in vitro* and then transplanted these into the CNS of host rats. Expression of transgene was detected for at least 5 months. Thus, adenovirus might constitute an alternative to retrovirus for transducing cells in preparation for transplantation into the brain. In many cases expression from retroviral vectors ceases after a few days to a few weeks; thus, it will be interesting to compare both systems vis-à-vis length of transgene expression.

Expression of transgenes after the administration of adenovirus recombinants into adult brains has been seen by some groups to last from 6 to 18 months. It also was reported that injection of recombinants into neonates has allowed expression to be sustained over periods up to a year. Long-term modulation of the immune response and the elucidation of its exact role in the regulation of long-term expression from adenoviral recombinants administered into the CNS will be crucial to achieve stable expression over a long period of time, which will be needed to implement gene therapy for chronic neurological disorders in humans. (Stratford-Perricaudet *et al.*, 1990; Kass-Eisler *et al.*, 1994; Geddes *et al.*, 1997; Thomas *et al.*, 2000a,b; Thomas *et al.*, 2000b, 2001a,b; Zermansky *et al.*, 2001).

Scant information is available on the interactions of adenovirus with the brain. Thus it will be of great importance to explore this field, e.g., adenovirus entry into brain cells, transport to the nucleus, and viral replication in neurons of permissive species. In addition, the effect of adenovirus recombinants on the electrophysiology of specific neuronal populations will have to be investigated. It also will be critical to assess the effects of adenoviral delivery into different brain regions on animal behavior. It is surprising that not much work on these topics has been published in the past 5 years.

## V. Adenoviral Recombinant Vectors: Applications to Neurological Gene Therapy

Essentially there are two types of applications for which vectors for gene transfer into the brain could be used. Short-term expression could be used, for example, to express cytotoxic products to kill tumor cells, or to provide drugs to block ischemia-induced neurotoxicity. Ideally, the area of brain tissue to be targeted should be focal, and therapeutic benefit should be predicted after short-term transgene expression (Table III). Long term expression would be needed to treat chronic neurodegenerative disorders, such as Parkinson's disease, Alzheimer's disease, or amyotrophic lateral dystrophy. In this case, the area of brain tissue will be larger, and the results of the therapeutic intervention may not be seen until after several years.

The availability of viral vectors that express short term at high levels of expression could lead to important new treatments for brain diseases. A case in point are brain tumors, which are now being treated experimentally with adenovirus recombinants expressing conditional cytotoxic gene products (Badie *et al.*, 1994; Boviatsis *et al.*, 1994; Brody *et al.*, 1994a,b; Chen *et al.*, 1994; Perez-Cruet *et al.*, 1994; Shewach *et al.*, 1994; Takamiya *et al.*, 1992; Barba *et al.*, 1994; Benedetti *et al.*, 1997; Ram *et al.*, 1997; Bansal and Engelhard, 2000; Ikeda *et al.*, 2000; Trask *et al.*, 2000; Jacobs *et al.*, 2001; Rainov and Kramm, 2001). Brain tumors constitute a good target disease for gene therapy for the following reasons: (1) the disease is focal; (2) the therapeutic objective—the destruction of tumor cells—needs to be achieved within the short term, and this can be done using cytotoxic or conditionally cytotoxic gene products; (3) the disease is life threatening within 6–12 months of diagnosis; (4) no effective treatments are available; and (5) adenovirus-particle-induced inflammation could be a beneficial adjunct to tumor cell elimination. Similarly, replication-deficient and replication-competent HSV1-derived vectors have also been also developed (Martuza *et al.*, 1991; Markert *et al.*, 1993) and these vectors have already been tested in Phase I–III clinical trials

(Martuza *et al.*, 1991; Markert *et al.*, 1993, 2000; Rampling *et al.*, 1998, 2000; Rainov, 2000).

Achieving long-term expression in brain following the administration of recombinant vectors will open up the development of gene therapy clinical trials for human neurological diseases where long-term expression is paramount, such as Alzheimer's disease, Parkinson's disease, Huntington's disease, and the metabolic and degenerative brain disorders (Amalfitano and Parks, 2002; Hsich *et al.*, 2002; Lowenstein, 2002; Lowenstein and Castro, 2002). What makes it difficult to treat such diseases is that they affect large areas of the brain, many extending throughout large anatomical regions of the human CNS, and they progress relatively slowly over years or decades. For Parkinson's diseases there are as yet no presymptomatic diagnostic methods available, so treatment can be instituted only when symptoms appear. In families suffering from Huntington's disease, presymptomatic diagnosis is now possible (Holloway *et al.*, 1994) and this might also soon become possible in Alzheimer's disease (Saunders *et al.*, 1993; Nalbantoglu *et al.*, 1994; Scinto *et al.*, 1994). The diffuse nature of these diseases has led to the development of other therapeutic interventions that could target genes and their products to different areas located throughout the brain, such as transplantation of genetically engineered cells (Fisher, 1993), the direct intracranial administration of neuronal growth factors (NGF), (Seiger *et al.*, 1993), or more recently the development of embryonic and or neural stem cells. HSV1 (Lokensgard *et al.*, 1994; Burton *et al.*, 2001; Kay *et al.*, 2001; Latchman and Coffin, 2001; Lilley *et al.*, 2001) and adeno-associated virus vectors (Kaplitt *et al.*, 1994; Sanlioglu *et al.*, 2001; Fu *et al.*, 2002; Muramatsu *et al.*, 2002) might also provide alternative vectors to achieve long-term transgene expression in brain for clinical applications in humans.

Although human brain infections caused by adenovirus are extremely rare, it has been reported that adenovirus can cause central nervous system disease, especially in children and immuncompromised individuals, although rare cases have also been described in nonimmunocompromised patients (Ginsberg and Prince, 1994; Engel, 1995; Chirmule *et al.*, 1999; Ginsberg, 1999). Meningoencephalitis, encephalitis, and cerebellar ataxia have been reported in children in association with adenovirus infection of the CNS, mostly adenovirus type 7, 1, 2, or 32 (Chou *et al.*, 1973; Kelsey, 1978; Davis *et al.*, 1988; Osamura *et al.*, 1993). Direct inoculation of adenovirus into the brain of rhesus monkeys has demonstrated that several adenovirus serotypes can cause neuropathology, although adenovirus did not appear to replicate to a great extent in the primate brain (Ginsberg and Prince, 1994; Engel, 1995; Chirmule *et al.*, 1999; Ginsberg, 1999). Also, neuropathology due to an anamnestic reponse of the immune system following the injection of the adenovirus type 5 into the primate brain was recently described (Davidson *et al.*, 1994).

The possibility that latent adenovirus infection might be localized to the human brain, and that adenovirus could enter the brain through infected macrophages from the periphery, has not yet been examined in enough detail, although widespread neuronal inclusions diagnostic of adenovirus particles have been seen in some fatal encephalitis cases (Chou *et al.*, 1973). However, all these data taken together with our observations that wild-type adenovirus can express endogenous proteins in neurons, and brain glial cells, strongly suggest that adenovirus could replicate, or at last express many of its genes after infecting human neurons or glial cells and that expression might proceed for a longer time than predicted, even in the absence of symptoms of encephalitis. If there is a risk of adenoviral encephalitis it will have to be carefully considered at a time adenovirus vector is directly administered into the brain. It is clear that many more studies will be required to clarify the origins of neurotoxicity after direct injections of adenovirus virions into the brain. The use of the completely deleted high-capacity helper-dependent adenoviral vectors will completely avoid these potential complications.

## VI. Immune Responses to Recombinant Adenoviral Vectors Delivered into the Brain

As stated above, recombinant adenoviruses are very attractive vectors for gene transfer and therapy within the CNS, however, they still cause acute inflammation, and any of their expressed genes can be the target of adaptive immune responses. Local inflammation is elicited in the brain in response to the injection of either a first-generation, or a high-capacity helper-dependent adenoviral vector. This acute inflammation is dose dependent, self-limited (disappears within 30 days), adenoviral-capsid dependent, and does not affect long-term adenoviral-mediated transgene expression (Byrnes *et al.*, 1995, 1996a; Kajiwara *et al.*, 1997; Gerdes *et al.*, 2000; Thomas *et al.*, 2000b, 2001a,b, 2002). Injection of adenoviral vectors into the brain leads to persistent expression (e.g., up to 18 months). This is thought to be due to the failure to stimulate an effective antiadenoviral T cell response following the careful injection of these vectors into the brain parenchyma (Byrnes *et al.*, 1996b; Wood *et al.*, 1996a; Kajiwara *et al.*, 1997; Gerdes *et al.*, 2000; Thomas *et al.* 2000b, 2001a,b, 2002). Activation of antiadenoviral immune response by peripheral immunization with adenovirus leads to massive lymphocyte infiltration of the brain parenchyma, macrophage/microglial activation, up-regulation of MHCI and II, and loss of vector-mediated transgene expression (Byrnes *et al.*, 1996a; Gerdes *et al.*, 2000; Thomas *et al.*, 2000b, 2001a,b, 2002). Mechanisms of loss of transgene expression have not been completely elucidated. Either cytotoxicity of transduced cells or downregulation of mRNA expression could explain it. Thus, loss of transgene expression from first-generation adenoviral vectors in the brain following systemic immunization against adenovirus limits their utility for long-term neurological gene therapy applications (Lowenstein and Castro, 2002).

Recent work from our laboratory indicates that high doses of E1/E3-deleted viral vectors ($>10^8$ IU) cause direct acute cytotoxicity, and chronic inflammation, which lead to reduced transgene expression and substantial tissue damage (Thomas *et al.*, 2001a,b). We thus reasoned that an absolute reduction in the dose of vectors would be necessary to eliminate both the acute cytotoxicity and chronic inflammation. The use of the powerful mCMV promoter in adenoviral vectors has recently uncovered the capacity to transfer transgenes into the brain, and achieving similar results in the levels of transgene expression while using 2–3 logs lower doses of vectors when compared to vectors employing the hCMV promoter. The important reduction in total viral dose needed thus allows transduction in the absence of acute cytotoxicity or chronic brain inflammation (Gerdes *et al.*, 2000).

Additionally, E1/E3-deleted adenoviruses do express adenoviral proteins encoded within their genomes (Yang *et al.*, 1994a,b). These can either generate specific immune responses or provide target epitopes recognized by the activated adaptive immune system. We thus studied transgene expression and inflammatory responses elicited by the intrastriatal injection of E1–E3-deleted or HC-Adv vectors in four different paradigms: (1) acutely in naive rats, (2) long term, in naive animals, (3) long term, in naive animals, followed by a peripheral immunization against adenovirus type 5, and (4) medium term, in animals immunized against adenovirus type 5 preceding intracranial vector injection (Thomas *et al.*, 2000b, 2001a,b, 2002).

These studies have shown the following. (1) Injection of $1\times10^7$ IU of either E1/E3-deleted vectors or HC-Adv into the brain causes an acute, transitory inflammatory response, with both cellular (e.g., activation of microglia) and molecular components (e.g., upregulation of MHCI-I). This acute inflammation does not affect long-term transgene expression (see Figs. 3–6). (2) Within 12 months the expression of a marker transgene from first generation vectors decreases. This decrease is less when expression is directed by an HC-Adv. (3) Following peripheral immunization against adenovirus type 5, transgene expression from an E1/E3-deleted adenovirus vector is completely abolished within 45 days, whereas expression from a HC-Adv remains unaffected (Fig. 4). (4) Downregulation of transgene expression appears to be mediated at the transcriptional level (specific reduction of transgene mRNA levels), rather than cytotoxicity (unpublished data). (5) Expression from an E1/E3-deleted adenovirus in brains of animals previously immunized against adenovirus type 5 is almost completely eliminated by 14 days postinjection, whereas expression from an HC-Adv is reduced to only 50%, and remains stable for at least 2 months (Thomas *et al.*, 2000b, 2001a,b, 2002; Lowenstein and Castro, 2002). The complexities, causes, and consequences of using viral vectors as gene transfer vectors in the brain have been explored elsewhere (Lowenstein, 2002).

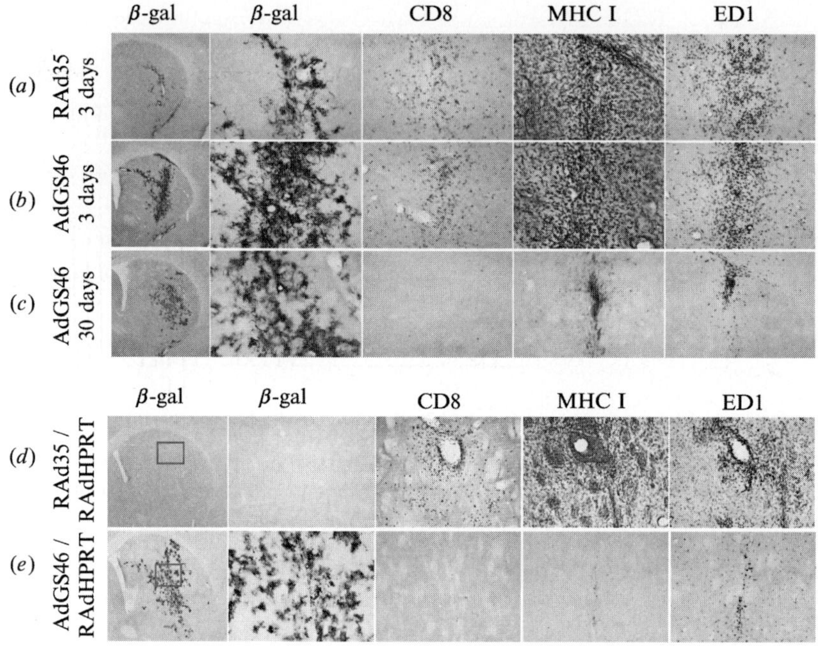


FIG. 3. Transgene expression ($\beta$-galactosidase) and inflammation in the brains of animals injected with either the E1/E3-deleted vector RAd35 or the HC-Adv AdGS46. Rows *a* and *b* show $\beta$-galactosidase expression (columns 1 and 2), $CD8^+$ cell infiltration (column 3), MHC class I upregulation (column 4), and microglial cell activation (ED1, column 5), 3 days after injection of $1 \times 10^7$ IU of RAd35 (row *a*) or AdGS46 (row *b*) into the brains of naive animals. Levels of transgene expression and inflammation mediated by the E1/E3-deleted, or the HC-Adv vector, were indistinguishable in naive animals. Transgene expression from $1 \times 10^7$ IU of both RAd35 and AdGS46 is stable for at least 30 days in naive animals and inflammation resolves within this time frame (shown in row *c* only for AdGS46). However, transgene expression from E1/E3-deleted vectors in the CNS is rapidly eliminated and accompanied by severe brain inflammation if animals receive a subsequent *peripheral* infection with adenovirus [row *d*; RAd35 injected in the brain at Day 0, RAdHPRT (Southgate *et al.*, 1999) injected in the skin at Day 60. See also the quanitative analysis of all markers, in panel *f*]. In contrast, transgene expression from HC-Adv remains stable, and no brain inflammation is elicited when animals are subsequently injected with RAdHPRT in the skin (row *e*; AdGS46 injected in the brain at Day 0, RAdHPRT injected in the skin at Day 60).

## VII. Summary: Targeting the Brain with Adenovirus-Derived Vectors

The early generations of replication-deficient adenovirus vectors, as well as the most recently developed high-capacity helper-dependent adenovirus vectors are extremely versatile vehicles for gene transfer into the brain, and have thus

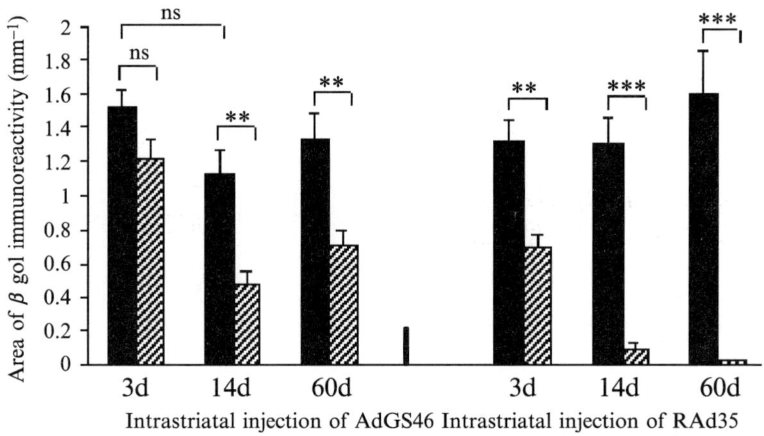


FIG. 4. Preimmunization against adenovirus eliminates rapidly transgene expression from E1/E3-deleted vectors, but not from high-capacity vectors. Quantification of the area within 40-$\mu$m-thick brain sections occupied by $\beta$-galactosidase and ED1. Error bars show the SEM value from the five animals in each experimental group. Student's *t* test was used to calculate the degree of significance of differences between levels of transgene expression and inflammation in the brains of nonimmunized animals (black bars) and immunized animals (hatched bars) after intrastriatal injection of RAd35 (right) or AdGS46 (left). The x-axis indicates the days after vector injection; vector was injected on day 0, and animals were immunized 14 days before the brain injection.

become invaluable reagents to neuroscientists. Current applications cover wide areas from helping to address basic neurobiological problems to clinical trials for the treatment of human diseases. The main advantages of adenovirus-derived vectors are as follows:

1. the existence of well-characterized methods to produce high titer GMP grade replication-defective vectors for human clinical trials, as well as methods to produce high titers of high-capacity helper-dependent vectors with contamination levels of helper virus <0.01%. Although these last ones have yet to be used in clinical trials, the production of high levels of such vectors is being pursued by several groups;
2. a capacity to infect, in the adult brain, both terminally differentiated neurons and dividing brain astrocytes, both *in vivo* and *in vitro*;
3. a capacity to transduce both immature neonatal and mature neurons;
4. the potential to achieve cell type-specific and regulated transgene expression using appropriate promoters and transcriptional regulatory elements;
5. an ability to achieve multiple infections of a single target cell with separate vector virions;
6. their use to elucidate neuronal morphology and connectivity;
7. long-term high-level expression *in vivo* in naive nonimmunized animals;

8. low incidence of adenovirus-induced brain infections in humans; and
9. large experience in the use of adenovirus to immunize humans against natural infections.

Although the results so far have been highly encouraging (Tables II and III), adenovirus vector systems continue to be optimized for application to somatic cell gene therapy and specifically for neurological gene transfer. Currently, progress in clinical neurological gene therapy is both dependent on and limited by the technology for gene transfer. Selection of the most appropriate vector system for each application is crucial. Adenoviral vectors clearly have distinct advantages for efficient gene delivery into nervous system both *in vitro* and *in vivo*. Vector development will undoubtedly further enhance the utility of these vectors and allow their implementation for gene therapy applications to treat neurological diseases.

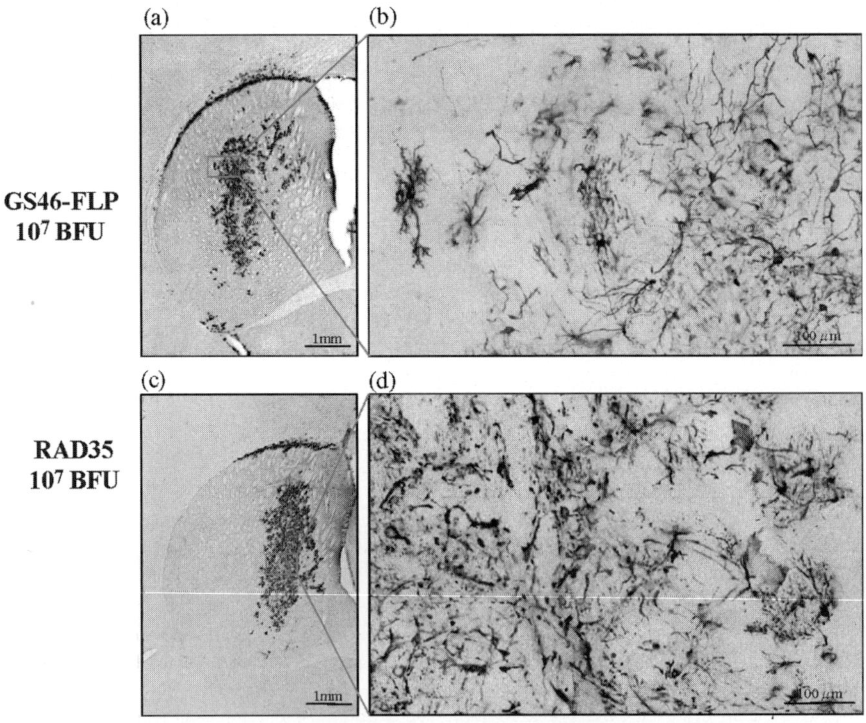


FIG. 5. *In vivo* $\beta$-galactosidase expression in rat brain infected with either GS46-FLP HC-Adv vector or E1/E3-deleted adenovirus (RAd35). $7 \times 10^7$ BFU of either virus (in 3 $\mu$l) was injected into the striatum of male Sprague–Dawley rats. Six days after virus inoculation, $\beta$-galactosidase expression was analyzed by immunohistochemistry. (a, b) GS46-FLP HD virus. (c, d) RAd35 virus. $\beta$-Galactosidase is expressed under the control of the hCMV promoter in both viruses. From Umana *et al.* (2001).

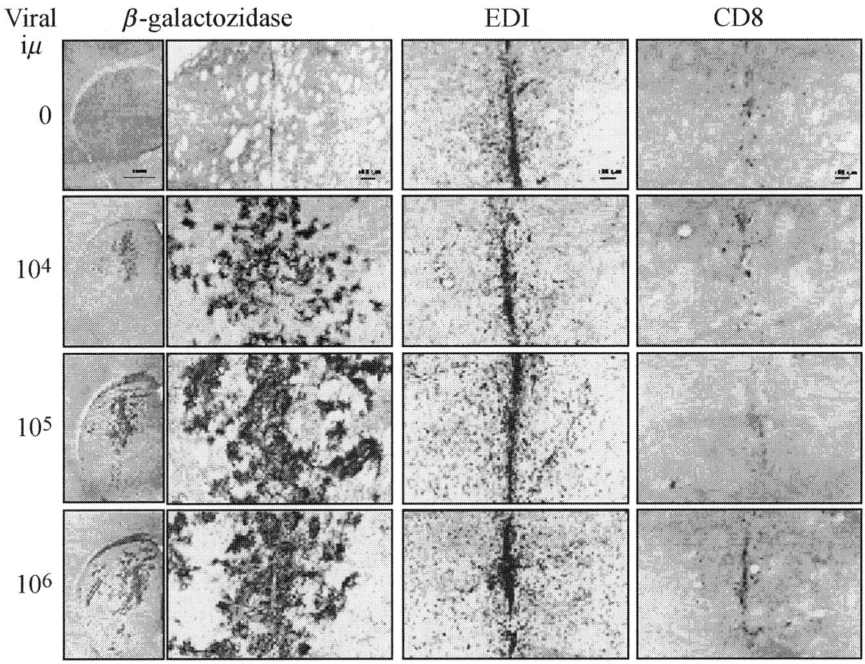


FIG. 6. The mCMV promoter allows high-level expression at low doses of vector, without inflammation. Increasing doses of RAd36 (mCMV-$\beta$gal) ($10^4$–$10^6$) were injected into the striatum of adult Sprague–Dawley rats in a total volume of 3 $\mu$l, and animals were perfused 5 days later. Brains were processed for immunohistochemistry to detect $\beta$-galactosidase (transgene expression; first two left hand side columns), ED1 (macrophages-microglial cells), and CD8 (infiltrating lymphocytes and NK cells). Low-power view of the striatum is shown in the left-hand side column, and a higher power view is shown in the column next to it. Scale bar shown in the upper left = 1 mm; all others = 100 $\mu$m. RAd36 allows strong transduction of the striatum, in the absence of recruitment of inflammatory cells, compared to animals injected with saline controls. This image illustrates three doses of animals injected with RAd36. The quantification of this experiment is shown in Fig. 4, above.

## Acknowledgments

Work in the Gene Therapeutics Research Institute at Cedars Sinai Medical Center is funded by R01 NS42893.01 and R01 NS44556.01 grants from the National Institutes of Health, National Institute for Neurological Disorders and Stroke. We also thank the funding our Institute receives from the Board of Governors at Cedars Sinai Medical Center and the encouragement and support of every one of its members. We also wish to thank Dr. Shlomo Melmed for unparalleled support and academic leadership. We are grateful to Ms. Cheryl Cathcart for her superb administrative and organizational skills and to Mr. Danny Malaniak for his encouragement, support, and commitment. Ms. Juanita Gutierrez was responsible for the editing and preparation of this manuscript for publication and we are grateful for her dedication and superb skills.

## References

Acsadi, G., Jani, A., *et al.* (1994a). Cultured human myoblasts and myotubes show markedly different transducibility by replication-defective adenovirus recombinants. *Gene Ther.* **1**(5), 338–340.

Acsadi, G., Jani, A., *et al.* (1994b). A differential efficiency of adenovirus-mediated in vivo gene transfer into skeletal muscle cells of different maturity. *Hum. Mol. Genet.* **3**(4), 579–584.

Ailles, L. E., and Naldini, L. (2002). HIV-1-derived lentiviral vectors. *Curr. Top. Microbiol. Immunol.* **261,** 31–52.

Akagi, K., Sandig, V., *et al.* (1997). Cre-mediated somatic site-specific recombination in mice. *Nucleic Acids Res.* **25**(9), 1766–1773.

Akli, S., Caillaud, C., *et al.* (1993). Transfer of a foreign gene into the brain using adenovirus vectors. *Nat. Genet.* **3**(3), 224–228.

Amalfitano, A., and Parks, R. J. (2002). Separating fact from fiction: Assessing the potential of modified adenovirus vectors for use in human gene therapy. *Curr. Gene Ther.* **2**(2), 111–133.

Ambar, B. B., Frei, K., *et al.* (1999). Treatment of experimental glioma by administration of adenoviral vectors expressing Fas ligand. *Hum. Gene Ther.* **10**(10), 1641–1648.

Babiss, L. E., Friedman, J. M., *et al.* (1986). Cellular promoters incorporated into the adenovirus genome: Effects of viral regulatory elements on transcription rates and cell specificity of albumin and beta-globin promoters. *Mol. Cell. Biol.* **6**(11), 3798–3806.

Badie, B., Hunt, K., *et al.* (1994). Stereotactic delivery of a recombinant adenovirus into a C6 glioma cell line in a rat brain tumor model. *Neurosurgery* **35**(5), 910–915; discussion 915–916.

Bain, D., Wilkinson, G. W., *et al.* (1994). Adenovirus vectors to transfer genes into neurones: Implications for gene therapy of neurological disorders. *Gene Ther.* **1**(Suppl. 1), S68.

Bajocchi, G., Feldman, S. H., *et al.* (1993). Direct in vivo gene transfer to ependymal cells in the central nervous system using recombinant adenovirus vectors. *Nat. Genet.* **3**(3), 229–234.

Bansal, K., and Engelhard, H. H. (2000). Gene therapy for brain tumors. *Curr. Oncol. Rep.* **2**(5), 463–472.

Barba, D., Hardin, J., *et al.* (1994). Development of anti-tumor immunity following thymidine kinase-mediated killing of experimental brain tumors. *Proc. Natl. Acad. Sci. USA* **91**(10), 4348–4352.

Bauknecht, T., and Meinhold-Heerlein, I. (2002). Gene therapy of ovarian cancer. *Curr. Womens Health Rep.* **2**(1), 39–46.

Benedetti, S., Dimeco, F., *et al.* (1997). Limited efficacy of the HSV-TK/GCV system for gene therapy of malignant gliomas and perspectives for the combined transduction of the interleukin-4 gene. *Hum. Gene Ther.* **8**(11), 1345–1353.

Bennett, J., Wilson, J., *et al.* (1994). Adenovirus vector-mediated in vivo gene transfer into adult murine retina. *Invest. Ophthalmol. Vis. Sci.* **35**(5), 2535–2542.

Bergelson, J. M., Cunningham, J. A., *et al.* (1997). Isolation of a common receptor for Coxsackie B viruses and adenoviruses 2 and 5. *Science* **275**(5304), 1320–1323.

Berkner, K. L. (1988). Development of adenovirus vectors for the expression of heterologous genes. *Biotechniques* **6**(7), 616–629.

Bessereau, J. L., Stratford-Perricaudet, L. D., *et al.* (1994). In vivo and in vitro analysis of electrical activity-dependent expression of muscle acetylcholine receptor genes using adenovirus. *Proc. Natl. Acad. Sci. USA* **91**(4), 1304–1308.

Bilang-Bleuel, A., Revah, F., *et al.* (1997). Intrastriatal injection of an adenoviral vector expressing glial-cell-line-derived neurotrophic factor prevents dopaminergic neuron degeneration and behavioral impairment in a rat model of Parkinson disease. *Proc. Natl. Acad. Sci. USA* **94**(16), 8818–8823.

Blomer, U., Naldini, L., *et al.* (1997). Highly efficient and sustained gene transfer in adult neurons with a lentivirus vector. *J. Virol.* **71**(9), 6641–6649.

Bohn, M. C. (2000). Parkinson's disease: A neurodegenerative disease particularly amenable to gene therapy. *Mol. Ther.* **1**(6), 494–496.

Bohn, M. C., Choi-Lundberg, D. L., *et al.* (1999). Adenovirus-mediated transgene expression in nonhuman primate brain. *Hum. Gene Ther.* **10**(7), 1175–1184.

Boucher, R. C., Knowles, M. R., *et al.* (1994). Gene therapy for cystic fibrosis using E1-deleted adenovirus: A phase I trial in the nasal cavity. The University of North Carolina at Chapel Hill. *Hum. Gene Ther.* **5**(5), 615–639.

Bout, A., Imler, J. L., *et al.* (1994a). In vivo adenovirus-mediated transfer of human CFTR cDNA to rhesus monkey airway epithelium: Efficacy, toxicity and safety. *Gene Ther.* **1**(6), 385–394.

Bout, A., Perricaudet, M., *et al.* (1994b). Lung gene therapy: In vivo adenovirus-mediated gene transfer to rhesus monkey airway epithelium. *Hum. Gene Ther.* **5**(1), 3–10.

Boviatsis, E. J., Chase, M., *et al.* (1994c). Gene transfer into experimental brain tumors mediated by adenovirus, herpes simplex virus, and retrovirus vectors. *Hum. Gene Ther.* **5**(2), 183–191.

Brody, S. L., Jaffe, H. A., *et al.* (1994a). Direct in vivo gene transfer and expression in malignant cells using adenovirus vectors. *Hum. Gene Ther.* **5**(4), 437–447.

Brody, S. L., Metzger, M., *et al.* (1994b). Acute responses of non-human primates to airway delivery of an adenovirus vector containing the human cystic fibrosis transmembrane conductance regulator cDNA. *Hum. Gene Ther.* **5**(7), 821–836.

Brooks, A. I., Stein, C. S., *et al.* (2002). Functional correction of established central nervous system deficits in an animal model of lysosomal storage disease with feline immunodeficiency virus-based vectors. *Proc. Natl. Acad. Sci. USA* **99**(9), 6216–6221.

Burcin, M. M., Schiedner, G., *et al.* (1999). Adenovirus-mediated regulable target gene expression in vivo. *Proc. Natl. Acad. Sci. USA* **96**(2), 355–360.

Burton, E. A., Bai, Q., *et al.* (2001). Targeting gene expression using HSV vectors. *Adv. Drug Deliv. Rev.* **53**(2), 155–170.

Byrnes, A. P., Rusby, J. E., *et al.* (1995). Adenovirus gene transfer causes inflammation in the brain. *Neuroscience* **66**(4), 1015–1024.

Byrnes, A. P., MacLaren, R. E., *et al.* (1996a). Immunological instability of persistent adenovirus vectors in the brain: Peripheral exposure to vector leads to renewed inflammation, reduced gene expression, and demyelination. *J. Neurosci.* **16**(9), 3045–3055.

Byrnes, A. P., Wood, M. J., *et al.* (1996b). Role of T cells in inflammation caused by adenovirus vectors in the brain. *Gene Ther.* **3**(7), 644–651.

Caillaud, C., Akli, S., *et al.* (1993). Adenoviral vector as a gene delivery system into cultured rat neuronal and glial cells. *Eur. J. Neurosci.* **5**(10), 1287–1291.

Card, J. P., Whealy, M. E., *et al.* (1991). Two alpha-herpesvirus strains are transported differentially in the rodent visual system. *Neuron* **6**(6), 957–969.

Card, J. P., Whealy, M. E., *et al.* (1992). Pseudorabies virus envelope glycoprotein gI influences both neurotropism and virulence during infection of the rat visual system. *J. Virol.* **66**(5), 3032–3041.

Chen, H. H., Mack, L. M., *et al.* (1997). Persistence in muscle of an adenoviral vector that lacks all viral genes. *Proc. Natl. Acad. Sci. USA* **94**(5), 1645–1650.

Chen, H. H., Mack, L. M., *et al.* (1999). DNA from both high-capacity and first-generation adenoviral vectors remains intact in skeletal muscle. *Hum. Gene. Ther.* **10**(3), 365–373.

Chen, S. H., Shine, H. D., *et al.* (1994). Gene therapy for brain tumors: Regression of experimental gliomas by adenovirus-mediated gene transfer in vivo. *Proc. Natl. Acad. Sci. USA* **91**(8), 3054–3057.

Chen, S. H., Kosai, K., *et al.* (1996). Combination suicide and cytokine gene therapy for hepatic metastases of colon carcinoma: Sustained antitumor immunity prolongs animal survival. *Cancer Res.* **56**(16), 3758–3762.

Chirmule, N., Propert, K., *et al.* (1999). Immune responses to adenovirus and adeno-associated virus in humans. *Gene Ther.* **6**(9), 1574–1583.

Choi-Lundberg, D. L., Lin, Q., *et al.* (1997). Dopaminergic neurons protected from degeneration by GDNF gene therapy. *Science* **275**(5301), 838–841.

Choi-Lundberg, D. L., Lin, Q., *et al.* (1998). Behavioral and cellular protection of rat dopaminergic neurons by an adenoviral vector encoding glial cell line-derived neurotrophic factor. *Exp. Neurol.* **154**(2), 261–275.

Chou, S. M., Roos, R., *et al.* (1973). Subacute focal adenovirus encephalitis. *J. Neuropathol. Exp. Neurol.* **32**(1), 34–50.

Connor, B., Kozlowski, D. A., *et al.* (1999). Differential effects of glial cell line-derived neurotrophic factor (GDNF) in the striatum and substantia nigra of the aged Parkinsonian rat. *Gene Ther.* **6**(12), 1936–1951.

Connor, B., Kozlowski, D. A., *et al.* (2001). Glial cell line-derived neurotrophic factor (GDNF) gene delivery protects dopaminergic terminals from degeneration. *Exp. Neurol.* **169**(1), 83–95.

Consiglio, A., Quattrini, A., *et al.* (2001). In vivo gene therapy of metachromatic leukodystrophy by lentiviral vectors: Correction of neuropathology and protection against learning impairments in affected mice. *Nat. Med.* **7**(3), 310–316.

Corti, O., Sabate, O., *et al.* (1999a). A single adenovirus vector mediates doxycycline-controlled expression of tyrosine hydroxylase in brain grafts of human neural progenitors. *Nat. Biotechnol.* **17**(4), 349–354.

Corti, O., Sanchez-Capelo, A., *et al.* (1999b). Long-term doxycycline-controlled expression of human tyrosine hydroxylase after direct adenovirus-mediated gene transfer to a rat model of Parkinson's disease. *Proc. Natl. Acad. Sci. USA* **96**(21), 12120–12125.

Cotten, M., Wagner, E., *et al.* (1993). Chicken adenovirus (CELO virus) particles augment receptor-mediated DNA delivery to mammalian cells and yield exceptional levels of stable transformants. *J. Virol.* **67**(7), 3777–3785.

Cregan, S. P., MacLaurin, J., *et al.* (2000). Helper-dependent adenovirus vectors: Their use as a gene delivery system to neurons. *Gene Ther.* **7**(14), 1200–1209.

Curran, M. A., and Nolan, G. P. (2002). Nonprimate lentiviral vectors. *Curr. Top. Microbiol. Immunol.* **261,** 75–105.

Davidson, B. L., Allen, E. D., *et al.* (1993). A model system for in vivo gene transfer into the central nervous system using an adenoviral vector. *Nat. Genet.* **3**(3), 219–223.

Davidson, B. L., Doran, S. E., *et al.* (1994). Expression of Escherichia coli beta-galactosidase and rat HPRT in the CNS of Macaca mulatta following adenoviral mediated gene transfer. *Exp. Neurol.* **125**(2), 258–267.

Davis, D., Henslee, P. J., *et al.* (1988). Fatal adenovirus meningoencephalitis in a bone marrow transplant patient. *Ann. Neurol.* **23**(4), 385–389.

De Palma, M., and Naldini, L. (2002). Transduction of a gene expression cassette using advanced generation lentiviral vectors. *Methods Enzymol.* **346,** 514–529.

DeFalco, J., Tomishima, M., *et al.* (2001). Virus-assisted mapping of neural inputs to a feeding center in the hypothalamus. *Science* **291**(5513), 2608–2613.

Deglon, N., and Aebischer, P. (2002). Lentiviruses as vectors for CNS diseases. *Curr. Top. Microbiol. Immunol.* **261,** 191–209.

Dewey, R. A., Morrissey, G., *et al.* (1999). Chronic brain inflammation and persistent herpes simplex virus 1 thymidine kinase expression in survivors of syngeneic glioma treated by adenovirus-mediated gene therapy: Implications for clinical trials. *Nat. Med.* **5**(11), 1256–1263.

Doronin, K., Toth, K., *et al.* (2000). Tumor-specific, replication-competent adenovirus vectors overexpressing the adenovirus death protein. *J. Virol.* **74**(13), 6147–6155.

Doronin, K., Kuppuswamy, M., *et al.* (2001). Tissue-specific, tumor-selective, replication-competent adenovirus vector for cancer gene therapy. *J. Virol.* **75**(7), 3314–3324.

Eloit, M., Gilardi-Hebenstreit, P., *et al.* (1990). Construction of a defective adenovirus vector expressing the pseudorabies virus glycoprotein gp50 and its use as a live vaccine. *J. Gen. Virol.* **71**(pt. 10), 2425–2431.

Engel, J. P. (1995). Viral upper respiratory infections. *Semin. Respir. Infect.* **10**(1), 3–13.

Engelhardt, J. F., Simon, R. H., *et al.* (1993). Adenovirus-mediated transfer of the CFTR gene to lung of nonhuman primates: Biological efficacy study. *Hum. Gene Ther.* **4**(6), 759–769.

Engelhardt, J. F., Ye, X., *et al.* (1994). Ablation of E2A in recombinant adenoviruses improves transgene persistence and decreases inflammatory response in mouse liver. *Proc. Natl. Acad. Sci. USA* **91**(13), 6196–6200.

Enquist, L. W., Husak, P. J., *et al.* (1998). Infection and spread of alphaherpesviruses in the nervous system. *Adv. Virus Res.* **51,** 237–347.

Fisher, L. J., and Gage, F. H. (1993). Grafting in the mammalian central nervous system. *Physiol. Rev.* **73**(3), 583–616.

Flint, S. J., Enquist, L. W., *et al.* (2000). "Principle of Virology: Molecular Biology, Pathogenesis, and Control." *Principles of Virology*. ASM Press, Washington, DC.

Fu, H., Samulski, R. J., *et al.* (2002). Neurological correction of lysosomal storage in a mucopolysaccharidosis IIIB mouse model by adeno-associated virus-mediated gene delivery. *Mol. Ther.* **5**(1), 42–49.

Galimi, F., and Verma, I. M. (2002). Opportunities for the use of lentiviral vectors in human gene therapy. *Curr. Top. Microbiol. Immunol.* **261,** 245–254.

Gallichan, W. S., Johnson, D. C., *et al.* (1993). Mucosal immunity and protection after intranasal immunization with recombinant adenovirus expressing herpes simplex virus glycoprotein B. *J. Infect. Dis.* **168**(3), 622–629.

Geddes, B. J., Harding, T. C., *et al.* (1997). Long-term gene therapy in the CNS: Reversal of hypothalamic diabetes insipidus in the Brattleboro rat by using an adenovirus expressing arginine vasopressin. *Nat. Med.* **3**(12), 1402–1404.

Gerdes, C. A., Castro, M. G., *et al.* (2000). Strong promoters are the key to highly efficient, noninflammatory and noncytotoxic adenoviral-mediated transgene delivery into the brain in vivo. *Mol. Ther.* **2**(4), 330–338.

Gilardi, P., Courtney, M., *et al.* (1990). Expression of human alpha 1-antitrypsin using a recombinant adenovirus vector. *FEBS Lett.* **267**(1), 60–62.

Ginsberg, H. S. (1999). The life and times of adenoviruses. *Adv. Virus Res.* **54,** 1–13.

Ginsberg, H. S., and Prince, G. A. (1994). The molecular basis of adenovirus pathogenesis. *Infect. Agents Dis.* **3**(1), 1–8.

Ginsberg, H. S., Moldawer, L. L., *et al.* (1991). A mouse model for investigating the molecular pathogenesis of adenovirus pneumonia. *Proc. Natl. Acad. Sci. USA* **88**(5), 1651–1655.

Glover, C. P., Bienemann, A. S., *et al.* (2002). Adenoviral-mediated, high-level, cell-specific transgene expression: A SYN1-WPRE cassette mediates increased transgene expression with no loss of neuron specificity. *Mol. Ther.* **5**(5, pt. 1), 509–516.

Grable, M., and Hearing, P. (1992). cis and trans requirements for the selective packaging of adenovirus type 5 DNA. *J. Virol.* **66**(2), 723–731.

Graham, F. L., Smiley, J., *et al.* (1977). Characteristics of a human cell line transformed by DNA from human adenovirus type 5. *J. Gen. Virol.* **36**(1), 59–74.

Greber, U. F., Willetts, M., *et al.* (1993). Stepwise dismantling of adenovirus 2 during entry into cells. *Cell* **75**(3), 477–486.

Grubb, B. R., Pickles, R. J., *et al.* (1994). Inefficient gene transfer by adenovirus vector to cystic fibrosis airway epithelia of mice and humans. *Nature* **371**(6500), 802–806.

Gupta, A., Ho, D. Y., *et al.* (2001). Neuroprotective effects of an adenoviral vector expressing the glucose transporter: A detailed description of the mediating cellular events. *Brain Res.* **908**(1), 49–57.

Hackett, N. R., Kaminsky, S. M., *et al.* (2000). Antivector and antitransgene host responses in gene therapy. *Curr. Opin. Mol. Ther.* **2**(4), 376–382.

Haddada, H., Lopez, M., *et al.* (1993). Efficient adenovirus-mediated gene transfer into human blood monocyte-derived macrophages. *Biochem. Biophys. Res. Commun.* **195**(3), 1174–1783.

Harding, T. C., Geddes, B. J., *et al.* (1998). Switching transgene expression in the brain using an adenoviral tetracycline-regulatable system. *Nat. Biotechnol.* **16**(6), 553–555.

Hardy, S., Kitamura, M., *et al.* (1997). Construction of adenovirus vectors through Cre-lox recombination. *J. Virol.* **71**(3), 1842–1849.

Hartigan-O'Connor, D., Barjot, C., *et al.* (2002). Generation and growth of gutted adenoviral vectors. *Methods Enzymol.* **346,** 224–246.

Hayashi, Y., DePaoli, A. M., *et al.* (1994). Expression of a thyroid hormone-responsive recombinant gene introduced into adult mice livers by replication-defective adenovirus can be regulated by endogenous thyroid hormone receptor. *J. Biol. Chem.* **269**(39), 23872–23875.

Herrlinger, U., Jacobs, A., *et al.* (2000). Helper virus-free herpes simplex virus type 1 amplicon vectors for granulocyte-macrophage colony-stimulating factor-enhanced vaccination therapy for experimental glioma. *Hum. Gene Ther.* **11**(10), 1429–1438.

Herz, J., and Gerard, R. D. (1993). Adenovirus-mediated transfer of low density lipoprotein receptor gene acutely accelerates cholesterol clearance in normal mice. *Proc. Natl. Acad. Sci. USA* **90**(7), 2812–2816.

Hida, H., Hashimoto, M., *et al.* (1999). Dopa-producing astrocytes generated by adenoviral transduction of human tyrosine hydroxylase gene: In vitro study and transplantation to hemiparkinsonian model rats. *Neurosci. Res.* **35**(2), 101–112.

Holloway, S., Mennie, M., *et al.* (1994). Predictive testing for Huntington disease: Social characteristics and knowledge of applicants, attitudes to the test procedure and decisions made after testing. *Clin. Genet.* **46**(2), 175–180.

Horellou, P., Vigne, E., *et al.* (1994). Direct intracerebral gene transfer of an adenoviral vector expressing tyrosine hydroxylase in a rat model of Parkinson's disease. *Neuroreport* **6**(1), 49–53.

Hsich, G., Sena-Esteves, M., *et al.* (2002). Critical issues in gene therapy for neurologic disease. *Hum. Gene Ther.* **13**(5), 579–604.

Huentelman, M. J., Reaves, P. Y., *et al.* (2002). Large-scale production of retroviral vectors for systemic gene delivery. *Methods Enzymol.* **346,** 562–573.

Ikeda, K., Wakimoto, H., *et al.* (2000). Complement depletion facilitates the infection of multiple brain tumors by an intravascular, replication-conditional herpes simplex virus mutant. *J. Virol.* **74**(10), 4765–4775.

Imperiale, M. J., Kao, H. T., *et al.* (1984). Common control of the heat shock gene and early adenovirus genes: Evidence for a cellular E1A-like activity. *Mol. Cell. Biol.* **4**(5), 867–874.

Ishibashi, S., Brown, M. S. I., *et al.* (1993). Hypercholesterolemia in low density lipoprotein receptor knockout mice and its reversal by adenovirus-mediated gene delivery. *J. Clin. Invest.* **92**(2), 883–893.

Jacobs, A., Tjuvajev, J. G., *et al.* (2001). Positron emission tomography-based imaging of transgene expression mediated by replication-conditional, oncolytic herpes simplex virus type 1 mutant vectors in vivo. *Cancer Res.* **61**(7), 2983–2995.

Jacobs, S. C., Stephenson, J. R., *et al.* (1992). High-level expression of the tick-borne encephalitis virus NS1 protein by using an adenovirus-based vector: Protection elicited in a murine model. *J. Virol.* **66**(4), 2086–2095.

Jomary, C., Piper, T. A., *et al.* (1994). Adenovirus-mediated gene transfer to murine retinal cells in vitro and in vivo. *FEBS Lett.* **347**(2–3), 117–122.

Kajiwara, K., Byrnes, A. P., *et al.* (1997). Immune responses to adenoviral vectors during gene transfer in the brain. *Hum. Gene Ther.* **8**(3), 253–265.

Kaplitt, M. G., Leone, P., *et al.* (1994). Long-term gene expression and phenotypic correction using adeno-associated virus vectors in the mammalian brain. *Nat. Genet.* **8**(2), 148–154.

Kass-Eisler, A., Falck-Pedersen, E., *et al.* (1993). Quantitative determination of adenovirus-mediated gene delivery to rat cardiac myocytes in vitro and in vivo. *Proc. Natl. Acad. Sci. USA* **90**(24), 11498–11502.

Kass-Eisler, A., Falck-Pedersen, E., *et al.* (1994). The impact of developmental stage, route of administration and the immune system on adenovirus-mediated gene transfer. *Gene Ther.* **1**(6), 395–402.

Kay, M. A., Glorioso, J. C., *et al.* (2001). Viral vectors for gene therapy: The art of turning infectious agents into vehicles of therapeutics. *Nat. Med.* **7**(1), 33–40.

Kelsey, D. S. (1978). Adenovirus meningoencephalitis. *Pediatrics* **61**(2), 291–293.

Kingston, P. A., Sinha, S., *et al.* (2001). Adenovirus-mediated gene transfer of a secreted transforming growth factor-beta type II receptor inhibits luminal loss and constrictive remodeling after coronary angioplasty and enhances adventitial collagen deposition. *Circulation* **104**(21), 2595–2601.

Kirn, D., Martuza, R. L., *et al.* (2001). Replication-selective virotherapy for cancer: Biological principles, risk management and future directions. *Nat. Med.* **7**(7), 781–787.

Kochanek, S., Clemens, P. R., *et al.* (1996). A new adenoviral vector: Replacement of all viral coding sequences with 28 kb of DNA independently expressing both full-length dystrophin and beta-galactosidase. *Proc. Natl. Acad. Sci. USA* **93**(12), 5731–5736.

Kozarsky, K. F., McKinley, D. R., *et al.* (1994). In vivo correction of low density lipoprotein receptor deficiency in the Watanabe heritable hyperlipidemic rabbit with recombinant adenoviruses. *J. Biol. Chem.* **269**(18), 13695–13702.

Kozlowski, D. A., Connor, B., *et al.* (2000). Delivery of a GDNF gene into the substantia nigra after a progressive 6-OHDA lesion maintains functional nigrostriatal connections. *Exp. Neurol.* **166**(1), 1–15.

Kozlowski, D. A., Bremer, E., *et al.* (2001). Quantitative analysis of transgene protein, mRNA, and vector DNA following injection of an adenoviral vector harboring glial cell line-derived neurotrophic factor into the primate caudate nucleus. *Mol. Ther.* **3**(2), 256–261.

Kramm, C. M., Chase, M., *et al.* (1997). Therapeutic efficiency and safety of a second-generation replication-conditional HSV1 vector for brain tumor gene therapy. *Hum. Gene Ther.* **8**(17), 2057–2068.

Latchman, D. S., and Coffin, R. S. (2001). Viral vectors for gene therapy in Parkinson's disease. *Rev. Neurosci.* **12**(1), 69–78.

Lawrence, M. S., Foellmer, H. G., *et al.* (1999). Inflammatory responses and their impact on beta-galactosidase transgene expression following adenovirus vector delivery to the primate caudate nucleus. *Gene Ther.* **6**(8), 1368–1379.

Le Gal La Salle, G., Robert, J. J., *et al.* (1993). An adenovirus vector for gene transfer into neurons and glia in the brain. *Science* **259**(5097), 988–990.

Legrand, V., Leissner, P., *et al.* (2002). Transductional targeting with recombinant adenovirus vectors. *Curr. Gene Ther.* **2**(3), 323–339.

Lemarchand, P., Jaffe, H. A., *et al.* (1992). Adenovirus-mediated transfer of a recombinant human alpha 1-antitrypsin cDNA to human endothelial cells. *Proc. Natl. Acad. Sci. USA* **89**(14), 6482–6486.

Li, T., Adamian, M., *et al.* (1994). In vivo transfer of a reporter gene to the retina mediated by an adenoviral vector. *Invest. Ophthalmol. Vis. Sci.* **35**(5), 2543–2549.

Lilley, C. E., Branston, R. H., *et al.* (2001). Herpes simplex virus vectors for the nervous system. *Curr. Gene Ther.* **1**(4), 339–358.

Lisovoski, F., Cadusseau, J., *et al.* (1994). In vivo transfer of a marker gene to study motoneuronal development. *Neuroreport* **5**(9), 1069–1072.

Logvinoff, C., and Epstein, A. L. (2001). A novel approach for herpes simplex virus type 1 amplicon vector production, using the Cre-loxP recombination system to remove helper virus. *Hum. Gene Ther.* **12**(2), 161–167.

Lokensgard, J. R., Bloom, D. C., *et al.* (1994). Long-term promoter activity during herpes simplex virus latency. *J. Virol.* **68**(11), 7148–7158.

Lowenstein, P. R. (1994). Molecular neurosurgery: Mending the broken brain. *Biotechnology* **12**(11), 1075–1080.

Lowenstein, P. R. (1995). Degenerative and inherited neurological disorders. *In* "Molecular and Cell Biology of Human Gene Therapeutics" (G. Dickson, Ed.), pp. 301–349. Chapman and Hall, London.

Lowenstein, P. R. (2002). Immunology of viral-vector-mediated gene transfer into the brain: An evolutionary and developmental perspective. *Trends Immunol.* **23**(1), 23–30.

Lowenstein, P. R., and Castro, M. G. (2001). Genetic engineering within the adult brain: Implications for molecular approaches to behavioral neuroscience. *Physiol. Behav.* **73**(5), 833–839.

Lowenstein, P. R., and Castro, M. G. (2002). Progress and challenges in viral vector-mediated gene transfer to the brain. *Curr. Opin. Mol. Ther.* **4**(4), 359–371.

Lowenstein, P. R., and Enquist, L. W. (1996). "Protocols for Gene Transfer into Neurons: Towards Gene Therapy of Neurological Disorders." John Wiley & Sons, Chichester, UK.

Lowenstein, P. R., Thomas, C. E., *et al.* (2002). High-capacity, helper-dependent, gutless adenoviral vectors for gene transfer into brain. *Methods Enzymol.* **346,** 292–311.

Macmillan, J. C., and Lowenstein, P. R. (1994). "Gene Therapy Approaches to the Treatment of Human Neurological Disease: The Future of Neurological Therapies?" *Protocols for Gene Transfer in Nueroscience: Towards Gene Therapy of Neurological Disorders.* Lowenstein, P. R., and Enquist, L. W., Eds., John Wiley and Sons, Chichester, UK.

Mahr, J. A., and Gooding, L. R. (1999). Immune evasion by adenoviruses. *Immunol. Rev.* **168,** 121–130.

Maione, D., Della Rocca, C., *et al.* (2001). An improved helper-dependent adenoviral vector allows persistent gene expression after intramuscular delivery and overcomes preexisting immunity to adenovirus. *Proc. Natl. Acad. Sci. USA* **98**(11), 5986–5991.

Markert, J. M., Malick, A., *et al.* (1993). Reduction and elimination of encephalitis in an experimental glioma therapy model with attenuated herpes simplex mutants that retain susceptibility to acyclovir. *Neurosurgery* **32**(4), 597–603.

Markert, J. M., Medlock, M. D., *et al.* (2000). Conditionally replicating herpes simplex virus mutant, G207 for the treatment of malignant glioma: Results of a phase I trial. *Gene Ther.* **7**(10), 867–874.

Martuza, R. L., Malick, A., *et al.* (1991). Experimental therapy of human glioma by means of a genetically engineered virus mutant. *Science* **252**(5007), 854–856.

Mastrangeli, A., Danel, C., *et al.* (1993). Diversity of airway epithelial cell targets for in vivo recombinant adenovirus-mediated gene transfer. *J. Clin. Invest.* **91**(1), 225–234.

Matyszak, M. K. (1998). Inflammation in the CNS: Balance between immunological privilege and immune responses. *Prog. Neurobiol.* **56**(1), 19–35.

Matyszak, M. K., and Perry, V. H. (1996). A comparison of leucocyte responses to heat-killed bacillus Calmette-Guerin in different CNS compartments. *Neuropathol. Appl. Neurobiol.* **22**(1), 44–53.

Mazarakis, N. D., Azzouz, M., *et al.* (2001). Rabies virus glycoprotein pseudotyping of lentiviral vectors enables retrograde axonal transport and access to the nervous system after peripheral delivery. *Hum. Mol. Genet.* **10**(19), 2109–2121.

Millecamps, S., Kiefer, H., *et al.* (1999). Neuron-restrictive silencer elements mediate neuron specificity of adenoviral gene expression. *Nat. Biotechnol.* **17**(9), 865–869.

Mitani, K., Graham, F. L., *et al.* (1995). Rescue, propagation, and partial purification of a helper virus-dependent adenovirus vector. *Proc. Natl. Acad. Sci. USA* **92**(9), 3854–3858.

Mittereder, N., Yei, S., *et al.* (1994). Evaluation of the efficacy and safety of in vitro, adenovirus-mediated transfer of the human cystic fibrosis transmembrane conductance regulator cDNA. *Hum. Gene Ther.* **5**(6), 717–729.

Morelli, A. E., Larregina, A. T., *et al.* (1999). Neuronal and glial cell type-specific promoters within adenovirus recombinants restrict the expression of the apoptosis-inducing molecule Fas ligand to predetermined brain cell types, and abolish peripheral liver toxicity. *J. Gen. Virol.* **80**(pt. 3), 571–583.

Morin, J. E., Lubeck, M. D., *et al.* (1987). Recombinant adenovirus induces antibody response to hepatitis B virus surface antigen in hamsters. *Proc. Natl. Acad. Sci. USA* **84**(13), 4626–4630.

Morral, N., O'Neal, W., *et al.* (1997). Immune responses to reporter proteins and high viral dose limit duration of expression with adenoviral vectors: Comparison of E2a wild type and E2a deleted vectors. *Hum. Gene Ther.* **8**(10), 1275–1286.

Morral, N., Parks, R. J., *et al.* (1998). High doses of a helper-dependent adenoviral vector yield supraphysiological levels of alpha1-antitrypsin with negligible toxicity. *Hum. Gene Ther.* **9**(18), 2709–2716.

Morsy, M. A., Gu, M., *et al.* (1998). An adenoviral vector deleted for all viral coding sequences results in enhanced safety and extended expression of a leptin transgene. *Proc. Natl. Acad. Sci. USA* **95**(14), 7866–7871.

Muchardt, C., Li, C., *et al.* (1990). CREB regulation of cellular cyclic AMP-responsive and adenovirus early promoters. *J. Virol.* **64**(9), 4296–4305.

Muramatsu, S., Fujimoto, K., *et al.* (2002). Behavioral recovery in a primate model of Parkinson's disease by triple transduction of striatal cells with adeno-associated viral vectors expressing dopamine-synthesizing enzymes. *Hum. Gene Ther.* **13**(3), 345–354.

Nalbantoglu, J., Gilfix, B. M., *et al.* (1994). Predictive value of apolipoprotein E genotyping in Alzheimer's disease: Results of an autopsy series and an analysis of several combined studies. *Ann. Neurol.* **36**(6), 889–895.

Naldini, L., Blomer, U., *et al.* (1996). In vivo gene delivery and stable transduction of nondividing cells by a lentiviral vector. *Science* **272**(5259), 263–267.

Nanda, D., Vogels, R., *et al.* (2001). Treatment of malignant gliomas with a replicating adenoviral vector expressing herpes simplex virus-thymidine kinase. *Cancer Res.* **61**(24), 8743–8750.

Ng, P., Beauchamp, C., *et al.* (2001). Development of a FLP/frt system for generating helper-dependent adenoviral vectors. *Mol. Ther.* **3**(5 pt. 1), 809–815.

Ng, P., Parks, R. J., *et al.* (2002). Preparation of helper-dependent adenoviral vectors. *Methods Mol. Med.* **69,** 371–388.

Nicklin, S. A., and Baker, A. H. (2002). Tropism-modified adenoviral and adeno-associated viral vectors for gene therapy. *Curr. Gene Ther.* **2**(3), 273–293.

Osamura, T., Mizuta, R., *et al.* (1993). Isolation of adenovirus type 11 from the brain of a neonate with pneumonia and encephalitis. *Eur. J. Pediatr.* **152**(6), 496–499.

Oualikene, W., Gonin, P., *et al.* (1994). Short and long term dissemination of deletion mutants of adenovirus in permissive (cotton rat) and non-permissive (mouse) species. *J. Gen. Virol.* **75**(pt. 10), 2765–2768.

Palu, G., Cavaggioni, A., *et al.* (1999). Gene therapy of glioblastoma multiforme via combined expression of suicide and cytokine genes: A pilot study in humans. *Gene Ther.* **6**(3), 330–337.

Papanastassiou, V., Rampling, R., *et al.* (2002). The potential for efficacy of the modified (ICP 34.5(-)) herpes simplex virus HSV1716 following intratumoural injection into human malignant glioma: A proof of principle study. *Gene Ther.* **9**(6), 398–406.

Parks, R. J., Chen, L., *et al.* (1996). A helper-dependent adenovirus vector system: Removal of helper virus by Cre-mediated excision of the viral packaging signal. *Proc. Natl. Acad. Sci. USA* **93**(24), 13565–13570.

Parks, R., Evelegh, C., *et al.* (1999a). Use of helper-dependent adenoviral vectors of alternative serotypes permits repeat vector administration. *Gene Ther.* **6**(9), 1565–1573.

Parks, R. J., Bramson, J. L., *et al.* (1999b). Effects of stuffer DNA on transgene expression from helper-dependent adenovirus vectors. *J. Virol.* **73**(10), 8027–8034.

Pechan, P. A., Herrlinger, U., *et al.* (1999). Combined HSV-1 recombinant and amplicon piggyback vectors: Replication-competent and defective forms, and therapeutic efficacy for experimental gliomas. *J. Gene Med.* **1**(3), 176–185.

Perez-Cruet, M. J., Trask, T. W., *et al.* (1994). Adenovirus-mediated gene therapy of experimental gliomas. *J. Neurosci. Res.* **39**(4), 506–511.

Perry, V. H. (1998). A revised view of the central nervous system microenvironment and major histocompatibility complex class II antigen presentation. *J. Neuroimmunol.* **90**(2), 113–121.

Phillips, M. I. (Ed.) (2002). Gene therapy methods. *Methods Enzymol.* **346,** 1–728.

Prevec, L., Schneider, M., *et al.* (1989). Use of human adenovirus-based vectors for antigen expression in animals. *J. Gen. Virol.* **70**(pt. 2), 429–434.

Quantin, B., Perricaudet, L. D., *et al.* (1992). Adenovirus as an expression vector in muscle cells in vivo. *Proc. Natl. Acad. Sci. USA* **89**(7), 2581–2584.

Rainov, N. G. (2000). A phase III clinical evaluation of herpes simplex virus type 1 thymidine kinase and ganciclovir gene therapy as an adjuvant to surgical resection and radiation in adults with previously untreated glioblastoma multiforme. *Hum. Gene Ther.* **11**(17), 2389–2401.

Rainov, N. G., and Kramm, C. M. (2001). Vector delivery methods and targeting strategies for gene therapy of brain tumors. *Curr. Gene Ther.* **1**(4), 367–383.

Ralph, G. S., Bienemann, A., *et al.* (2000). Targeting of tetracycline-regulatable transgene expression specifically to neuronal and glial cell populations using adenoviral vectors. *Neuroreport* **11**(9), 2051–2055.

Ram, Z., Culver, K. W., *et al.* (1993). In situ retroviral-mediated gene transfer for the treatment of brain tumors in rats. *Cancer Res.* **53**(1), 83–88.

Ram, Z., Culver, K. W., *et al.* (1997). Therapy of malignant brain tumors by intratumoral implantation of retroviral vector-producing cells. *Nat. Med.* **3**(12), 1354–1361.

Rampling, R., Cruickshank, G., *et al.* (1998). Therapeutic replication-competent herpes virus. *Nat. Med.* **4**(2), 133.

Rampling, R., Cruickshank, G., *et al.* (2000). Toxicity evaluation of replication-competent herpes simplex virus (ICP 34.5 null mutant 1716) in patients with recurrent malignant glioma. *Gene Ther.* **7**(10), 859–866.

Raper, S. E., Yudkoff, M., *et al.* (2002). A pilot study of in vivo liver-directed gene transfer with an adenoviral vector in partial ornithine transcarbamylase deficiency. *Hum. Gene Ther.* **13**(1), 163–175.

Ridoux, V., Robert, J. J., *et al.* (1994a). Adenoviral vectors as functional retrograde neuronal tracers. *Brain Res.* **648**(1), 171–175.

Ridoux, V., Robert, J. J., *et al.* (1994). The use of adenovirus vectors for intracerebral grafting of transfected nervous cells. *Neuroreport* **5**(7), 801–804.

Roelvink, P. W., Mi Lee, G., *et al.* (1999). Identification of a conserved receptor-binding site on the fiber proteins of CAR-recognizing adenoviridae. *Science* **286**(5444), 1568–1571.

Roessler, B. J., Allen, E. D., *et al.* (1993). Adenoviral-mediated gene transfer to rabbit synovium in vivo. *J. Clin. Invest.* **92**(2), 1085–1092.

Rosenfeld, M. A., Siegfried, W., *et al.* (1991). Adenovirus-mediated transfer of a recombinant alpha 1-antitrypsin gene to the lung epithelium in vivo. *Science* **252**(5004), 431–434.

Rosenfeld, M. A., Yoshimura, K., *et al.* (1992). In vivo transfer of the human cystic fibrosis transmembrane conductance regulator gene to the airway epithelium. *Cell* **68**(1), 143–155.

Ross, D., and Ziff, E. (1992). Defective synthesis of early region 4 mRNAs during abortive adenovirus infections in monkey cells. *J. Virol.* **66**(5), 3110–3117.

Rubin, B. A., and Rorke, L. B. (1994). Adenovirus vaccines.. *In* "Vaccines" (S. A. a. M. Plotkin, Jr. , *et al.*, Ed.), pp. 475–502. W. B. Saunders Company, Philadelphia.

Rutanen, J., Markkanen, J., *et al.* (2002). Gene therapy for restenosis: current status. *Drugs* **62**(11), 1575–1585.

Sambrook, J., Fritsch, E. F., and Maniatis, T. (1989). "Molecular Cloning." Cold Spring Harbor Publishers, Cold Spring Harbor, NY.

Sandig, V., Youil, R., *et al.* (2000). Optimization of the helper-dependent adenovirus system for production and potency in vivo. *Proc. Natl. Acad. Sci. USA* **97**(3), 1002–1007.

Sandmair, A. M., Loimas, S., *et al.* (2000). Thymidine kinase gene therapy for human malignant glioma, using replication-deficient retroviruses or adenoviruses. *Hum. Gene Ther.* **11**(16), 2197–2205.

Sanlioglu, S., Monick, M. M., *et al.* (2001). Rate limiting steps of AAV transduction and implications for human gene therapy. *Curr. Gene Ther.* **1**(2), 137–147.

Saunders, A. M., Strittmatter, W. J., *et al.* (1993). Association of apolipoprotein E allele epsilon 4 with late-onset familial and sporadic Alzheimer's disease. *Neurology* **43**(8), 1467–1472.

Schadeck, E. B., Partidos, C. D., *et al.* (1999). CTL epitopes identified with a defective recombinant adenovirus expressing measles virus nucleoprotein and evaluation of their protective capacity in mice. *Virus Res.* **65**(1), 75–86.

Schiedner, G., Morral, N., *et al.* (1998). Genomic DNA transfer with a high-capacity adenovirus vector results in improved in vivo gene expression and decreased toxicity. *Nat. Genet.* **18**(2), 180–183.

Schwartz, A. R., Togo, Y., *et al.* (1974). Clinical evaluation of live, oral types 1, 2, and 5 adenovirus vaccines. *Annu. Rev. Respir. Dis.* **109**(2), 233–239.

Scinto, L. F., Daffner, K. R., *et al.* (1994). A potential noninvasive neurobiological test for Alzheimer's disease. *Science* **266**(5187), 1051–1054.

Segerman, A., Mei, Y. F., *et al.* (2000). Adenovirus types 11p and 35p show high binding efficiencies for committed hematopoietic cell lines and are infective to these cell lines. *J. Virol.* **74**(3), 1457–1467.

Seiger, A., Nordberg, A., *et al.* (1993). Intracranial infusion of purified nerve growth factor to an Alzheimer patient: The first attempt of a possible future treatment strategy. *Behav. Brain Res.* **57**(2), 255–261.

Shayakhmetov, D. M., Papayannopoulou, T., *et al.* (2000). Efficient gene transfer into human CD34(+) cells by a retargeted adenovirus vector. *J. Virol.* **74**(6), 2567–2583.

Shenk, T. (2001). Adenoviridae: The viruses and their replication. *In* "Virology" (D. Krupe and D. M. Howley, Eds.). Raven Press, New York.

Shering, A. F., Bain, D., *et al.* (1997). Cell type-specific expression in brain cell cultures from a short human cytomegalovirus major immediate early promoter depends on whether it is inserted into herpesvirus or adenovirus vectors. *J. Gen. Virol.* **78**(pt. 2), 445–459.

Shewach, D. S., Zerbe, L. K., *et al.* (1994). Enhanced cytotoxicity of antiviral drugs mediated by adenovirus directed transfer of the herpes simplex virus thymidine kinase gene in rat glioma cells. *Cancer Gene Ther.* **1**(2), 107–112.

Shinoura, N., Ohashi, M., *et al.* (2000). Adenovirus-mediated overexpression of Fas induces apoptosis of gliomas. *Cancer Gene Ther.* **7**(2), 224–232.

Shinoura, N., Yoshida, Y., *et al.* (2000). Adenovirus-mediated transfer of p53 and Fas ligand drastically enhances apoptosis in gliomas. *Cancer Gene Ther.* **7**(5), 732–738.

Shono, T., Tofilon, P. J., *et al.* (2002). Apoptosis induced by adenovirus-mediated p53 gene transfer in human glioma correlates with site-specific phosphorylation. *Cancer Res.* **62**(4), 1069–1076.

Simon, R. H., Engelhardt, J. F., *et al.* (1993). Adenovirus-mediated transfer of the CFTR gene to lung of nonhuman primates: toxicity study. *Hum. Gene Ther.* **4**(6), 771–780.

Smith, T. A., Mehaffey, M. G., *et al.* (1993). Adenovirus mediated expression of therapeutic plasma levels of human factor IX in mice. *Nat. Genet.* **5**(4), 397–402.

Smith-Arica, J. R., Morelli, A. E., *et al.* (2000). Cell-type-specific and regulatable transgenesis in the adult brain: Adenovirus-encoded combined transcriptional targeting and inducible transgene expression. *Mol. Ther.* **2**(6), 579–587.

Smith-Arica, J. R., Williams, J. C., *et al.* (2001). Switching on and off transgene expression within lactotrophic cells in the anterior pituitary gland in vivo. *Endocrinology* **142**(6), 2521–2532.

Southgate, T. D., Bain, D., Fairbanks, L. D., Morelli, A. E., Larrengina, A. T., Simmonds, H. A., Castro, M. G., Lowenstein, P. R. (1999). Adenoviruses encoding HPRT correct biochemical abnormalities of HPRT-deficient cells and allow their survival in negative selection medium. *Metab. Brain Dis.* **14**(4), 205–21.

Southgate, T. D., Stone, D., *et al.* (2001). Long-term transgene expression within the anterior pituitary gland in situ: Impact on circulating hormone levels, cellular and antibody-mediated immune responses. *Endocrinology* **142**(1), 464–476.

Spergel, J. M., and Chen-Kiang, S. (1991). Interleukin 6 enhances a cellular activity that functionally substitutes for E1A protein in transactivation. *Proc. Natl. Acad. Sci. USA* **88**(15), 6472–6476.

Stratford-Perricaudet, L. D., Levrero, M., *et al.* (1990). Evaluation of the transfer and expression in mice of an enzyme-encoding gene using a human adenovirus vector. *Hum. Gene Ther.* **1**(3), 241–256.

Stratford-Perricaudet, L. D., Makeh, I., *et al.* (1992). Widespread long-term gene transfer to mouse skeletal muscles and heart. *J. Clin. Invest.* **90**(2), 626–630.

Suhr, S. T., and Gage, F. H. (1999). Gene therapy in the central nervous system: The use of recombinant retroviruses. *Arch. Neurol.* **56**(3), 287–292.

Takamiya, Y., Short, M. P., *et al.* (1992). Gene therapy of malignant brain tumors: A rat glioma line bearing the herpes simplex virus type 1-thymidine kinase gene and wild type retrovirus kills other tumor cells. *J. Neurosci. Res.* **33**(3), 493–503.

Tamura, K., Tamura, M., *et al.* (2001). Eradication of murine brain tumors by direct inoculation of concentrated high titer-recombinant retrovirus harboring the herpes simplex virus thymidine kinase gene. *Gene Ther.* **8**(3), 215–222.

Tate, M. C., Shear, D. A., *et al.* (2002). Fibronection promotes survival and migration of primary neural stem cells transplanted into the traumatically injured mouse brain. *Cell Transplant* **11**(3), 283–295.

Thomas, C. E., and Abordo-Adesida, E., and Maleniak, T. C., and Stone, D., and Gerdes, G., and Lowenstein, P. R. (2000a). Gene transfer into rat brain using adenoviral vectors.. *In* "Current Protocols in Neuroscience" (J. N. Gerfen, R. McKay, M. A. Rogawski, D. R. Sibley, and P. Skolnick, Eds.), pp. 4.23.1–4.23.40. John Wiley & Sons, New York.

Thomas, C. E., Schiedner, G., *et al.* (2000b). Peripheral infection with adenovirus causes unexpected long-term brain inflammation in animals injected intracranially with first-generation, but not with high-capacity, adenovirus vectors: Toward realistic long-term neurological gene therapy for chronic diseases. *Proc. Natl. Acad. Sci. USA* **97**(13), 7482–7487.

Thomas, C. E., Birkett, D., *et al.* (2001). Acute direct adenoviral vector cytotoxicity and chronic, but not acute, inflammatory responses correlate with decreased vector-mediated transgene expression in the brain. *Mol. Ther.* **3**(1), 36–46.

Thomas, C. E., Schiedner, G., *et al.* (2001). Preexisting antiadenoviral immunity is not a barrier to efficient and stable transduction of the brain, mediated by novel high-capacity adenovirus vectors. *Hum. Gene Ther.* **12**(7), 839–846.

Thomas, C. E., Edwards, P., *et al.* (2002). Adenovirus binding to the coxsackievirus and adenovirus receptor or integrins is not required to elicit brain inflammation but is necessary to transduce specific neural cell types. *J. Virol.* **76**(7), 3452–3460.

Tominaga-Yoshino, K. (2001). Neurotoxic and neuroprotective effects of glutamate are enhanced by introduction of amyloid precursor protein cDNA. *Brain Res.* **918**(2), 121–130.

Tomko, R. P., Xu, R., *et al.* (1997). HCAR and MCAR: The human and mouse cellular receptors for subgroup C adenoviruses and group B coxsackieviruses. *Proc. Natl. Acad. Sci. USA* **94**(7), 3352–3356.

Top, F. H., Jr., Grossman, R. A., *et al.* (1971a). Immunization with live types 7 and 4 adenovirus vaccines. I. Safety, infectivity, antigenicity, and potency of adenovirus type 7 vaccine in humans. *J. Infect. Dis.* **124**(2), 148–154.

Top, F. H., Jr., Buescher, E. L., *et al.* (1971b). Immunization with live types 7 and 4 adenovirus vaccines. II. Antibody response and protective effect against acute respiratory disease due to adenovirus type 7. *J. Infect. Dis.* **124**(2), 155–160.

Trask, T. W., Trask, R. P., *et al.* (2000). Phase I study of adenoviral delivery of the HSV-tk gene and ganciclovir administration in patients with current malignant brain tumors. *Mol. Ther.* **1**(2), 195–203.

Trotman, L. C., Mosberger, N., *et al.* (2001). Import of adenovirus DNA involves the nuclear pore complex receptor CAN/Nup214 and histone H1. *Nat. Cell Biol.* **3**(12), 1092–1100.

Ugolini, G., Kuypers, H. G., *et al.* (1989). Transneuronal transfer of herpes virus from peripheral nerves to cortex and brainstem. *Science* **243**(4887), 89–91.

Umana, P., Gerdes, C. A., *et al.* (2001). Efficient FLPe recombinase enables scalable production of helper-dependent adenoviral vectors with negligible helper-virus contamination. *Nat. Biotechnol.* **19**(6), 582–585.

Wade-Martins, R., Smith, E. R., *et al.* (2001). An infectious transfer and expression system for genomic DNA loci in human and mouse cells. *Nat. Biotechnol.* **19**(11), 1067–1070.

Wadsworth, S. C., Zhou, H., *et al.* (1997). Adenovirus vector-infected cells can escape adenovirus antigen-specific cytotoxic T-lymphocyte killing in vivo. *J. Virol.* **71**(7), 5189–5196.

Wang, S., Fraefel, C., *et al.* (2002). HSV-1 amplicon vectors. *Methods Enzymol.* **346,** 593–603.

Welsh, M. J., Smith, A. E., *et al.* (1994). Cystic fibrosis gene therapy using an adenovirus vector: In vivo safety and efficacy in nasal epithelium. *Hum. Gene Ther.* **5**(2), 209–219.

Wickham, T. J. (2000). Targeting adenovirus. *Gene Ther.* **7**(2), 110–114.

Wickham, T. J., Mathias, P., *et al.* (1993). Integrins alpha v beta 3 and alpha v beta 5 promote adenovirus internalization but not virus attachment. *Cell* **73**(2), 309–319.

Wickham, T. J., Filardo, E. J., *et al.* (1994). Integrin alpha v beta 5 selectively promotes adenovirus mediated cell membrane permeabilization. *J. Cell Biol.* **127**(1), 257–264.

Wilkinson, G. W. G. (1994a). Gene therapy and viral vaccination. *Rev. Med. Microbiol.* **5,** 97–106.

Wilkinson, G. W. G. (1994b). Viral vectors for gene therapy. *In* "From Genetics to Gene Therapy" (D. Latchman, Ed.), pp. 161–192. BioScientific Publishers, London.

Wilkinson, G. W., and Borysiewicz, L. K. (1995). Gene therapy and viral vaccination: The interface. *Br. Med. Bull.* **51**(1), 205–216.

Wilson, J. M., Engelhardt, J. F., *et al.* (1994). Gene therapy of cystic fibrosis lung disease using E1 deleted adenoviruses: A phase I trial. *Hum. Gene Ther.* **5**(4), 501–519.

Wold, W. S., Hermiston, T. W., *et al.* (1994). Adenovirus proteins that subvert host defenses. *Trends Microbiol.* **2**(11), 437–443.

Wold, W. S., Doronin, K., *et al.* (1999). Immune responses to adenoviruses: Viral evasion mechanisms and their implications for the clinic. *Curr. Opin. Immunol.* **11**(4), 380–386.

Wood, M. J., Charlton, H. M., *et al.* (1996a). Immune responses to adenovirus vectors in the nervous system. *Trends Neurosci.* **19**(11), 497–501.

Wood, M. J. A., and Byrnes, A. P., and McMenamin, M., and Kajiwara, K., and Vine, A., and Gordon, I., and Lang, J., and Wood, K. J., and Charlton, H. M. (1996b). Immune responses to viruses: Practical implications for the use of viruses as vectors for experimental and clinical gene therapy. *In* "Protocols for Gene Transfer in Neuroscience" (P. R. Lowenstein, *et al.*, Ed.). John Wiley & Sons, New York.

Yang, Y., Nunes, F. A., *et al.* (1994a). Cellular immunity to viral antigens limits E1-deleted adenoviruses for gene therapy. *Proc. Natl. Acad. Sci. USA* **91**(10), 4407–4411.

Yang, Y., Nunes, F. A., *et al.* (1994b). Inactivation of E2a in recombinant adenoviruses improves the prospect for gene therapy in cystic fibrosis. *Nat. Genet.* **7**(3), 362–369.

Yei, S., Bachurski, C. J., *et al.* (1994a). Adenoviral-mediated gene transfer of human surfactant protein B to respiratory epithelial cells. *Am. J. Respir. Cell Mol. Biol.* **11**(3), 329–336.

Yei, S., Mittereder, N., *et al.* (1994b). In vivo evaluation of the safety of adenovirus-mediated transfer of the human cystic fibrosis transmembrane conductance regulator cDNA to the lung. *Hum. Gene Ther.* **5**(6), 731–744.

Zabner, J., Couture, L. A., *et al.* (1994). Correction of cAMP-stimulated fluid secretion in cystic fibrosis airway epithelia: Efficiency of adenovirus-mediated gene transfer in vitro. *Hum. Gene Ther.* **5**(5), 585–593.

Zabner, J., Chillon, M., *et al.* (1999). A chimeric type 2 adenovirus vector with a type 17 fiber enhances gene transfer to human airway epithelia. *J. Virol.* **73**(10), 8689–8695.

Zemanick, M. C., Strick, P. L., *et al.* (1991). Direction of transneuronal transport of herpes simplex virus 1 in the primate motor system is strain-dependent. *Proc. Natl. Acad. Sci. USA* **88**(18), 8048–8051.

Zermansky, A. J., Bolognani, F., *et al.* (2001). Towards global and long-term neurological gene therapy: Unexpected transgene dependent, high-level, and widespread distribution of HSV-1 thymidine kinase throughout the CNS. *Mol. Ther.* **4**(5), 490–498.

Zhou, H., Zhao, T., *et al.* (2001). A Cre-expressing cell line and an E1/E2a double-deleted virus for preparation of helper-dependent adenovirus vector. *Mol. Ther.* **3**(4), 613–622.

Zhou, H., Pastore, L., *et al.* (2002). Helper-dependent adenoviral vectors. *Methods Enzymol.* **346,** 177–198.

# ADENO-ASSOCIATED VIRAL VECTORS

E. Lehtonen and L. Tenenbaum

Free University of Brussels
Laboratory of Experimental Neurosurgery and Interdisciplinary Research Institute (IRIBHM)
B-1070 Brussels Belgium



## I. Biology of Adeno-Associated Virus and Its Recombinant Vectors

### A. Introduction

In an earlier edition of this book (Latchmann, 1996), the chapter on adeno-associated virus (AAV) vectors mentioned only one published application of AAV vectors in the central nervous system (CNS) (Kaplitt *et al.*, 1994). Since this pioneering work, an extraordinary expansion of this field has taken place. Numerous applications of recombinant adeno-associated virus (rAAV) vectors have been proposed, in particular in the CNS (Fig. 1), and these have boosted not only the field of gene therapy but also fundamental research on the biology of this






0074-7742/03 $35.00

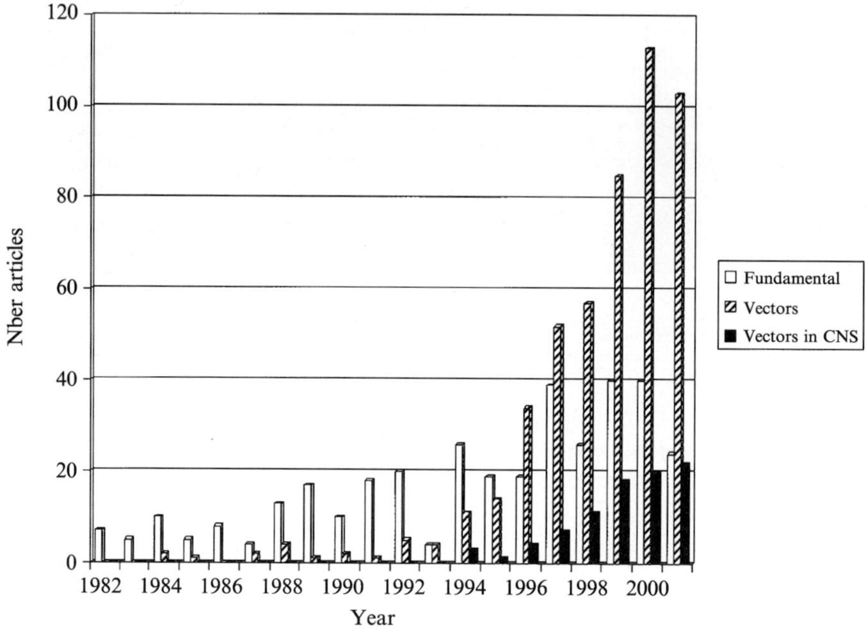


FIG. 1. Evolution of AAV literature. Number of articles concerning AAV published per year from 1982 to 2001, sorted using the Medline (NIH). Open bars, fundamental studies about AAV and AAV vectors biology; hatched bars, AAV vectors technology and applications in animal and cellular models for all diseases; solid bars, AAV vectors applications in the CNS.

defective human virus. Hence, the number of fundamental publications about AAV (Fig. 1) has continuously increased since the obtainment in 1982 of the first molecular clones of AAV (Samulski *et al.*, 1982; Laughlin *et al.*, 1983) and the subsequent construction of the first vectors (Hermonat and Muzyczka, 1984; Tratschin *et al.*, 1985; McLaughlin *et al.*, 1988). Even more striking is the exponential increase in the number of publications concerning rAAV vectors (Fig. 1). Numbers of them describe the development of improved/alternative vectors and production/purification methods, which could have been developed mainly as a result of the increasing knowledge of the biology of the virus. There is no doubt that AAV vectors will continue to play a key role in preclinical gene therapy studies for a number of diseases, including diseases of the brain.

## B. Infectiology

AAVs have many appealing characteristics that make its recombinant derivatives very attractive tools for human gene therapy. AAV is classified as a member of the mammalian Parvovirus family, which is among the smallest and structurally simplest of the DNA viruses. The Parvovirus family has been divided into three genera: parvovirus, densovirus, and dependovirus (Siegl *et al.*, 1985). AAVs constitute the latter genus, so called because of the requirement for a helper virus coinfection to enable productive infection to occur (Berns and Linden, 1995; Tenenbaum and Hooghe-Peters, 1996). Adeno-associated viruses specific for species ranging from chicken to human have been isolated (Berns and Linden, 1995). A total of eight different serotypes of AAV (AAV1–8) has been isolated to date of which the serotype 2 is the most characterized as its infectious clone was the first one made (Samulski *et al.*, 1982). More recently the full sequences for the other serotypes have also been determined (Xiao *et al.*, 1999; Muramatsu *et al.*, 1996; Chiorini *et al.*, 1997, 1999; Rutledge *et al.*, 1998, Gao *et al.*, 2002). It appears that AAV1 and AAV6 are serologically indistinguishable and that the AAV6 is actually a naturally occurring hybrid of AAV1 and AAV2 formed by homologous recombination (Xiao *et al.*, 1999; Rabinowitz *et al.*, 2002).

The natural tissue tropism of wild-type (wt) AAV has been proposed to be for lung epithelial cells (Malik *et al.*, 2000), although a latent infection of AAV has been discovered in other tissues such as muscle (Dutheil *et al.*, 2000). The rAAV vector has been reported to infect various other cells and tissues, including skeletal muscle (Xiao *et al.*, 1996), blood progenitor cells (Miller *et al.*, 1994), brain (Kaplitt *et al.*, 1994; McCown *et al.*, 1996; Bartlett *et al.*, 1998; Klein *et al.*, 1998), liver (Koeberl *et al.*, 1997; Snyder *et al.*, 1997; Xiao *et al.*, 1998), retina (Ali *et al.*, 1996), and gut epithelium (During *et al.*, 1998). The ability to mediate highly efficient gene transfer into several clinically relevant, nondividing, terminally differentiated cell types makes rAAV vectors very appealing tools for human gene therapy approaches in which the target tissue is, e.g., brain or muscle.

### C. Genome Organization

The ligation of the double-stranded form of AAV into pBR322 plasmid led to a concomitant isolation of two different molecular clones of the AAV type 2 and opened the door for the genetic analysis of this virus (Samulski *et al.*, 1982; Laughlin *et al.*, 1983). The wtAAV genome is a simple 4.7-kb linear single-stranded DNA molecule. It contains two large nonoverlapping open reading frames (ORFs), cap and rep, occupying the right and left halves of the genome, respectively. The AAV1–4 and AAV6-8 genomes show about 80% identity whereas serotype 5 seems to be the most distantly related member of the AAV family (Chiorini *et al.*, 1999, Gao *et al.*, 2002).

All eight serotypes of AAV share the same basic genome organization. The genetic map of AAV is presented in Fig. 2. Three promoters have been identified and named according to their approximate map positions using a scale of 100

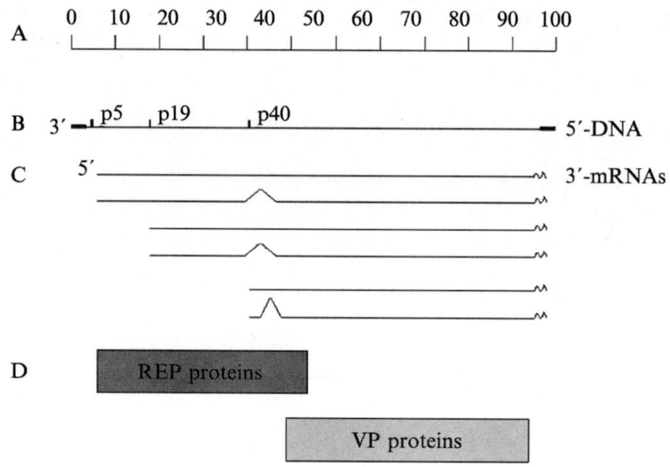


FIG. 2. Wild-type AAV genome organization. (A) AAV genome divided into 100 arbitrary map units. (B) Position of the inverted terminal repeats (ITR; bold lines) and of the three AAV promoters, p5, p19 driving the expression of Rep proteins, and p40 driving the expression of Cap proteins. (C) Messenger RNAs terminate at a common polyadenylation site; p5 promoter initiates a transcript, which unspliced codes for Rep78 and spliced, codes for Rep68. Promoter p19 initiates a transcript that is either translated intact, yielding Rep52, or spliced, yielding Rep40. Promoter p40 initiates all three capsid proteins mRNA species. VP1 is produced by an alternative splicing event. VP2 in produced from the same mRNA as VP3 but it initates at an unusual starting codon ACG. (D) Overlapping Rep and Cap proteins.

map units: p5, p19, and p40 (Fig. 2B). The p5 promoter region shows some divergence from serotype to serotype but shares common critical regulatory elements such as repeated YY1 sites (Chiorini *et al.*, 1997, 1999; Xiao *et al.*, 1999). The two other promoters seem to be more conserved among the different serotypes. Transcripts initiating at each of these promoters have been identified and they share a common intron (Berns and Bohenzky, 1987). In addition, all transcripts terminate at the same polyadenylation site. Promoters p5 and p19 drive the transcription of the regulatory proteins, designated as Rep78, Rep68, Rep52, and Rep40, according to their molecular masses (Berns and Giraud, 1996). Rep78 and 68 are translated from p5 as unspliced and spliced forms, respectively (Fig 2C). Similarly, Rep52 and 40 are translated from the p19 promoter as unspliced and spliced forms, respectively. The rep ORFs of AAV1–4 and AAV6 are approximately 90% identical whereas that of AAV5 shows only 67% homology as compared to AAV2 (Chiorini *et al.*, 1999).

The structural mRNAs encoding VP1, VP2, and VP3 capsid proteins are translated from a spliced p40 transcript. The amino acid sequence of the most abundant capsid protein VP3 is contained within the two larger capsid proteins

VP1 and VP2. The VP1 is produced by an alternative splicing event whereas VP2 is synthesized from the same mRNA as VP3, but VP2 initiates at an unusual starting codon ACG (Berns and Giraud, 1996). The genomic sequences of the capsid proteins include several regions with significant variations between the serotypes, presumably playing an important role in serotype-specific functions, such as antigen specificity and binding to the host cell receptors.

The AAV genome also contains two inverted terminal repeats (ITR), whose overall features are conserved among the eight serotypes. The length varies between 143 and 146 bases according to the serotype. The first 125 bases constitute an overall palindrome, which is interrupted by two smaller ones located symmetrically on either side of the overall axis (Berns and Giraud, 1996). Within the ITR region, there are two elements, which are central to the functions of the ITR: a GAGC repeat motif and a terminal resolution site (TRS) (Chiorini *et al.*, 1997). The repeat motif has been shown to bind Rep78/68 protein and is thus termed the Rep78/68 binding site (RBS). It serves to position Rep proteins for site-specific cleavage at the TRS. This cleavage permits the replication of the ends of the linear genome (Chiorini *et al.*, 1996). The TRS motif is well conserved among the other serotypes except AAV5, which shows only weak homology (Chiorini *et al.*, 1999). When folded on itself, ITR forms a T-shaped hairpin structure, which is involved in the initiation of AAV DNA replication, serving as the origin (ori) of replication and the primer of initiation (see Fig. 3). The ITRs have been shown to be the only *cis*-acting element needed for replication and encapsidation of the single-stranded viral DNA, integration into host genome, and rescue from plasmids or cellular genomes (McLaughlin *et al.*, 1988). This observation enabled the generation of recombinant vectors.

To make recombinant AAV (rAAV) vectors, the entire AAV genome containing *rep* and *cap* genes is deleted, leaving only the ITR sequences of the parent virus. A heterologous gene of interest is then cloned in place of the *rep* and *cap* genes between the ITRs, generating the rAAV vector substrate (Fig. 4). Typically, the foreign DNA encodes a transgene cassette with the promoter/enhancer and polyadenylation signals designed to express from an ectopic location. There is a size limit for the inserts as only genomes within approximately 5% of the wild-type AAV (wtAAV) genome, i.e., 5 kb, will be efficiently packaged (Muzyczka, 1992).

### D. Life Cycle

#### 1. *Wild-Type AAV*

Adeno-associated virus has a unique biphasic life cycle as illustrated in Fig. 5. AAV can be propagated as a lytic virus in the presence of a helper virus coinfection or can establish latency by physically recombining its DNA into the host cell

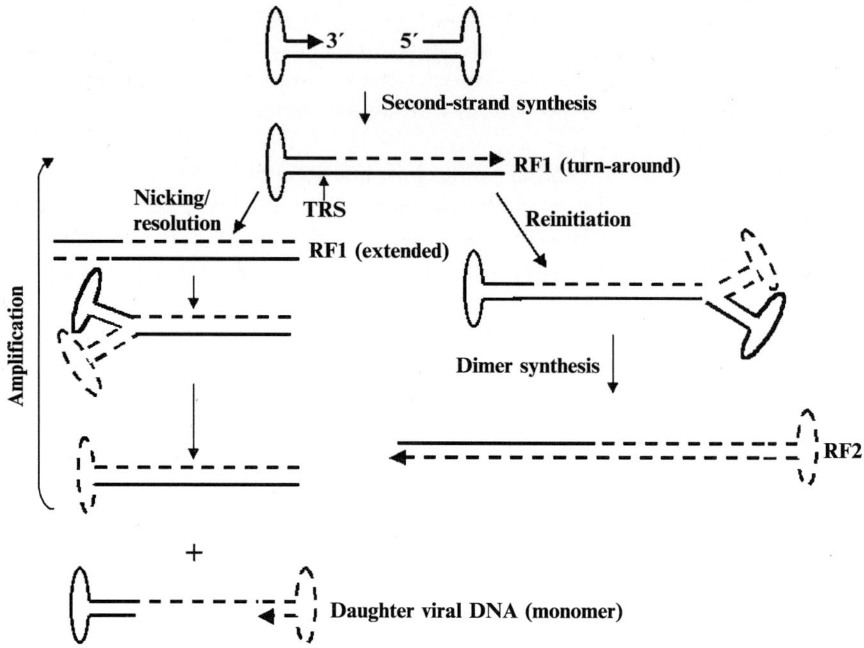


FIG. 3. Model for AAV DNA replication. The conversion of single-stranded viral DNA into the double-stranded form occurs by using the 3′-end hairpin formed by AAV terminal palindromes as a primer for replication resulting in a T-shaped double-stranded intermediate (RF1 "turn-around" form). To replicate the 3′-end ITR, a nick is made by the Rep proteins at the terminal resolution site (TRS), generating a new primer to finish the replication of the 3′-ITR. This duplex replicative form (RF1 "extended form") serves as a template for strand-displacement synthesis generating a new RF1 turn-around replicative form and a single-stranded AAV genome that is packaged to produce progeny virions. A dimeric replicative form (RF2) is also observed presumably resulting from a second round of replication on monomeric duplexes (RF1), which have not been resolved. RF2 constitute a "dead end" replicative form as single-stranded genomes generated by strand-displacement synthesis would be too large to be encapsidated. According to Berns and Bohenzky (1987). Solid lines represent parental DNA molecules; dotted lines represent newly synthesized DNA molecules.

genome and persisting as an integrated provirus in the absence of a helper virus. AAV is the only known eukaryotic virus capable of undergoing site-specific integration into the human genome (Berns and Giraud, 1996). However, even if AAV is strongly oriented to latency, in cell culture AAV DNA can be rescued from its integrated state by adenovirus superinfection leading to DNA replication and productive infection.

At the molecular level, when AAV infects a healthy dividing cell, there is only limited expression of Rep78/68 proteins. This has several biological consequences: AAV gene expression is repressed, DNA replication to a great extent is inhibited, and integration of AAV DNA into the host cell genome occurs

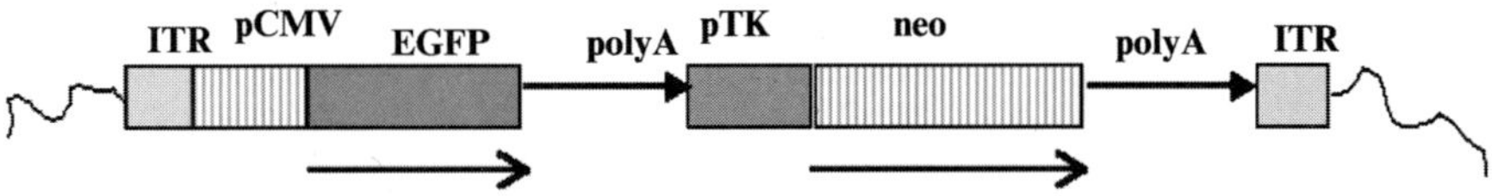


FIG. 4. Schematic presentation of an rAAV vector plasmid. The vector plasmid contains the transgenic cassette(s) flanked by the AAV noncoding inverted terminal repeats (ITRs; map position 1–145 in AAV sequence, gene bank accession number J01901).

(Linden *et al.*, 1996b). When AAV–cellular junction DNA sequences were isolated and used as a probe to screen latently infected cell lines, it was demonstrated that AAV integrated in a sequence-specific manner at the human chromosome 19. The exact physical location of AAV provirus integration has been identified at q13.3-ter of chromosome 19 (ch-19). It was also found that even at a specific site of integration, the actual junctions between AAV and cellular DNA sequences could vary by several hundred nucleotides (Samulski *et al.*, 1991; Kotin *et al.*, 1992).

The integration locus, termed AAVS1, has been cloned and sequenced (Kotin *et al.*, 1992). It covers 8.2 kb, although, the critical site has been narrowed down to a fragment of 33 bp that is sufficient for site-specific recombination (Linden *et al.*, 1996a). This fragment has two interesting features: First, there is a sequence corresponding to RBS of the AAV ITR showing a direct link between Rep proteins and targeted integration (Berns and Giraud, 1996; Linden *et al.*, 1996a). Rep78/68 has actually been shown to bind to this cellular sequence

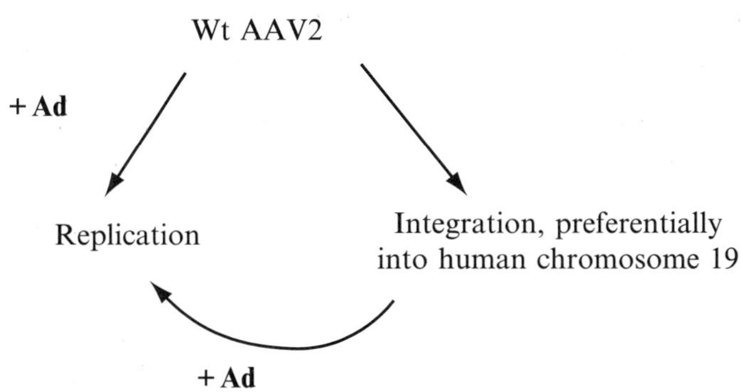


FIG. 5. Adeno-associated virus life cycle. The productive replicative cycle of adeno-associated virus is dependent on the presence of a helper virus (often of the adenovirus family). In the absence of a helper, AAV can penetrate the cell but it cannot replicate. Instead, its genome integrates into the cellular genome at a specific location in human chromosome 19. In the latent state, AAV genes are repressed and the genome remains integrated. However, in case of a helper virus superinfection AAV genome is mobilized and replicates, its genes are expressed, and progeny virions are produced.

(Weitzman *et al.*, 1994) and to initiate replication on ch-19 *in vitro* (Urcelay *et al.*, 1995). Second, a sequence similar to the AAV TRS, which is a specific cleavage site for bound Rep78/68, appeared to be correctly positioned relative to the RBS (Berns and Giraud, 1996; Linden *et al.*, 1996a). Both of these sequences seem to play an integral role in site-specific integration (Linden *et al.*, 1996a). Recently, the spacer element between the RBS and TRS was also shown to be critical as alterations in the distance between these two motifs completely abolished site-specific integration (Meneses *et al.*, 2000).

Establishment of wtAAV latency involves tandem array of head-to-tail oriented viral structures, with rearrangements, amplifications, or inversions of virus and/or cellular DNA at the integration site (Samulski *et al.*, 1991; Linden *et al.*, 1996b; Yang *et al.*, 1997). Even if the exact mechanism of the site-specific integration is not yet clear, models based on the characteristics of the AAV–cellular junction DNA both *in vitro* and *in vivo* describing this process have been proposed (Linden *et al.*, 1996b; Young *et al.*, 2001). The current model requires Rep proteins *in trans*, ch-19 RBS and TRS sites *in cis*, and the AAV ITR elements (Young *et al.*, 2001). In this model, Rep initiates replication on ch-19 by unwinding the ch-19 RBS and thus creating a replication bubble followed by site- and strand-specific nicking at the ch-19 TRS. Nicking creates a free 3′-OH group that can be used to initiate replication (Ni *et al.*, 1998). Additional cellular machinery is required for DNA replication and repair of noncomplementary strands (Linden *et al.*, 1996b; Yang *et al.*, 1997). It is possible that this kind of unscheduled DNA replication would generate substrates that become recombination "hot spots" (Young *et al.*, 2001). Because Rep is covalently attached to the displaced strand, it is likely that this strand remains unprocessed resulting in observed amplification and rearrangements of the chromosomal region. Once integrated, the AAV genome can persist indefinitely in a latent state until rescued and mobilized by helper superinfection.

The requirement for coinfection with a helper virus to establish a productive infection in cell cultures renders AAV naturally defective. Helper virus can be adenovirus (Ad), herpes simplex virus (HSV), or vaccinia virus (Bartlett and Samulski, 1996; Tenenbaum and Hooghe-Peters, 1996). AAV6 seems to use HSV rather than Ad as the natural helper virus (Chiorini *et al.*, 1999). However, the helper functions have been best characterized for Ad coinfection. It has been shown that the lytic phase of the AAV life cycle requires the expression of Ad early gene products (Richardson and Westphal, 1981), including E1A and E1B (Richardson and Westphal, 1984), E2A (Jay *et al.*, 1979), VA (West *et al.*, 1987), and E4 ORF6 (Richardson and Westphal, 1981). Although E4 ORF6 alone is sufficient to induce AAV genome replication, it has been suggested that E1 strengthens the induction effect based on the observations that Ad mutants that are deleted in E1 but are expressing ORF6 enhance transduction in 293 cells, which express E1, but not in HeLa cells, which do not express the constitutive

E1 product. E1 is also known to induce host cellular gene expression and stimulate DNA synthesis through regulation of the cell cycle (Fisher *et al.*, 1996).

When HSV functions as the helper, it seems that HSV enzymes, directly involved in viral DNA synthesis, provide the helper functions (Berns and Linden, 1995). It has also been suggested that host cellular responses induced by the helper virus, rather than viral functions, are responsible for the helper effect (Tenenbaum and Hooghe-Peters, 1996). This hypothesis is supported by the observations that cytotoxic cell culture treatments render the cells permissive for AAV infection. These treatments include, e.g., UV light (Yakobson *et al.*, 1989), chemical carcinogens, heat shock (Yalkinoglu *et al.*, 1988), and hydroxyurea (Yakobson *et al.*, 1987). It has been hypothesized that because most of these treatments result in DNA damage, the induction of a DNA damage repair system could be the mechanism that allows the conversion of a single-stranded AAV genome to double-stranded transcriptionally active templates in a manner analogous to filling a gapped region of chromosomal DNA (Samulski *et al.*, 1999).

2. *Recombinant AAV*

Due to deletion of the *rep* gene, rAAV does not integrate site specifically at AAVS1 like the wtAAV (Linden *et al.*, 1996b; Ponnazhagan *et al.*, 1997; Yang *et al.*, 1997). It rather persists as unintegrated episomes (Afione *et al.*, 1996; Duan *et al.*, 1998) or as randomly integrated genomes (Ponnazhagan *et al.*, 1997; Yang *et al.*, 1997; Wu *et al.*, 1998). Efficient circularization of rAAV genomes has recently been shown to occur in a time-dependent fashion after delivery to muscle (Duan *et al.*, 1998; Vincent-Lacaze *et al.*, 1999). The conversion of monomeric to multimeric high-molecular-weight species occurred over time and was associated with long-term persistence of rAAV genomes. These high-molecular-weight forms can be either head-to-tail concatamers or interlocked circular forms (Vincent-Lacaze *et al.*, 1999), probably arising from intermolecular recombination events between independent monomers (Yang *et al.*, 1999) or a rolling-circle replicative mechanism (Vincent-Lacaze *et al.*, 1999). These circular genomes could act as preintegration intermediates (Duan *et al.*, 1998). Although it is not always clear whether integration or episomal vector genomes are truly responsible for the long-term gene expression and persistence, the presence of integrated rAAV genomes has been demonstrated *in vivo* in the brain (Wu *et al.*, 1998) and muscle (Duan *et al.*, 1998; Dutheil *et al.*, 2000).

Integration of rAAV into host genome occurs at a low frequency, as monomers or low copy number concatamers (Malik *et al.*, 2000). The integration process *in vitro* seems to involve substantial rearrangements such as deletions and insertions within the integration site and even translocations between two individual chromosomes (Miller *et al.*, 2002). It has also recently been suggested that *in vitro* integrated rAAV vector sequences could spontaneously mobilize by an excision-type mechanism (Musatov *et al.*, 2002). The unpredictable nature of the integration process, provirus structure, and chromosomal alterations have consequences for gene therapy and should be addressed as these phenomena may influence both transgene expression and gene expression within the host cells.

### E. Physical Properties

The nonenveloped virion of AAV has a diameter of 20–24 nm and is icosahedral in shape (Berns and Bohenzky, 1987). The three-dimensional structure of the AAV2 capsid has recently been established (xie *et al.*, 2002). The virion is composed solely of three structural proteins called VP1, VP2, and VP3. The VP3 is the most abundant comprising about 90% of the total virion protein with VP1 and VP2 each accounting for 5% of the total virion protein (Bartlett and Samulski, 1996). AAV viral particles have been shown to be very stable in a wide pH range from 3 to 9 (Berns and Bohenzky, 1987). In addition, they are resistant to many harsh physical and chemical factors such as various detergents, certain proteases, repeated freezing/thawing, lyophilization and heat denaturation up to 60°C (Muzyczka, 1992) thereby permitting the AAV viral particles to be purified and concentrated for rAAV vector production.

## II. AAV Vector Production

Traditional protocols for producing AAV-based vectors involve cotransfection of rAAV vector plasmid and a packaging plasmid that provides *in trans* the AAV *rep* and *cap* genes into adenovirus-infected cells, usually the human transformed 293 cell line (Samulski *et al.*, 1989). Initial vector-producing systems resulted in low productivity and a mixture of rAAV vector particles and adenovirus particles (Samulski *et al.*, 1989; Muzyczka, 1992). Furthermore, homologous and nonhomologous recombination events between the vector plasmid and complementing plasmids generate replication-competent AAV (rcAAV) particles (Allen *et al.*, 1997).

The increasing interest in the use of rAAV vectors, especially for clinical trials, demands large quantities of vectors and hence a need to improve vector production. First, the specific productivity needed to be improved. Second, adenovirus or wild-type AAV must be removed because even marginal amounts of Ad or Ad viral proteins can result in host cellular immune responses (Monahan *et al.*, 1998), including cytotoxic T-lymphocytes capable of eliminating transduced cells following transgene expression (Xiao *et al.*, 1996), while wild-type AAV poses a risk of mobilization of the rAAV vector in the patient. Third, transient DNA transfection is not costeffective for large-scale commercial production. Fourth, the traditional method of purification by CsCl density gradient is not desirable for commercial manufacturing.

### A. Helper Virus-Free Vector Production

The discovery of Ferrari *et al.* (1996) that helper viral functions can be provided by transfection rather than by adenovirus infection led to the development of vector manufacturing protocols producing rAAV2 stocks totally free of helper virus. A major advantage was achieved with the development of plasmids containing only the parts of the adenovirus genome that are necessary for rAAV production. Proposed strategies include, e.g., triple or double plasmid protocols. In the triple plasmid approach two plasmids, the adenoviral plasmid and another plasmid containing the AAV *rep/cap* genes, are cotransfected together with the vector plasmid to generate rAAV2 particles (Xiao *et al.*, 1998). In the double plasmid approach, as illustrated in Fig. 6, these two separate plasmids are replaced by one plasmid named pDG, carrying both the necessary Ad genes and the AAV2

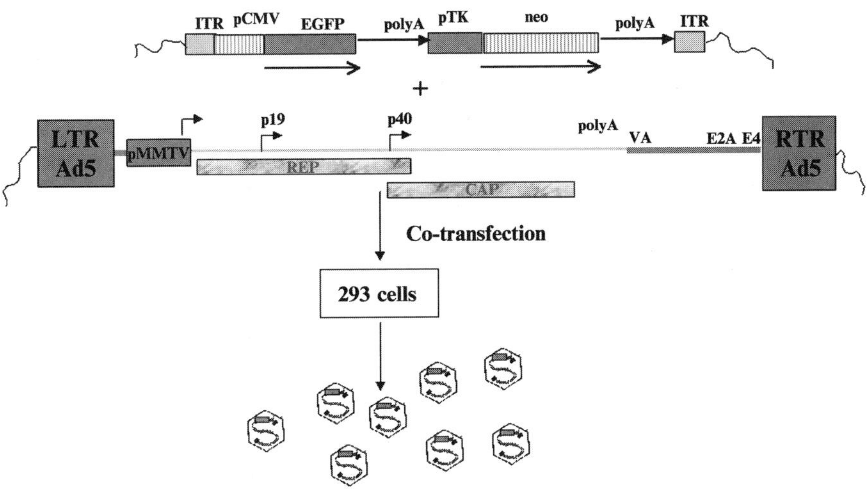


Fig. 6. Production of rAAV by the cotransfection method. Recombinant AAV virus is obtained by cotransfection of adenovirus-immortalized human embryonic kidney (HEK-293) cells with the vector plasmid together with a helper/packaging plasmid (e.g., pDG) expressing AAV viral genes and the adenoviral genes required for AAV replication and encapsidation (Grimm *et al.*, 1998). The rAAV vector stock, devoid of adenovirus, is purified by cesium chloride gradient (Snyder *et al.*, 1996) or by iodixanol gradient and heparin agarose column chromatography (Zolotukhin *et al.*, 2000). (See Color Insert.)

*rep/cap* genes (Grimm *et al.*, 1998). A similar double plasmid approach was reported by Matsushita *et al.* (1998), although in their protocol the AAV2 *rep* and *cap* genes and vector sequences were provided on the same plasmid, whereas adenoviral genes were provided on a separate construct. In general, all of these approaches increase the specific productivity compared to earlier systems and an important advance in vector safety is provided by the fact that resulting vector stocks are free of detectable Ad, wild-type AAV, and rcAAV.

Recently, an increasing use of serotypes other than AAV2 has created a novel demand for efficient production schemes specific for these serotypes. Considering the differences in their genome sequence, each serotype requires their corresponding *rep/cap* genes provided *in trans* using the triple plasmid protocol to create recombinant virus.

### B. Producer Cell Lines

Recent development of rAAV packaging cell lines might eliminate the requirement for transient plasmid transfection. In this approach the *rep* and *cap* complementing systems, and sometimes the vector genome as well, are incorporated into the same cell line. Although packaging cell lines for AAV production have been reported (Clark *et al.*, 1995; Tamayose *et al.*, 1996), their yields were not better than those prepared by the transfection method, suggesting that gene expression is the rate-limiting step for efficient vector production. To improve these systems relying on overexpression of viral gene products, the problems of Rep protein toxicity (Yang *et al.*, 1994) as well as the regulated rep/cap expression levels must be solved. Inoue and Russell (1998) proposed inducible amplification of AAV helper and vector constructs using a tetracycline-controlled transactivator system. Infection of the producer cell line with Ad after activation with doxycycline resulted in vector production with titers 10 times higher than those obtained using the transfection method. Although the producer cell lines are less likely to generate rcAAV (Clark *et al.*, 1996; Inoue and Russell, 1998), they still require the helper adenovirus infection.

### C. Purification

AAV2 stocks have traditionally been purified from cell lysates followed by ultracentrifugation in a CsCl density gradient to concentrate and purify the vector particles and to separate adenovirus particles (Snyder *et al.*, 1996). This still remains the basis of many laboratory-scale vector purification schemes but it is unacceptable for commercial production. Tamayose *et al.* (1996) reported that rAAV2 could be purified and concentrated by sulfonated cellulose column chromatography. Alternative procedures followed and include ion-exchange chromatography using ceramic hydroxyapatite and diethylaminoethyl (DEAE)-Sepharosein combination with sulfonated cellulose (O'Riordan *et al.*, 2000), analytical cation-exchange high-performance liquid chromatography (HPLC) utilizing a TSKgel SP-NPR strong cation-exchange column (Debelak *et al.*, 2000), HPLC using two ion-exchange columns in tandem (Drittanti *et al.*, 2001), or affinity chromatography with an antibody A20 that recognizes assembled AAV particles and not free AAV capsid proteins (Grimm *et al.*, 1998).

Since the discovery of heparin sulfate proteoglycan as a receptor for AAV2 (Summerford and Samulski, 1998), smaller scale purification using heparin sulfate column chromatography in combination with ultracentrifugation through a discontinuous gradient of iodixanol has been suggested (Zolotukhin *et al.*, 1999). Also rAAV3 binds efficiently to heparin (Rabinowitz *et al.*, 2002) and can thus be purified with the same method as rAAV2. This is not the case for rAAV1, rAAV4, or rAAV5 (Rabinowitz *et al.*, 2002) because instead of binding to heparin, rAAV4 and rAAV5 have been shown to bind to sialic acid (Kaludov *et al.*, 2001; Walters *et al.*, 2001). Recently, a purification method for AAV5 using single-step gravity flow columns filled with a sialic acid-rich protein called mucin has been proposed (Auricchio *et al.*, 2001). This method results in vector preparations that are highly infectious and more pure than those purified with the conventional CsCl gradient.

## III. Recombinant AAV-Mediated Transduction in the CNS

Recombinant AAV-mediated transgene expression in different areas of the rat brain has been studied (Kaplitt *et al.*, 1994; McCown *et al.*, 1996; Xiao *et al.*, 1997; Bartlett *et al.*, 1998; Klein *et al.*, 1998; Mandel *et al.*, 1999; Davidson *et al.*, 2000; Tenenbaum *et al.*, 2000). There are striking differences in transduction efficiency between different regions of the brain. In addition, distinct serotypes seem to have different cellular tropism.

### A. Transduction in Different Brain Regions

Most of the *in vivo* studies have been performed by using AAV2 vectors, which have became the best characterized so far, containing the strong cytomegalovirus (CMV) immediate-early promoter (see Table I). These vectors have been used to deliver marker and therapeutic genes into different regions of the brain in order to develop new treatment strategies for neurological diseases. Recombinant AAV2 vectors have generated positive results following injection into

TABLE I
PATTERN OF rAAV2-CMV-EGFP-MEDIATED TRANSDUCTION IN THE CNS

| Time post infusion | Striatum | Hippocampus | Substantia nigra | Globus pallidus | Internal capsule |
|---|---|---|---|---|---|
| 1 week | + | ++++ | Not shown | ++++ | NT |
| 1 month | +++ | ++++ | ++++ | ++++ | ++ |
| 3 months | ++ | + | ++ | NT | NT |
| 1 year | + | − | NT | NT | NT |
| Cell type | Interneurons | Gabaergic neurons | Dopaminergic neurons | Gabaergic neurons | Glial cells |
| Reference | McCown *et al.* (1996); Kirik *et al.* (2000); Tenenbaum *et al.* (2000) | McCown *et al.* (1996); Klein *et al.* (1998) | McCown *et al.* (1996); Mandel *et al.* (1997); Klein *et al.* (1998) | Tenenbaum *et al.* (2000) | Tenenbaum *et al.* (2000) |

hippocampus, substantia nigra, caudate, piriform and lateral cortex, olfactory tubercle, cerebellum, inferior colliculus, and globus pallidus (Kaplitt *et al.*, 1994; McCown *et al.*, 1996; Mandel *et al.*, 1997; Xiao *et al.*, 1997; Bartlett *et al.*, 1998; Klein *et al.*, 1998; Davidson *et al.*, 2000; Tenenbaum *et al.*, 2000). Transduction efficiencies, however, vary markedly from one region to the other (see Table I). In a systematic study using AAV2, efficiency of the transgene expression in the brain was demonstrated with the following order: hippocampus, inferior colliculus, piriform cortex > olfactory tubercle > caudate (McCown *et al.*, 1996).

In addition to the differences in efficiency, transduction in the rat brain seems to result in variable duration of gene expression depending on the site of the injection and the promoter used to drive the transgene. Although long-term gene expression up to 1 year from the CMV-based vector has been reported (Mandel *et al.*, 1999; Tenenbaum *et al.*, 2000), reductions in transgene expression levels over time in different brain regions have been described (Kaplitt *et al.*, 1994; McCown *et al.*, 1996; Klein *et al.*, 1998; Mandel *et al.*, 1998). The most striking difference has been seen between hippocampus and inferior colliculus (McCown *et al.*, 1996; Klein *et al.*, 1998). Both areas showed strong transgene expression 1 week postinjection. However, 3 months postinjection, a dramatic reduction in the number of transduced neurons (up to 96%) was observed in hippocampus, whereas in inferior colliculus no significant loss occurred. Other areas, such as caudate, did not show either strong initial transduction or substantial reduction in the number of positive cells over time (McCown *et al.*, 1996; Tenenbaum *et al.*, 2000). Similarly, when another viral promoter, the Rous sarcoma virus (RSV) long terminal promoter, was used in the mouse striatum, transgene expression in the striatum was completely lost 15 weeks postinfection (Davidson *et al.*, 2000). Observed decline in transgene expression has been suggested to occur because of (1) silencing of the CMV or RSV promoters by hypermethylation (Prosch *et al.*, 1996), (2) loss of transduced neurons by apoptosis (Xiao *et al.*, 1997), or (3) loss of vector episomes prior to integration (Xiao *et al.*, 1997). There is no evidence that loss of transduced cells occurs by immune rejection (Lo *et al.*, 1999).

Most diseases affecting the brain progress over long periods of time and thus potential therapeutic proteins will need to be expressed stably for months or even years. This may require the use of neuron-specific promoters. Indeed, when an rAAV2 vector containing the neuron-specific enolase (NSE) promoter was injected into the rat hippocampus or substantia nigra, both the efficiency and the persistence of transgene expression were shown to substantially improve as compared to CMV (Klein *et al.*, 1998) (see Table II). Regardless of the promoter used, transduction efficiency can be increased using posttranscriptional regulatory elements such as the Woodchuck hepatitis virus element (WPRE), which has been shown to increase the steady-state level of messenger RNA and the efficiency of translation, thus resulting in higher levels of gene expression (Loeb *et al.*, 1999). In a comparative study,Paterna *et al.* (2000) injected rAAV2 vectors containing either CMV or platelet-derived growth factor (PDGF)-$\beta$ promoter in conjunction with WPRE in the substantia nigra. The vector containing the PDGF-$\beta$ promoter and WPRE resulted in more efficient and widespread transduction than the PDGF-$\beta$ promoter alone or the two corresponding vectors containing the CMV promoter. Accordingly, Xu *et al.* (2001) have reported that WPRE boosted the NSE-driven transgene expression by 13-fold in the striatum and by 35-fold in the hippocampus. Furthermore, WPRE seemed to prolong the duration of transgene expression (Xu *et al.*, 2001).

Altogether, these results show that rAAV-mediated transgene expression in the brain can be extremely durable, although little is known about the mechanism behind this phenomenon. Integration in the host genome remains one possibility (Wu *et al.*, 1998).

TABLE II
Comparison of the Performances of rAAV2 Vectors Containing CMV and NSE Promoters

| Time postinfusion | Hippocampus | | Substantia nigra | | Striatum | |
|---|---|---|---|---|---|---|
| | CMV | NSE | CMV | NSE | CMV | NSE |
| 3 days | − | + | NT | NT | NT | NT |
| 3 weeks | +++ | ++++ | +++ | ++++ | +++ | ++++ |
| 3 months | ++ | ++++ | ++ | ++++ | ++ | ++++ |
| 9 months | − | ++++ | − | ++++ | + | NT |
| Reference | Klein *et al.* (1998) | | Klein *et al.* (1999) | | Xu *et al.* (2001) | |

## B. Cellular Tropism

Several studies converged to conclude that rAAV2-based vectors using the CMV promoter almost exclusively transduce neurons when delivered in the rat CNS in regions containing mixed neuronal-glial cell populations (Kaplitt *et al.*, 1994; McCown *et al.*, 1996; Klein *et al.*, 1998; Tenenbaum *et al.*, 2000) (see Fig. 7A,C–E), transduction in astroglial cells occuring only very infrequently (McCown *et al.*, 1996; Klein *et al.*, 1998). It should be noted however, that not all types of neurons are susceptible; for example, dendate granule cells of the hippocampus do not exhibit transgene expression (McCown *et al.*, 1996).

In accordance with these empirical observations, studies using virus with fluorophore-labeled viral capsids demonstrated that binding and uptake

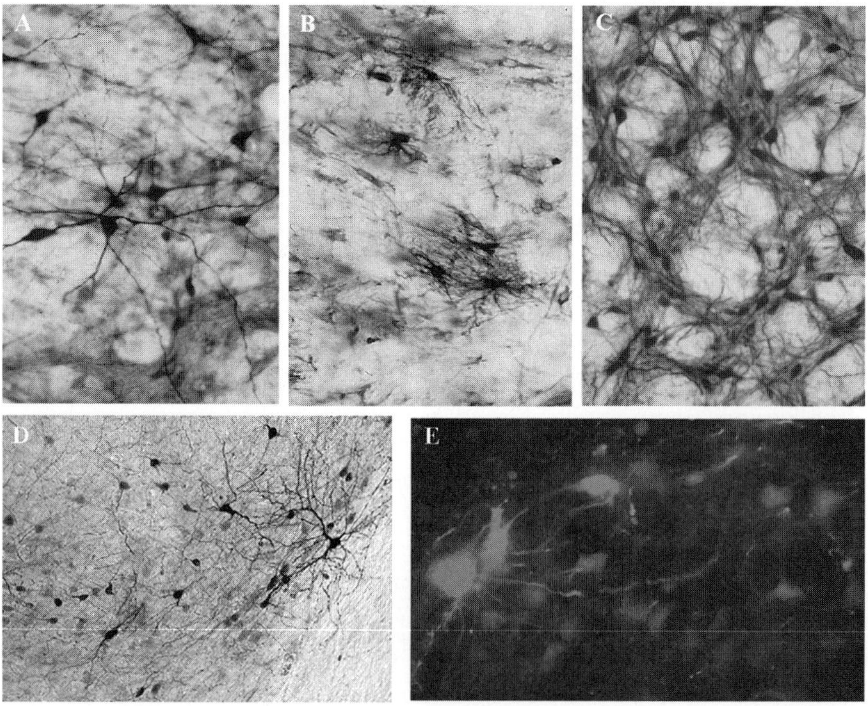


Fig. 7. Tropism of rAAV vectors for specific cells in the brain. A purified viral suspension of rAAV2 expressing gfp-marker gene under the control of the CMV promoter was stereotaxically injected in various regions of the brain. One month after injection, transgene expression was detected either by observing native gfp fluorescence or by immunohistochemistry using peroxidase staining. Almost exclusively neurons are transduced in the striatum (A), globus pallidus (C), cortex (D), and substantia nigra pars compacta (E). In white matter, e.g., in the internal capsule (B), glial cells are efficiently transduced. (See Color Insert.)

preferentially occur in neurons (Bartlett *et al.*, 1998). However, there seems to be no absolute obstacle to glial cell transduction. Indeed, cultured glial cells of fetal (Keir *et al.*, 1997) or tumoral (Tenenbaum *et al.*, 1994) origin can be efficiently infected by wt and rAAV. Furthermore, *in vivo*, in a region devoid of neurons, the internal capsule, glial cells were transduced by an rAAV2 vector using the CMV promoter (Tenenbaum *et al.*, 2000) (Fig. 7B).

The choice of the promoter may nevertheless affect the cellular specificity of gene transfer. As expected, when the NSE promoter is used, transduction seems to occur primarily in neurons. However, within a brain region devoid of neurons, the corpus callosum, oligodendrocytes were also transduced, although at a low efficiency (Klein *et al.*, 1998). This low level of gene transfer in cells in which the NSE promoter is normally not active possibly reflects interference with promoter/enhancer elements present in the viral ITRs (Flotte and Carter, 1995; Haberman *et al.*, 2000). In contrast, when a vector containing the myelin basic protein (MBP) promoter, specifically active in oligodendrocytes, was injected in the corpus callosum a high level of transgene expression was observed for up to 3 months (Chen *et al.*, 1998, 1999).

Injections of rAAV2 vectors containing the astrocyte-specific glial fibrillary acidic protein (GFAP) promoter resulted in transduction of almost exclusively neurons in the spinal cord and in the hippocampus and of a majority of neurons with only 5% astrocytes in the striatum (Peel and Klein, 2000; Xu *et al.*, 2001). Similarly to the unspecific activity of the MBP promoter (see above) transcriptional elements present in the ITRs might complicate the interpretation of studies with cell type-specific promoters.

The GFAP promoter is known to be upregulated in reactive astrocytes in response to brain injury (Eng *et al.*, 2000). Interestingly, in lesioned brain and spinal cord up to 15–30% of astrocytes can be transduced (Peel and Klein, 2000; Xu *et al.*, 2001). It remains to be determined whether these data reflect a modified pattern of virus binding and uptake in injured brain.

### C. Vectors Based on Alternative AAV Serotypes

In a recent study where rAAV vectors based on serotypes 4 and 5 were used, it was observed that these vectors not only show distinct regional patterns of transduction but are also capable of transducing cell types different than rAAV2 (Davidson *et al.*, 2000). Indeed, rAAV5-mediated marker gene delivery in the striatum resulted in a greater number of transduced cells, including both neurons and a significant proportion of astrocytes, over a larger volume of tissue than AAV2. Transgene-expressing cells were also observed at a distance from the delivery site in the septal region and neocortex. In contrast, when injected into the striatum, rAAV4 appears to transduce almost exclusively ependymal cells (Davidson *et al.*,

2000). Because rAAV2 and rAAV4 particles contain nearly identical ITRs (Chiorini *et al.*, 1997), differences between their preferential target cells must be accounted for by the variations in their capsids. The authors predicted that rAAV4 and rAAV5 have different receptor requirements for cell entry compared to rAAV2.

### D. *Ex Vivo* Gene Delivery

Cell replacement therapies such as transplantation of fetal mesencephalon in Parkinson's disease could be advantageously combined with gene therapy in two ways. First, grafted cells are submitted to oxidative stress and apoptosis and neuroprotective gene transfer could improve the survival and functionality of the graft. Second, protection of remaining host cells could be achieved by the means of *ex vivo* gene delivery using the graft as a cellular vector. We have shown that rAAV2 vector mediates the expression of the EGFP reporter gene under the control of the CMV promoter in organotypic cultures of freshly explanted solid fragments of human and rat fetal mesencephalic tissue from 3 days to at least 6 weeks postinfection. Furthermore, gfp expression was stable *in vivo* in the rat striatum until at least 3 months posttransplantation (Lehtonen *et al.*, 2002) (Fig. 8). These promising data suggest that AAV vectors expressing neurotrophic factors could be used to genetically modify the fetal tissue prior to transplantation in order to promote graft survival and integration. The graft has been recently shown to be maintained for at least 10 years in patients (Piccini *et al.*, 1999) is furthermore an excellent cellular vector to deliver neurotrophic factors to the host in order to provide long-term neuroprotection and possibly halt the progression of the disease.

Current trends are focused on neuronal precursors or stem cells that might provide an alternate source of cellular vectors able to differentiate in response to epigenetic stimuli and to integrate in the environment of the host brain. They could also provide ideal cellular vectors to deliver appropriate neuroprotective genes. However, these cells are not yet stable *in vivo*. Cell replacement therapies using gene transfer to influence the differentiation of stem cells are currently under development. It has been recently shown that AAV2 vectors mediate efficient and stable gene expression in human neural progenitors both *in vitro* and *in vivo* after transplantation in the spinal cord (Wu *et al.*, 2002).

## IV. Factors Limiting Efficient Transduction

Although AAV is known to have a very broad host range, there are striking differences in *in vivo* transduction efficiency of rAAV vectors in various tissues and cells. The inability of the vector to enter specific cells due to a lack of its

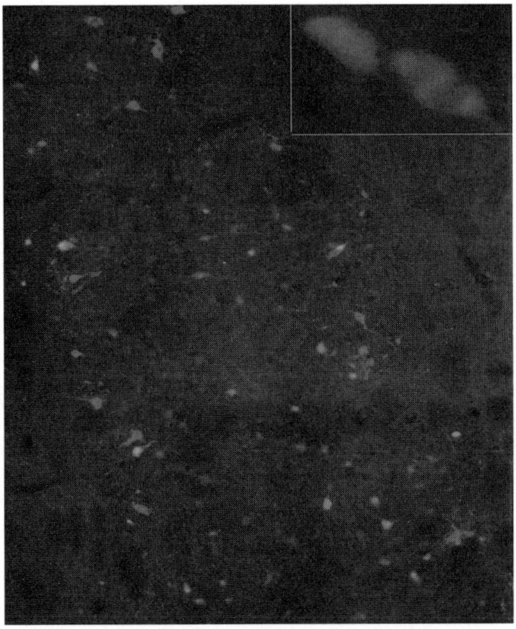

FIG. 8. *Ex vivo* gene delivery to embryonic dopaminergic grafts using rAAV2 vectors. rAAV2–CMV–gfp and transplanted into the striatum of 6-hydroxydopamine-lesioned rats. Three months posttransplantation, brains were analyzed by double immunofluorescence for gfp (in green) and tyrosine hydroxylase (in red). Dopaminergic neurons are located at the border between the graft and the host striatum whereas gfp-positive cells are present all over the graft. (See Color Insert.)

receptors that mediate virus binding and entry (Summerford and Samulski, 1998; Summerford *et al.*, 1999; Qing *et al.*, 1999; Qiu *et al.*, 1999) is a limiting step for efficient gene delivery. If virus can actually enter the cell, subsequent intracellular trafficking and nuclear uptake (Duan *et al.*, 1999; Bartlett *et al.*, 2000; Hansen *et al.*, 2000, 2001a, 2001b) can be impaired and thus rate limiting. In addition, because AAV is a single-stranded DNA virus, the target cell must allow the conversion of the single-stranded viral genome into a transcriptionally active double-stranded intermediate (Ferrari *et al.*, 1996; Fisher *et al.*, 1996).

### A. Viral Receptors

Viral receptors are often involved in defining the host range and specific tissue or cell type tropism of a virus. Studies in which fluorophore-labeled AAV2 particles were injected into the rat inferior colliculus or the hippocampus showed rapid binding and uptake of the virus into neurons, supporting a receptor-based neuronal tropism (Bartlett *et al.*, 1998). A cell surface heparan sulfate proteoglycan (HSPG) serves as the primary attachment receptor for AAV (Summerford and Samulski, 1998). Recently other coreceptors have been proposed, such as fibroblast growth factor receptor 1 (FGFR1) (Qing *et al.*, 1999) together with $\alpha_V\beta_5$ integrin (Summerford *et al.*, 1999). Although the expression of these receptors is high in neurons (Nurcombe *et al.*, 1993; Pinkstaff *et al.*, 1999) it is noteworthy that glial cells can also express a wide range of HSPGs as well as FGFR1 and $\alpha_V\beta_5$ integrin (Tawil *et al.*, 1994; Bansal and Pfeiffer, 1997; Takami *et al.*, 1998).

On the basis of the cap ORF, it was likely and subsequently demonstrated that other serotypes of AAV, such as AAV4 and AAV5, mediate infection through attachment to different cellular receptors. Indeed, binding to sialic acid has recently been shown to function as a receptor for serotypes 4 and 5 (Kaludov *et al.*, 2001; Walters *et al.*, 2001).

### B. Virus Internalization and Intracellular Trafficking

Internalization of rAAV particles into the target cells and subsequent intracellular trafficking might constitute a rate-limiting step for efficient transgene expression in the brain. Recent *in vitro* experiments using HeLa cells (Bartlett *et al.*, 2000) as well as a real-time imaging of single AAV molecules (Seisenberger *et al.*, 2001) have demonstrated that progressing through all the steps from the cell surface to nuclear entry may be as short as 15 min. Following attachment to the HSPGs, virus particles enter the cell through receptor-mediated endocytosis (Bartlett *et al.*, 2000). It seems that both AAV2 entry and infection are regulated by a cytosolic GTP-hydrolyzing protein dynamin (Duan *et al.*, 1999).

The efficiency of viral infection relies to a great extent on the targeting of the viral genome to the host cell nucleus following infection. In less permissive mouse fibroblasts, the virus fails to transport to the nucleus due to altered endocytic processing (Hansen *et al.*, 2000, 2001a). There is also evidence that ubiquitination of the AAV2 capsid proteins plays a role as a barrier for intracellular trafficking by rerouting the virus particles to a nonexpressible compartment or by promoting degradation of the incoming virions by proteasomes (Duan *et al.*, 2000; Yan *et al.*, 2002). When proteasome function is inhibited, a substantial increase in gene transfer has been shown to occur.

Relatively little is known about the intracellular site of viral uncoating and transport of the virion across the nuclear envelope. It has recently been shown that AAV2 can efficiently enter intact nuclei isolated from both permissive and less permissive cell types (Hansen *et al.*, 2001b). Furthermore, entry into the nucleus is time and temperature dependent, but is not saturable and seems to occur independently of the nuclear pore complex. Hansen *et al.* (2001b) also demonstrated that purified nuclei contain all of the machinery necessary for uncoating and viral second-strand DNA synthesis even in the absence of a helper virus.

## C. Second-Strand Synthesis

Because AAV has a single-stranded DNA genome, it has to be converted to a transcriptionally active double-stranded form so that transgene expression can occur. Thus, the rate at which the genome is converted to double-stranded form is limiting efficient transduction (Ferrari *et al.*, 1996; Fisher *et al.*, 1996).

Experiments *in vitro* correlating the induction of transgene expression with the conversion of the single-stranded viral genome to duplex form have been performed (Ferrari *et al.*, 1996; Fisher *et al.*, 1996). DNA from rAAV-infected cells was analyzed on denaturing agarose gels and single-strand conversion to duplex form was observed as the appearance of monomeric and dimeric double-stranded replicative-form intermediates (RFm and RFd, respectively). These experiments revealed that this rate-limiting step is greatly enhanced by expression of Ad genes E1 and E4 ORF6. Data also indicate that expression of the ORF6 protein is alone sufficient and necessary to increase the efficiency of transduction by an rAAV vector in a way that is dependent on second-strand synthesis.

Recently, a 53-kDa host cellular tyrosine phosphoprotein that appears to inhibit second-strand synthesis of the rAAV genome has been identified (Qing *et al.*, 1997, 1998). This protein is designated the single-stranded D-sequence binding protein (ssD-BP) for its specific interaction with the single-stranded D(−) sequence within the AAV 3′ ITR. The ssD-BP is phosphorylated at tyrosine, serine, and threonine residues. Interestingly, it appears to inhibit second-strand synthesis of the rAAV genome when phosphorylated at tyrosine residues. It complexes with the D(−) sequence in the AAV ITR and thus blocks the initiation of DNA replication from the 3′-OH end. Furthermore, in the presence of cellular protein tyrosine kinase inhibitors, as well as the Ad E4 ORF6 gene product, genistein, or hydroxyurea, the ssD-BP converts to a dephosphorylated state, which leads to some type of a conformational change. This results in the accessibility of the 3′-OH end to act as a primer for the second-strand synthesis leading to significant augmentation of transgene expression from rAAV (Qing *et al.*, 1997). It appears that the ratio of phosphorylated to unphosphorylated ssD-BP differs between cell types. The amount of unphosphorylated ssD-BP correlates well with the efficiency by which AAV can transduce cells (Qing *et al.*, 1998).

The exact mechanisms by which dephosphorylation of ssD-BP facilitates rAAV second-strand synthesis is not completely clear. It has been suggested that ssD-BP might itself possess a DNA-polymerase-like activity or that its dephosphorylation could activate a cellular DNA polymerase(s) necessary for host DNA synthesis or DNA repair pathway (Qing *et al.*, 1998). It also appears that this protein undergoes autophosphorylation followed by autodephosphorylation but the significance of this phenomenon is still not clear. Manipulation of the phosphorylation state of ssD-BP may be exploitable as a strategy to improve the transduction efficiency of rAAV vectors.

## V. New Vectors

The small packaging capability limits the use of rAAV vectors for some applications. Thus strategies are being developed to expand the size of transgenes that can be delivered. These strategies include the development of hybrid viral capsid structures (e.g., parvovirus B19/AAV (Srivastava *et al.*, 1989), packaging rAAV genomes into adenovirus capsids (Goncalves *et al.*, 2001), the use of expression cassettes with small promoters including the AAV p5 promoter, or promoter activity derived from the ITR sequence (Monahan *et al.*, 2000). In addition to these, two novel approaches that rely on either homologous recombination between overlapping regions in two independent vectors (Duan *et al.*, 2001) or heterodimerization of two AAV vectors when forming concatamers (Sun *et al.*, 2000; Yan *et al.*, 2000; Duan *et al.*, 2001) have recently been described. In the *trans*-splicing approach a large candidate gene is split into two segments and separately packaged with appropriate engineered intron splice donor and acceptor sites into two individual AAV vectors. AAV forms concatamers by fusing the 5′ and 3′ segments of the split gene and thus reconstitutes the gene of interest (Sun *et al.*, 2000). This approach should allow the use of original full-sized promoters, as well as packaging of large genes without concern for size limitation.

The conversion of the single-stranded viral genome to a double-stranded form causes a delay of several days or even weeks in the transduction. An interesting approach to overcome this problem is the use of self-complementary (sc) AAV vectors (McCarty *et al.*, 2001). McCarty *et al.* (2001) exploited the occurrence of dimeric intermediate replicative forms during the productive cycle (see Fig. 3) by designing vectors having half (or less) of the wtAAV size. Indeed, during the production of recombinant virus, the dimeric forms, whose size is within the limits of packaging capacity, are encapsidated. The resulting viral stock contains viral particles with both monomeric forms (one copy of single-stranded viral DNA) and dimeric forms consisting of two inverted complementary copies of single-stranded DNA, which can fold to form a double-stranded vector DNA.

Upon infection of a target cell with self-complementary vectors, onset of trangene expression is immediate as these vectors do not require cellular factors to undergo second-strand synthesis. Furthermore, scAAV vectors could allow us to transduce cell types in which second-strand synthesis was a limiting factor for transduction by full size vectors. The restricted insert size capacity (2.3 kb) will nevertheless limit the usefulness of these vectors to small genes and promoters.

The broad host range of AAV is a disadvantage for some applications, because it reduces the specificity of the gene transfer. However, AAV vectors can be targeted to a specific cell population by circumventing the normal virus receptors. This can be achieved by coupling of different ligands, such as antibodies (Bartlett *et al.*, 1999), to the viral capsid. Efficient retargeting has also been achieved by incorporating into the AAV *cap*-sequence peptide epitopes recognizing receptors different than HSPG (Girod *et al.*, 1999; Shi *et al.*, 2001), tumor-targeting sequences (Grifman *et al.*, 2001), or immunoglobulin binding domains (Ried *et al.*, 2002). Although applications of the above-described new vectors in the CNS have not been described yet, these will probably open new routes for efficient gene transfer in brain regions and cell types that are poorly transduced or not transduced by traditional rAAV2 vectors.

## VI. Regulation of Transgene Expression

In some circumstances, the level and duration of transgene expression need to be adjusted in order to minimize undesirable effects. For example, motor disturbances resulting from inappropriate expression of genes coding for neurotrophic factors in the brain have been described (Kirik *et al.*, 2000). In this study rAAV-mediated expression of GDNF in the right striatum of healthy animals resulted in asymmetrical rotational behavior, presumably reflecting an excess of dopamine in the injected side.

Exogenously regulated gene expression systems are usually based on two constructs, with the first one containing the gene of interest under the control of an inducible/repressible promoter, and the second one expressing a transactivator. None of the available systems combines absence of toxicity of the inducer and/or the transactivator, low level of residual expression in the absence of inducer, efficient and stable induced expression, absence of nonspecific effects such as influence of the inducer on endogenous transcription, or induction of immune response against the transactivator. Nevertheless some potentially useful AAV vectors with regulatable promoters have been described. Those are regulated by the antibiotic tetracycline or its analogs (Haberman *et al.*, 1998; Rendahl *et al.*, 1998; Fitzsimons *et al.*, 2001; McGee Sanftner *et al.*, 2001; Chtarto *et al.*, 2003) or by rapamycin (Ye *et al.*, 1999).

Disease-inducible cellular promoters could constitute a promising approach. For example, a recently described nerve growth factor-inducible promoter, pre-protachykinin-A promoter (Harrison *et al.*, 1999), could be used to treat lesions, which induce nerve growth factor upregulation.

## VII. Biosafety

Three of the eight serotypes of AAV, AAV2, AAV3, and AAV5, have been isolated from humans (Chiorini *et al.*, 1999). Although 80–90% of adults are seropositive for antibodies against AAV, AAV is not known to be an etiological agent of any disease in humans (Berns and Linden, 1995). Seroconversion to AAV positive occurs during childhood and is usually concomitant with Ad infection. AAV infection might actually be beneficial or even protective, as, e.g., women with cervical carcinomas are seronegative for AAV antibodies (Georg-Fries *et al.*, 1984).

It is tempting to assume that rAAV vectors have the same nonpathogenic behavior as the wtAAV. Indeed, initial safety results from, e.g., a clinical trial where patients with hemophilia were treated intramuscularly with rAAV encoding factor IX, are very optimistic (Monahan and Samulski, 2000). None of the three patients treated has experienced any toxicity, none has developed antifactor IX antibodies, and none had vector in the semen (Kay *et al.*, 2000). However, in a subsequent clinical trial where rAAV encoding factor IX was delivered via the hepatic artery, polymerase chain reaction (PCR) signals were detected in the semen of two patients. Although, the risk of germ line transmission might be raised by these data, in the two patients tested, the vector was cleared from the semen within a few weeks. Furthermore, at the time points where the PCR signal was positive, no vector sequences were found in the germ cells (Marshall, 2001).

The biodistribution of rAAV delivered by various routes in animal models is only poorly documented. Conrad *et al.* (1996) observed vector scattering to the systemic circulation and distant organs (heart, liver, kidney, lymph nodes, spleen, pancreas, and even brain) following rAAV vector delivery to the lung of nonhuman primates. After intramuscular injection, vector sequences were present in body fluids at short term, as well as in peripheral blood mononucleated cells (PBMC) for several months and in lymph nodes and liver up to at least 18 months (Favre *et al.*, 2001). However, none of the animals had detectable rAAV sequences in their gonads. The biodistribution of rAAV sequences after delivery of recombinant virus to the brain has not yet been assessed.

Additional safety concerns are raised by the occurrence of integration of rAAV sequences at several uncharacterized integration sites in the absence of Rep proteins (Yang *et al.*, 1997). Indeed, insertional mutagenesis could theoretically occur and, in addition, the presence of transcriptionally active sequences within the ITRs (Flotte *et al.*, 1993; Haberman *et al.*, 2000) could result in an activation of otherwise silent genes. Consistent with the consequence of insertional mutagenesis or of oncogene activation, metastatic liver tumors were observed in 6 of 59 rAAV-treated mice after 18 months of treatment in a mouse model of mucopolysaccharidosis Type VII ($\beta$-glucoronidase deficiency) (Donsante *et al.*, 2001). These data were surprising as several previous studies performed in the same model failed to show any tumor formation. Furthermore, no correlation between the tumor formation and the presence of rAAV sequences was shown. Nevertheless, this issue should be further addressed. The former study should be regarded keeping in mind that in the biodistribution studies (Conrad *et al.*, 1996; Favre *et al.*, 2001), no tissue abnormalities were observed until 18 months. However, the consequences of long-term persistence of rAAV sequences in PBMC, lymph nodes, and liver are still not predictable.

Another safety issue has arisen from recently established correlations between the occurrence of male infertility and the presence of AAV DNA in human semen (Rohde *et al.*, 1999) on one hand and the occurrence of miscarriage and the presence of infectious AAV in embryonic material as well as in the cervical epithelium (Walz *et al.*, 1997; Tobiasch *et al.*, 1998) on the other hand. It should be noted that when documented, the pathogenicity of parvoviruses is related to the Rep proteins, which are absent in recombinant AAV vectors. Nevertheless, the concomitant presence of wt AAV and helper viruses such as herpes simplex virus or human papilloma virus (Walz *et al.*, 1997) in the genital tract raises the possibility that rAAV, if present in the same cells, could be horizontally or even vertically transmitted.

Finally, in view of the potential toxicity of the rAAV viral capsid proteins and immune reactions against the capsid as well as against the transgene product, the amount of vector that could be safely administered as well as the site and route of delivery require further examination before rAAV can be considered as a completely safe gene delivery vector. In these aspects, the immune privilege of the brain is an advantage over other organs. Indeed, it has been shown that immune responses against rAAV2 vary as a function of the route of administration. When injected intraperitoneally, intravenously, or subcutaneously, rAAV2 vectors induce cytotoxic T-lymphocytes and antibodies against the transgene product as well as antibodies to AAV2 capsids (Brockstedt *et al.*, 1999). In contrast, upon intramuscular injection, the animals developed only a humoral response against the transgene and capsid proteins but no cytotoxic T cells (Brockstedt *et al.*, 1999). Accordingly, readministration of rAAV vectors in the periphery requires transient immunosuppression (Halbert *et al.*, 1997; Manning *et al.*, 1998). In the brain, antibodies to AAV2 capsid and transgene product appeared at low levels at 2–4 months after injection but did not correlate with a reduction in transgene expression and did not prevent readministration of vector (Lo *et al.*, 1999).

A more detailed evaluation of the safety profile of rAAV vectors in the context of clinical uses has recently appeared (Monahan *et al.*, 2002).

## References

Afione, S. A., Conrad, C. K., Kearns, W. G., Chunduru, S., Adams, R., Reynolds, T. C., Guggino, W. B., Cutting, G. R., Carter, B. J., and Flotte, T. R. (1996). In vivo model of adeno-associated virus vector persistence and rescue. *J. Virol.* **70,** 3235–3241.

Ali, R. R., Reichel, M. B., Thrasher, A. J., Levinsky, R. J., Kinnon, C., Kanuga, N., Hunt, D. M., and Bhattacharya, S. S. (1996). Gene transfer into the mouse retina mediated by an adeno-associated viral vector. *Hum. Mol. Genet.* **5,** 591–594.

Allen, J. M., Debelak, D. J., Reynolds, T. C., and Miller, A. D. (1997). Identification and elimination of replication-competent adeno-associated virus (AAV) that can arise by nonhomologous recombination during AAV vector production. *J. Virol.* **71,** 6816–6822.

Auricchio, A., O'Connor, E., Hildinger, M., and Wilson, J. M. (2001). A single-step affinity column for purification of serotype-5 based adeno-associated viral vectors. *Mol. Ther.* **4,** 372–374.

Bansal, R., and Pfeiffer, S. E. (1997). FGF-2 converts mature oligodendrocytes to a novel phenotype. *J. Neurosci. Res.* **50,** 215–228.

Bartlett, J. S., and Samulski, R. J. (1996). Adeno-associated virus vectors for gene therapy. *In* "Gene Therapy" (N. R. Lemoine and D. N. Cooper Eds.), BIOS Scientific Publishers Ltd., Oxford.

Bartlett, J. S., Samulski, R. J., and McCown, T. J. (1998). Selective and rapid uptake of adeno-associated virus type 2 in brain. *Hum. Gene Ther.* **9,** 1181–1186.

Bartlett, J. S., Kleinschmidt, J., Boucher, R. C., and Samulski, R. J. (1999). Targeted adeno-associated virus vector transduction of nonpermissive cells mediated by a bispecific F(ab′gamma)2 antibody. *Nat. Biotechnol.* **17,** 181–186.

Bartlett, J. S., Wilcher, R., and Samulski, R. J. (2000). Infectious entry pathway of adeno-associated virus and adeno-associated virus vectors. *J. Virol.* **74,** 2777–2785.

Berns, K. I., and Bohenzky, R. A. (1987). Adeno-associated viruses: An update. *Adv. Virus Res.* **32,** 243–306.

Berns, K. I., and Giraud, C. (1996). Biology of adeno-associated virus. *Curr. Top. Microbiol. Immunol.* **218,** 1–23.

Berns, K. I., and Linden, R. M. (1995). The cryptic life style of adeno-associated virus. *BioEssays* **17,** 237–245.

Brockstedt, D. G., Podsakoff, G. M., Fong, L., Kurtzman, G., Mueller-Ruchholtz, W., and Engleman, E. G. (1999). Induction of immunity to antigens expressed by recombinant adeno-associated virus depends on the route of administration. *Clin. Immunol.* **92,** 67–75.

Chen, H., McCarty, D. M., Bruce, A. T., Suzuki, K., and Suzuki, K. (1998). Gene transfer and expression in oligodendrocytes under the control of myelin basic protein transcriptional control region mediated by adeno-associated virus. *Gene Ther.* **5,** 50–58.

Chen, H., McCarty, D. M., Bruce, A. T., and Suzuki, K. (1999). Oligodendrocyte-specific gene expression in mouse brain: Use of a myelin-forming cell type-specific promoter in an adeno-associated virus. *J. Neurosci. Res.* **55,** 504–513.

Chiorini, J. A., Wiener, S. M., Yang, L., Smith, R. H., Safer, B., Kilcoin, N. P., Liu, Y., Urcelay, E., and Kotin, R. M. (1996). The roles of AAV Rep proteins in gene expression and targeted integration. *Curr. Top. Microbiol. Immunol.* **218,** 25–33.

Chiorini, J. A., Yang, L., Liu, Y., Safer, B., and Kotin, R. M. (1997). Cloning of adeno-associated virus type 4 (AAV4) and generation of recombinant AAV4 particles. *J. Virol.* **71,** 6823–6833.

Chiorini, J. A., Kim, F., Yang, L., and Kotin, R. M. (1999). Cloning and characterization of adeno-associated virus type 5. *J. Virol.* **73,** 1309–1319.

Chtarto, A., Bender, H. V., Haneman, C. O., Kemp, T., Lentonen, E., Levivier, M., Brotchi, J., Velu, T., and Tenenbaum, L. (2003). Tetracycline-inducible transgene expression mediated by a single AAV vector. *Gene Ther.* **10,** 84–94.

Clark, K. R., Voulgaropoulou, F., Fraley, D. M., and Johnson, P. R. (1995). Cell lines for the production of recombinant adeno-associated virus. *Hum. Gene Ther.* **6,** 1329–1341.

Clark, K. R., Voulgaropoulou, F., and Johnson, P. R. (1996). A stable cell line carrying adenovirus-inducible rep and cap genes allows for infectivity titration of adeno-associated virus vectors. *Gene Ther.* **3,** 1124–1132.

Conrad, C. K., Allen, S. S., Afione, S. A., Reynolds, T. C., Beck, S. E., Fee-Maki, M., Barrazza-Ortiz, X., Adams, R., Askin, F. B., Carter, B. J., Guggino, W. B., and Flotte, T. R. (1996). Safety of single-dose administration of an adeno-associated virus (AAV)-CFTR vector in the primate lung. *Gene Ther.* **3,** 658–668.

Davidson, B. L., Stein, C. S., Heth, J. A., Martins, I., Kotin, R. M., Derksen, T. A., Zabner, J., Ghodsi, A., and Chiorini, J. A. (2000). Recombinant adeno-associated virus type 2, 4, and 5 vectors: Transduction of variant cell types and regions in the mammalian central nervous system. *Proc. Natl. Acad. Sci. USA* **97,** 3428–3432.

Debelak, D., Fisher, J., Iuliano, S., Sesholtz, D., Sloane, D. L., and Atkinson, E. M. (2000). Cation-exchange high-performance liquid chromatography of recombinant adeno-associated virus type 2. *J. Chromatogr. B Biomed. Sci. Appl.* **740,** 195–202.

Donsante, A., Vogler, C., Muzyczka, N., Crawford, J. M., Barker, J., Flotte, T., Campbell-Thompson, M., Daly, T., and Sands, M. S. (2001). Observed incidence of tumorigenesis in long-term rodent studies of rAAV vectors. *Gene Ther.* **8,** 1343–1346.

Drittanti, L., Jenny, C., Poulard, K., Samba, A., Manceau, P., Soria, N., Vincent, N., Danos, O., and Vega, M. (2001). Optimised helper virus-free production of high-quality adeno-associated virus vectors. *J. Gene Med.* **3,** 59–71.

Duan, D., Sharma, P., Yang, J., Yue, Y., Dudus, L., Zhang, Y., Fisher, K. J., and Engelhardt, J. F. (1998). Circular intermediates of recombinant adeno-associated virus have defined structural characteristics responsible for long-term episomal persistence in muscle tissue. *J. Virol.* **72,** 8568–8577.

Duan, D., Li, Q., Kao, A. W., Yue, Y., Pessin, J. E., and Engelhardt, J. F. (1999). Dynamin is required for recombinant adeno-associated virus type 2 infection. *J. Virol.* **73,** 10371–10376.

Duan, D., Yue, Y., Yan, Z., Yang, J., and Engelhardt, J. F. (2000). Endosomal processing limits gene transfer to polarized airway epithelia by adeno-associated virus. *J. Clin. Invest.* **105,** 1573–1587.

Duan, D., Yue, Y., and Engelhardt, J. F. (2001). Expanding AAV packaging capacity with trans-splicing or overlapping vectors: A quantitative comparison. *Mol. Ther.* **4,** 383–391.

During, M. J., Xu, R., Young, D., Kaplitt, M. G., Sherwin, R. S., and Leone, P. (1998). Peroral gene therapy of lactose intolerance using an adeno-associated virus vector. *Nat. Med.* **4,** 1131–1135.

Dutheil, N., Shi, F., Dupressoir, T., and Linden, R. M. (2000). Adeno-associated virus site-specifically integrates into a muscle-specific DNA region. *Proc. Natl. Acad. Sci. USA* **97,** 4862–4866.

Eng, L. F., Ghirnikar, R. S., and Lee, Y. L. (2000). Glial fibrillary acidic protein: GFAP—thirty-one years (1969–2000). *Neurochem. Res.* **25,** 1439–1451.

Favre, D., Provost, N., Blouin, V., Blancho, G., Cherel, Y., Salvetti, A., and Moullier, P. (2001). Immediate and long-term safety of recombinant adeno-associated virus injection into the nonhuman primate muscle. *Mol. Ther.* **4,** 559–566.

Ferrari, F. K., Samulski, T., Shenk, T., and Samulski, R. J. (1996). Second-strand synthesis is a rate-limiting step for efficient transduction by recombinant adeno-associated virus vectors. *J. Virol.* **70,** 3227–3234.

Fisher, K. J., Gao, G. P., Weitzman, M. D., DeMatteo, R., Burda, J. F., and Wilson, J. M. (1996). Transduction with recombinant adeno-associated virus for gene therapy is limited by leading-strand synthesis. *J. Virol.* **70,** 520–532.

Fitzsimons, H. L., Mckenzie, J. M., and During, M. J. (2001). Insulators coupled to a minimal bidirectional tet cassette for tight regulation of rAAV-mediated gene transfer in the mammalian brain. *Gene Ther.* **8,** 1675–1681.

Flotte, T. R., and Carter, B. J. (1995). Adeno-associated virus vectors for gene therapy. *Gene Ther.* **2,** 357–362.

Flotte, T. R., Afione, S. A., Solow, R., Drumm, M. L., Markakis, D., Guggino, W. B., Zeitlin, P. L., and Carter, B. J. (1993). Expression of the cystic fibrosis transmembrane conductance regulator from a novel adeno-associated virus promoter. *J. Biol. Chem.* **268,** 3781–3790.

Gao, G. P., Alvica, M. C., Wang, L., Calcedo, R., Johnston, J., and Wilson, J. M. (2002). Novel adeno-associated viruses from rhesus monkeys as vectors for human gene therapy. *Proc. Natl. Acad. Sci. USA* **99,** 11854–11859.

Georg-Fries, B., Biederlack, S., Wolf, J., and zur Hausen, H. (1984). Analysis of proteins, helper dependence, and seroepidemiology of a new human parvovirus. *Virology* **134,** 64–71.

Girod, A., Ried, M., Wobus, C., Lahm, H., Leike, K., Kleinschmidt, J., Deleage, G., and Hallek, M. (1999). Genetic capsid modifications allow efficient re-targeting of adeno-associated virus type 2. *Nat. Med.* **5,** 1052–1056.

Goncalves, M. A., Pau, M. G., de Vries, A. A., and Valerio, D. (2001). Generation of a high-capacity hybrid vector: Packaging of recombinant adenoassociated virus replicative intermediates in adenovirus capsids overcomes the limited cloning capacity of adenoassociated virus vectors. *Virology* **288,** 236–246.

Grifman, M., Trepel, M., Speece, P., Gilbert, L. B., Arap, W., Pasqualini, R., and Weitzman, M. D. (2001). Incorporation of tumor-targeting peptides into recombinant adeno-associated virus capsids. *Mol. Ther.* **3,** 964–975.

Grimm, D., Kern, A., Rittner, K., and Kleinschmidt, J. A. (1998). Novel tools for production and purification of recombinant adenoassociated virus vectors. *Hum. Gene Ther.* **9,** 2745–2760.

Haberman, R. P., McCown, T. J., and Samulski, R. J. (1998). Inducible long-term gene expression in brain with adeno-associated virus gene transfer. *Gene Ther.* **5,** 1604–1611.

Haberman, R. P., McCown, T. J., and Samulski, R. J. (2000). Novel transcriptional regulatory signals in the adeno-associated virus terminal repeat A/D junction element. *J. Virol.* **74,** 8732–8739.

Halbert, C. L., Standaert, T. A., Aitken, M. L., Alexander, I. E., Russell, D. W., and Miller, A. D. (1997). Transduction by adeno-associated virus vectors in the rabbit airway: Efficiency, persistence, and readministration. *J. Virol.* **71,** 5932–5941.

Hansen, J., Qing, K., and Srivastava, A. (2001a). Adeno-associated virus type 2-mediated gene transfer: Altered endocytic processing enhances transduction efficiency in murine fibroblasts. *J. Virol.* **75,** 4080–4090.

Hansen, J., Qing, K., and Srivastava, A. (2001b). Infection of purified nuclei by adeno-associated virus 2. *Mol. Ther.* **4,** 289–296.

Hansen, J., Qing, K., Kwon, H. J., Mah, C., and Srivastava, A. (2000). Impaired intracellular trafficking of adeno-associated virus type 2 vectors limits efficient transduction of murine fibroblasts. *J. Virol.* **74,** 992–996.

Harrison, P. T., Dalziel, R. G., Ditchfield, N. A., and Quinn, J. P. (1999). Neuronal-specific and nerve growth factor-inducible expression directed by the preprotachykinin-A promoter delivered by an adeno-associated virus vector. *Neuroscience* **94,** 997–1003.

Hermonat, P. L., and Muzyczka, N. (1984). Use of adeno-associated virus as a mammalian DNA cloning vector: Transduction of neomycin resistance into mammalian tissue culture cells. *Proc. Natl. Acad. Sci. USA* **81,** 6466–6470.

Inoue, N., and Russell, D. W. (1998). Packaging cells based on inducible gene amplification for the production of adeno-associated virus vectors. *J. Virol.* **72,** 7024–7031.

Jay, F. T., de la Maza, L. M., and Carter, B. J. (1979). Parvovirus RNA transcripts containing sequences not present in mature mRNA: A method for isolation of putative mRNA precursor sequences. *Proc. Natl. Acad. Sci. USA* **76,** 625–629.

Kaludov, N., Brown, K. E., Walters, R. W., Zabner, J., and Chiorini, J. A. (2001). Adeno-associated virus serotype 4 (AAV4) and AAV5 both require sialic acid binding for hemagglutination and efficient transduction but differ in sialic acid linkage specificity. *J. Virol.* **75,** 6884–6893.

Kaplitt, M. G., Leone, P., Samulski, R. J., Xiao, X., Pfaff, D. W., O'Malley, K. L., and During, M. J. (1994). Long-term gene expression and phenotypic correction using adeno-associated virus vectors in the mammalian brain. *Nat. Genet.* **8,** 148–154.

Kay, M. A., Manno, C. S., Ragni, M. V., Larson, P. J., Couto, L. B., McClelland, A., Glader, B., Chew, A. J., Tai, S. J., Herzog, R. W., Arruda, V., Johnson, F., Scallan, C., Skarsgard, E., Flake, A. W., and High, K. A. (2000). Evidence for gene transfer and expression of factor IX in haemophilia B patients treated with an AAV vector. *Nat. Genet.* **24,** 257–261.

Keir, S. D., Miller, J., Yu, G., Hamilton, R., Samulski, R. J., Xiao, X., and Tornatore, C. (1997). Efficient gene transfer into primary and immortalized human fetal glial cells using adeno-associated virus vectors: Establishment of a glial cell line with a functional CD4 receptor. *J. Neurovirol.* **3,** 322–330.

Kirik, D., Rosenblad, C., Bjorklund, A., and Mandel, R. J. (2000). Long-term rAAV-mediated gene transfer of GDNF in the rat Parkinson's model: Intrastriatal but not intranigral transduction promotes functional regeneration in the lesioned nigrostriatal system. *J. Neurosci.* **20,** 4686–4700.

Klein, R. L., Meyer, E. M., Peel, A. L., Zolotukhin, S., Meyers, C., Muzyczka, N., and King, M. A. (1998). Neuron-specific transduction in the rat septohippocampal or nigrostriatal pathway by recombinant adeno-associated virus vectors. *Exp. Neurol.* **150,** 183–194.

Koeberl, D. D., Alexander, I. E., Halbert, C. L., Russell, D. W., and Miller, A. D. (1997). Persistent expression of human clotting factor IX from mouse liver after intravenous injection of adeno-associated virus vectors. *Proc. Natl. Acad. Sci. USA* **94,** 1426–1431.

Kotin, R. M., Linden, R. M., and Berns, K. I. (1992). Characterization of a preferred site on human chromosome 19q for integration of adeno-associated virus DNA by non-homologous recombination. *EMBO J.* **11,** 5071–5078.

Laughlin, C. A., Tratschin, J. D., Coon, H., and Carter, B. J. (1983). Cloning of infectious adeno-associated virus genomes in bacterial plasmids. *Gene* **23,** 65–73.

Lehtonen, E., Bonnaud, F., Melas, C., Lubansu, A., Malgrange, B., Chtarto, A., Velu, T., Brotchi, J., Levivier, M., Peschanski, M., and Tenenbaum, L. (2002). AAV2 vectors mediate efficient and sustained transduction of rat embryonic ventral mesencephalon. *Neurosport.* 13, 1503–1507.

Linden, R. M., Winocour, E., and Berns, K. I. (1996a). The recombination signals for adeno-associated virus site-specific integration. *Proc. Natl. Acad. Sci. USA* **93,** 7966–7972.

Linden, R. M., Ward, P., Giraud, C., Winocour, E., and Berns, K. I. (1996b). Site-specific integration by adeno-associated virus. *Proc. Natl. Acad. Sci. USA* **93,** 11288–11294.

Lo, W. D., Qu, G., Sferra, T. J., Clark, R., Chen, R., and Johnson, P. R. (1999). Adeno-associated virus-mediated gene transfer to the brain: duration and modulation of expression. *Hum. Gene Ther.* **10,** 201–213.

Loeb, J. E., Cordier, W. S., Harris, M. E., Weitzman, M. D., and Hope, T. J. (1999). Enhanced expression of transgenes from adeno-associated virus vectors with the woodchuck hepatitis virus posttranscriptional regulatory element: Implications for gene therapy. *Hum. Gene Ther.* **10,** 2295–2305.

Malik, A. K., Monahan, P. E., Allen, D. L., Chen, B. G., Samulski, R. J., and Kurachi, K. (2000). Kinetics of recombinant adeno-associated virus-mediated gene transfer. *J. Virol.* **74,** 3555–3565.

Mandel, R. J., Spratt, S. K., Snyder, R. O., and Leff, S. E. (1997). Midbrain injection of recombinant adeno-associated virus encoding rat glial cell line-derived neurotrophic factor protects nigral neurons in a progressive 6-hydroxydopamine-induced degeneration model of Parkinson's disease in rats. *Proc. Natl. Acad. Sci. USA* **94,** 14083–14088.

Mandel, R. J., Rendahl, K. G., Spratt, S. K., Snyder, R. O., Cohen, L. K., and Leff, S. E. (1998). Characterization of intrastriatal recombinant adeno-associated virus-mediated gene transfer of human tyrosine hydroxylase and human GTP-cyclohydrolase I in a rat model of Parkinson's disease. *J. Neurosci.* **18,** 4271–4284.

Mandel, R. J., Snyder, R. O., and Leff, S. E. (1999). Recombinant adeno-associated viral vector-mediated glial cell line-derived neurotrophic factor gene transfer protects nigral dopamine neurons after onset of progressive degeneration in a rat model of Parkinson's disease. *Exp. Neurol.* **160,** 205–214.

Manning, W. C., Zhou, S., Bland, M. P., Escobedo, J. A., and Dwarki, V. (1998). Transient immunosuppression allows transgene expression following readministration of adeno-associated viral vectors. *Hum. Gene Ther.* **9,** 477–485.

Marshall, E. (2001). Gene therapy. Panel reviews risks of germ line changes. *Science* **294,** 2268–2269.

Matsushita, T., Elliger, S., Podsakoff, G., Villarreal, L., Kurtzman, G. J., Iwaki, Y., and Lolosi, D. (1998). Adeno-associated virus vectors can be efficiently produced without helper virus. *Gene Ther.* **5,** 938–945 .

McCarty, D. M., Monahan, P. E., and Samulski, R. J. (2001). Self-complementary recombinant adeno-associated virus (scAAV) vectors promote efficient transduction independently of DNA synthesis. *Gene Ther.* **8,** 1248–1254.

McCown, T. J., Xiao, X., Li, J., Breese, G. R., and Samulski, R. J. (1996). Differential and persistent expression patterns of CNS gene transfer by an adeno-associated virus (AAV) vector. *Brain Res.* **713,** 99–107.

McGee Sanftner, L. H., Rendahl, K. G., Quiroz, D., Coyne, M., Ladner, M., Manning, W. C., and Flannery, J. G. (2001). Recombinant AAV-mediated delivery of a tet-inducible reporter gene to the rat retina. *Mol. Ther.* **3,** 688–696.

McLaughlin, S. K., Collis, P., Hermonat, P. L., and Muzyczka, N. (1988). Adeno-associated virus general transduction vectors: Analysis of proviral structures. *J. Virol.* **62,** 1963–1973.

Meneses, P., Berns, K. I., and Winocour, E. (2000). DNA sequence motifs which direct adeno-associated virus site-specific integration in a model system. *J. Virol.* **74,** 6213–6216.

Miller, D. G., Rutledge, E. A., and Russell, D. W. (2002). Chromosomal effects of adeno-associated virus vector integration. *Nat. Genet.* **30,** 147–148.

Miller, J. L., Donahue, R. E., Sellers, S. E., Samulski, R. J., Young, N. S., and Nienhuis, A. W. (1994). Recombinant adeno-associated virus (rAAV)-mediated expression of a human gamma-globin gene in human progenitor-derived erythroid cells. *Proc. Natl. Acad. Sci. USA* **91,** 10183–10187.

Monahan, P. E., and Samulski, R. J. (2000). AAV vectors: Is clinical success on the horizon? *Gene Ther.* **7,** 24–30.

Monahan, P. E., Samulski, R. J., Tazelaar, J., Xiao, X., Nichols, T. C., Bellinger, D. A., Read, M. S., and Walsh, C. E. (1998). Direct intramuscular injection with recombinant AAV vectors results in sustained expression in a dog model of hemophilia. *Gene Ther.* **5,** 40–49.

Monahan, P. E., Jooss, K., and Sands, M. S. (2002). Safety of adeno-associated virus gene therapy vectors: Current evaluation. *Expert Opin. Drug Saf.* (in press).

Muramatsu, S., Mizukami, H., Young, N. S., and Brown, K. E. (1996). Nucleotide sequencing and generation of an infectious clone of adeno-associated virus 3. *Virology* **221,** 208–217.

Musatov, S. A., Dudus, L., Parrish, C. M., Scully, T. A., and Fisher, K. J. (2002). Spontaneous mobilization of integrated recombinant adenoassociated virus in a cell culture model of virus latency. *Virology* **294,** 151–169.

Muzyczka, N. (1992). Use of adeno-associated virus as a general transduction vector for mammalian cells. *Curr. Top. Microbiol. Immunol.* **158,** 97–129.

Ni, T. H., McDonald, W. F., Zolotukhin, I., Melendy, T., Waga, S., Stillman, B., and Muzyczka, N. (1998). Cellular proteins required for adeno-associated virus DNA replication in the absence of adenovirus coinfection. *J. Virol.* **72,** 2777–2787.

Nurcombe, V., Ford, M. D., Wildschut, J. A., and Bartlett, P. F. (1993). Developmental regulation of neural response to FGF-1 and FGF-2 by heparan sulfate proteoglycan. *Science* **260,** 103–106.

O'Riordan, C. R., Lachapelle, A. L., Vincent, K. A., and Wadsworth, S. C. (2000). Scaleable chromatographic purification process for recombinant adeno-associated virus (rAAV). *J. Gene Med.* **2,** 444–454.

Paterna, J. C., Moccetti, T., Mura, A., Feldon, J., and Bueler, H. (2000). Influence of promoter and WHV post-transcriptional regulatory element on AAV-mediated transgene expression in the rat-brain. *Gene Ther.* **7,** 1304–1311.

Peel, A. L., and Klein, R. L. (2000). Adeno-associated virus vectors: Activity and applications in the CNS. *J. Neurosci. Methods.* **98,** 95–104.

Piccini, P., Brooks, D. J., Bjorklund, A., Gunn, R. N., Grasby, P. M., Rimoldi, O., Brundin, P., Hagell, P., Rehncrona, S., Widner, H., and Lindvall, O. (1999). Dopamine release from nigral transplants visualized in vivo in a Parkinson's patient. *Nat. Neurosci.* **2,** 1137–1140.

Pinkstaff, J. K., Detterich, J., Lynch, G., and Gall, C. (1999). Integrin subunit gene expression is regionally differentiated in adult brain. *J. Neurosci.* **19,** 1541–1556.

Ponnazhagan, S., Erikson, D., Kearns, W. G., Zhou, S. Z., Nahreini, P., Wang, X. S., and Srivastava, A. (1997). Lack of site-specific integration of the recombinant adeno-associated virus 2 genomes in human cells. *Hum. Gene Ther.* **8,** 275–284.

Prosch, S., Stein, J., Staak, K., Liebenthal, C., Volk, H. D., and Kruger, D. H. (1996). Inactivation of the very strong HCMV immediate early promoter by DNA CpG methylation in vitro. *Biol. Chem. Hoppe Seyler* **377,** 195–201.

Qing, K., Wang, X. S., Kube, D. M., Ponnazhagan, S., Bajpai, A., and Srivastava, A. (1997). Role of tyrosine phosphorylation of a cellular protein in adeno-associated virus 2-mediated transgene expression. *Proc. Natl. Acad. Sci. USA* **94,** 10879–10884.

Qing, K., Khuntirat, B., Mah, C., Kube, D. M., Wang, X. S., Ponnazhagan, S., Zhou, S., Dwarki, V. J., Yoder, M. C., and Srivastava, A. (1998). Adeno-associated virus type 2-mediated gene transfer: Correlation of tyrosine phosphorylation of the cellular single-stranded D sequence-binding protein with transgene expression in human cells in vitro and murine tissues in vivo. *J. Virol.* **72,** 1593–1599.

Qing, K., Mah, C., Hansen, J., Zhou, S., Dwarki, V., and Srivastava, A. (1999). Human fibroblast growth factor receptor 1 is a co-receptor for infection by adeno-associated virus 2. *Nat. Med.* **5,** 71–77.

Qiu, J., Mizukami, H., and Brown, K. E. (1999). Adeno-associated virus 2 co-receptors? *Nat. Med.* **5,** 467–468.

Rabinowitz, J. E., Rolling, F., Li, C., Conrath, H., Xiao, W., Xiao, X., and Samulski, R. J. (2002). Cross-packaging of a single adeno-associated virus (AAV) type 2 vector genome into multiple AAV serotypes enables transduction with broad specificity. *J. Virol.* **76,** 791–801.

Rendahl, K. G., Leff, S. E., Otten, G. R., Spratt, S. K., Bohl, D., van Roey, M., Donahue, B. A., Cohen, L. K., Mandel, R. J., Danos, O., and Snyder, R. O. (1998). Regulation of gene expression in vivo following transduction by two separate rAAV vectors. *Nat. Biotechnol.* **16,** 757–761.

Richardson, W. D., and Westphal, H. (1981). A cascade of adenovirus early functions is required for expression of adeno-associated virus. *Cell* **27,** 133–141.

Richardson, W. D., and Westphal, H. (1984). Requirement for either early region 1a or early region 1b adenovirus gene products in the helper effect for adeno-associated virus. *J. Virol.* **51,** 404–410.

Ried, M. U., Girod, A., Leike, K., Buning, H., and Hallek, M. (2002). Adeno-associated virus capsids displaying immunoglobulin-binding domains permit antibody-mediated vector retargeting to specific cell surface receptors. *J. Virol.* **76,** 4559–4566.

Rohde, V., Erles, K., Sattler, H. P., Derouet, H., Wullich, B., and Schlehofer, J. R. (1999). Detection of adeno-associated virus in human semen: Does viral infection play a role in the pathogenesis of male infertility? *Fertil. Steril.* **72,** 814–816.

Rutledge, E. A., Halbert, C. L., and Russell, D. W. (1998). Infectious clones and vectors derived from adeno-associated virus (AAV) serotypes other than AAV type 2. *J. Virol.* **72,** 309–319.

Samulski, R. J., Berns, K. I., Tan, M., and Muzyczka, N. (1982). Cloning of adeno-associated virus into pBR322: Rescue of intact virus from the recombinant plasmid in human cells. *Proc. Natl. Acad. Sci. USA* **79,** 2077–2081.

Samulski, R. J., Chang, L. S., and Shenk, T. (1989). Helper-free stocks of recombinant adeno-associated viruses: Normal integration does not require viral gene expression. *J. Virol.* **63,** 3822–3828.

Samulski, R. J., Zhu, X., Xiao, X., Brook, J. D., Housman, D. E., Epstein, N., and Hunter, L. A. (1991). Targeted integration of adeno-associated virus (AAV) into human chromosome 19. *EMBO J.* **10,** 3941–3950.

Samulski, R. J., Sally, M., and Muzyczka, N. (1999). Adeno-associated viral vectors. *In* "The Development of Human Gene Therapy" (T. Friedman, Ed.), pp. 131–172. Cold Spring Harbor Laboratory Press, Cold Spring Harbor, NY.

Seisenberger, G., Ried, M. U., Endress, T., Buning, H., Hallek, M., and Brauchle, C. (2001). Real-time single-molecule imaging of the infection pathway of an adeno-associated virus. *Science* **294,** 1929–1932.

Shi, W., Arnold, G. S., and Bartlett, J. S. (2001). Insertional mutagenesis of the adeno-associated virus type 2 (AAV2) capsid gene and generation of AAV2 vectors targeted to alternative cell-surface receptors. *Hum. Gene Ther.* **12,** 1697–1711.

Siegl, G., Bates, R. C., Berns, K. I., Carter, B. J., Kelly, D. C., Kurstak, E., and Tattersall, P. (1985). Characteristics and taxonomy of Parvoviridae. *Intervirology* **23,** 61–73.

Snyder, R. O., Xiao, X., and Samulski, R. J. (1996). Production of recombinant adeno-associated viral vectors. *In* "Current Protocols in Human Genetics" (N. Dracopoli, Ed.), pp. 12.1.1–12.1.23. John Wiley & Son, New York.

Snyder, R. O., Miao, C. H., Patijn, G. A., Spratt, S. K., Danos, O., Nagy, D., Gown, A. M., Winther, B., Meuse, L., Cohen, L. K., Thompson, A. R., and Kay, M. A. (1997). Persistent and therapeutic concentrations of human factor IX in mice after hepatic gene transfer of recombinant AAV vectors. *Nat. Genet.* **16,** 270–276.

Srivastava, C. H., Samulski, R. J., Lu, L., Larsen, S. H., and Srivastava, A. (1989). Construction of a recombinant human parvovirus B19: Adeno-associated virus 2 (AAV) DNA inverted terminal repeats are functional in an AAV-B19 hybrid virus. *Proc. Natl. Acad. Sci USA* **86,** 8078–8082.

Summerford, C., and Samulski, R. J. (1998). Membrane-associated heparan sulfate proteoglycan is a receptor for adeno-associated virus type 2 virions. *J. Virol.* **72,** 1438–1445.

Summerford, C., Bartlett, J. S., and Samulski, R. J. (1999). AlphaVbeta5 integrin: A co-receptor for adeno-associated virus type 2 infection. *Nat. Med.* **5,** 78–82.

Sun, L., Li, J., and Xiao, X. (2000). Overcoming adeno-associated virus vector size limitation through viral DNA heterodimerization. *Nat. Med.* **6,** 599–602.

Takami, K., Matsuo, A., Terai, K., Walker, D. G., McGeer, E. G., and McGeer, P. L. (1998). Fibroblast growth factor receptor-1 expression in the cortex and hippocampus in Alzheimer's disease. *Brain Res.* **802,** 89–97.

Tamayose, K., Hirai, Y., and Shimada, T. (1996). A new strategy for large-scale preparation of high-titer recombinant adeno-associated virus vectors by using packaging cell lines and sulfonated cellulose column chromatography. *Hum. Gene Ther.* **7,** 507–513.

Tawil, N. J., Wilson, P., and Carbonetto, S. (1994). Expression and distribution of functional integrins in rat CNS glia. *J. Neurosci. Res.* **39,** 436–447.

Tenenbaum, L., and Hooghe-Peters, E. (1996). Gene delivery using adeno-associated virus. *In* "Genetic Manipulation of the Nervous System" (D. S. Latchman, Ed.), pp. 73–97. Academic Press, San Diego.

Tenenbaum, L., Darling, J. L., and Hooghe-Peters, E. (1994). Adeno-associated virus (AAV) as a vector for gene transfer into glial cells of the human central nervous system. *Gene Ther* **1**(Suppl. 1), S80.

Tenenbaum, L., Jurysta, F., Stathopoulos, A., Puschban, Z., Melas, C., Hermens, W. T., Verhaagen, J., Pichon, B., Velu, T., and Levivier, M. (2000). Tropism of AAV-2 vectors for neurons of the globus pallidus. *Neuroreport* **11,** 2277–2283.

Tobiasch, E., Burguete, T., Klein-Bauernschmitt, P., Heilbronn, R., and Schlehofer, J. R. (1998). Discrimination between different types of human adeno-associated viruses in clinical samples by PCR. *J. Virol. Methods* **71,** 17–25.

Tratschin, J. D., Miller, I. L., Smith, M. G., and Carter, B. J. (1985). Adeno-associated virus vector for high-frequency integration, expression, and rescue of genes in mammalian cells. *Mol. Cell. Biol.* **5,** 3251–3260.

Urcelay, E., Ward, P., Wiener, S. M., Safer, B., and Kotin, R. M. (1995). Asymmetric replication in vitro from a human sequence element is dependent on adeno-associated virus Rep protein. *J. Virol.* **69,** 2038–2046.

Vincent-Lacaze, N., Snyder, R. O., Gluzman, R., Bohl, D., Lagarde, C., and Danos, O. (1999). Structure of adeno-associated virus vector DNA following transduction of the skeletal muscle. *J. Virol.* **73,** 1949–1955.

Walters, R. W., Yi, S. M., Keshavjee, S., Brown, K. E., Welsh, M. J., Chiorini, J. A., and Zabner, J. (2001). Binding of adeno-associated virus type 5 to 2,3-linked sialic acid is required for gene transfer. *J. Biol. Chem.* **276,** 20610–20616.

Walz, C., Deprez, A., Dupressoir, T., Durst, M., Rabreau, M., and Schlehofer, J. R. (1997). Interaction of human papillomavirus type 16 and adeno-associated virus type 2 co-infecting human cervical epithelium. *J. Gen. Virol.* **78**(pt. 6), 1441–1452.

Weitzman, M. D., Kyostio, S. R., Kotin, R. M., and Owens, R. A. (1994). Adeno-associated virus (AAV) Rep proteins mediate complex formation between AAV DNA and its integration site in human DNA. *Proc. Natl. Acad. Sci. USA* **91,** 5808–5812.

West, M. H., Trempe, J. P., Tratschin, J. D., and Carter, B. J. (1987). Gene expression in adeno-associated virus vectors: the effects of chimeric mRNA structure, helper virus, and adenovirus VA1 RNA. *Virology* **160,** 38–47.

Wu, P., Phillips, M. I., Bui, J., and Terwilliger, E. F. (1998). Adeno-associated virus vector-mediated transgene integration into neurons and other nondividing cell targets. *J. Virol.* **72,** 5919–5926.

Wu, P., Ye, Y., and Svendsen, C. N. (2002). Transduction of human neural progenitor cells using recombinant adeno-associated viral vectors. *Gene Ther.* **9,** 245–255.

Xiao, W., Berta, S. C., Lu, M. M., Moscioni, A. D., Tazelaar, J., and Wilson, J. M. (1998). Adeno-associated virus as a vector for liver-directed gene therapy. *J. Virol.* **72,** 10222–10226.

Xiao, W., Chirmule, N., Berta, S. C., McCullough, B., Gao, G., and Wilson, J. M. (1999). Gene therapy vectors based on adeno-associated virus type 1. *J. Virol.* **73,** 3994–4003.

Xiao, X., Li, J., and Samulski, R. J. (1996). Efficient long-term gene transfer into muscle tissue of immunocompetent mice by adeno-associated virus vector. *J. Virol.* **70,** 8098–8108.

Xiao, X., Li, J., McCown, T. J., and Samulski, R. J. (1997). Gene transfer by adeno-associated virus vectors into the central nervous system. *Exp. Neurol.* **144,** 113–124.

Xiao, X., Li, J., and Samulski, R. J. (1998). Production of high-titer recombinant adeno-associated virus vectors in the absence of helper adenovirus. *J. Virol.* **72,** 2224–2232.

Xie, Q., Bu, W., Bhatia, S., Hare, J., Somasundaram, T., Azzi, A., and Chapman, M. S. (2002). The atomic structure of adeno-associated virus (AAV-2), a vector for human gene therapy. *Proc. Natl. Acad. Sci. USA* **99,** 10405–10410.

Xu, R., Janson, C. G., Mastakov, M., Lawlor, P., Young, D., Mouravlev, A., Fitzsimons, H., Choi, K. L., Ma, H., Dragunow, M., Leone, P., Chen, Q., Dicker, B., and During, M. J. (2001). Quantitative comparison of expression with adeno-associated virus (AAV-2) brain-specific gene cassettes. *Gene Ther.* **8,** 1323–1332.

Yalkinoglu, A. O., Heilbronn, R., Burkle, A., Schlehofer, J. R., and zur Hausen, H. (1988). DNA amplification of adeno-associated virus as a response to cellular genotoxic stress. *Cancer Res.* **48,** 3123–3129.

Yakobson, B., Koch, T., and Winocour, E. (1987). Replication of adeno-associated virus in synchronized cells without the addition of a helper virus. *J. Virol.* **61,** 972–981.

Yakobson, B., Hrynko, T. A., Peak, M. J., and Winocour, E. (1989). Replication of adeno-associated virus in cells irradiated with UV light at 254 nm. *J. Virol.* **63,** 1023–1030.

Yan, Z., Zhang, Y., Duan, D., and Engelhardt, J. F. (2000). From the cover: Trans-splicing vectors expand the utility of adeno-associated virus for gene therapy. *Proc. Natl. Acad. Sci. USA* **97,** 6716–6721.

Yan, Z., Zak, R., Luxton, G. W., Ritchie, T. C., Bantel-Schaal, U., and Engelhardt, J. F. (2002). Ubiquitination of both adeno-associated virus type 2 and 5 capsid proteins affects the transduction efficiency of recombinant vectors. *J. Virol.* **76,** 2043–2053.

Yang, C. C., Xiao, X., Zhu, X., Ansardi, D. C., Epstein, N. D., Frey, M. R., Matera, A. G., and Samulski, R. J. (1997). Cellular recombination pathways and viral terminal repeat hairpin structures are sufficient for adeno-associated virus integration in vivo and in vitro. *J. Virol.* **71,** 9231–9247.

Yang, J., Zhou, W., Zhang, Y., Zidon, T., Ritchie, T., and Engelhardt, J. F. (1999). Concatamerization of adeno-associated virus circular genomes occurs through intermolecular recombination. *J. Virol.* **73,** 9468–9477.

Yang, Q., Chen, F., and Trempe, J. P. (1994). Characterization of cell lines that inducibly express the adeno-associated virus Rep proteins. *J. Virol.* **68,** 4847–4856.

Ye, X., Rivera, V. M., Zoltick, P., Cerasoli, F., Jr., Schnell, M. A., Gao, G., Hughes, J. V., Gilman, M., and Wilson, J. M. (1999). Regulated delivery of therapeutic proteins after in vivo somatic cell gene transfer. *Science* **283,** 88–91.

Young, S. M., Jr., and Samulski, R. J. (2001). Adeno-associated virus (AAV) site-specific recombination does not require a Rep-dependent origin of replication within the AAV terminal repeat. *Proc. Natl. Acad. Sci. USA* **98,** 13525–13530.

Zolotukhin, S., Byrne, B. J., Mason, E., Zolotukhin, I., Potter, M., Chesnut, K., Summerford, C., Samulski, R. J., and Muzyczka, N. (1999). Recombinant adeno-associated virus purification using novel methods improves infectious titer and yield. *Gene Ther.* **6,** 973–985.

# PROBLEMS IN THE USE OF HERPES SIMPLEX VIRUS AS A VECTOR

Lawrence T. Feldman

Department of Microbiology, Immunology, and Molecular Genetics
UCLA School of Medicine
Los Angeles, California 90095



## I. Introduction

Herpes simplex virus (HSV) is an attractive candidate for use as a viral vector to express foreign genes. It has a large, double-stranded DNA genome of 152 kb. Several viral genes can be deleted, allowing for insertion of considerably more foreign DNA than is possible with many other viral vectors. Because HSV is neurotropic, it may be particularly suitable for transfer and expression of genes in the nervous system. Not only can HSV enter neurons eciently, it can also establish a life-long latent infection within peripheral neurons. It is this property that has attracted a number of investigators to develop the virus as a neuronal vector.

The challenges of using HSV as a vector have been many. The wild-type virus is very cytotoxic. Considerable eort has been made to reduce its cytotoxicity by the alteration of appropriate viral genes. Neal DeLuca and his colleagues at the University of Pittsburg, including Joseph Glorioso, have made a number of interesting improvements in this area, which will be discussed below. A second general method for reducing the toxicity of HSV has been to make amplicons. HSV will package any plasmid DNA that contains an origin of replication and the packaging signals. These amplicons do not contain any of the essential viral genes, so they are propagated with the aid of wild-type or conditional mutants of HSV, which provide the necessary proteins for complementation. Gene therapy investigators initially used this system to package plasmids that contained a suitable promoter-gene insert between the packaging signals and origin. Such






0074-7742/03 $35.00

mixed viral stocks were rather toxic, and below I will discuss improvements in this system.

In addition to viral toxicity, another general problem has been how to get long-term expression of the foreign gene in the latent state in neurons. There has been some success on that front in peripheral neurons; however, long-term expression in the central nervous system has not been as good.

In this article I will explain the general features of the lytic and latent phases of the viral life cycle, focusing on molecular aspects as they relate to either expression or toxicity. I will then explain the molecular aspects of amplicons and how they work. Finally, I will discuss recent eorts at understanding the molecular basis of long-term expression in neuronal latency. A recent article by Glorioso and colleagues discusses other aspects of HSV vectors such as applications to neuronal disease and applications to the use of HSV and the HSV viral thymidine kinase to target cancer cells for destruction (Burton *et al.*, 2001).

## II. General Aspects of Lytic Cycle Replication and Latent Infections of Neurons *in Vivo*

### A. Immediate-Early Genes

HSV has evolved a strategy for high titer replication in epithelial cells based on the sequential transcription of the three classes of viral genes, which have been termed immediate-early, early, and late genes (Honess and Roizman, 1974, 1975). There are five immediate-early genes; each of them is marked by the presence of one or more TAATGARAT motifs in the enhancer region (Mackem and Roizman, 1982). These motifs are the DNA binding sites for a combination of regulatory proteins. Oct-1 is a cellular transcription factor from the POU family of proteins (He *et al.*, 1989). VP16 is a viral-encoded protein found in the tegument of the virion (Batterson and Roizman, 1983). Upon infection, VP16 is brought into the nucleus, and these two proteins combine with a host cell factor to assemble and bind to the TAATGARAT motifs of the immediate-early genes, stimulating their transcription (O'Hare and Goding, 1988; Preston *et al.*, 1983). Of the IE gene products, only ICP4 and ICP27 are essential for viral replication (DeLuca *et al.*, 1985; Sacks *et al.*, 1985; Dixon and Schaer, 1980).

### B. Three Waves of Viral Transcripts Are Made in the Lytic Cycle

Two immediate-early genes, ICP0 and ICP4, encode transcriptional activators (Fig. 1). These two proteins induce the transcription of the entire set of early viral genes, most of which are proteins needed for viral DNA synthesis

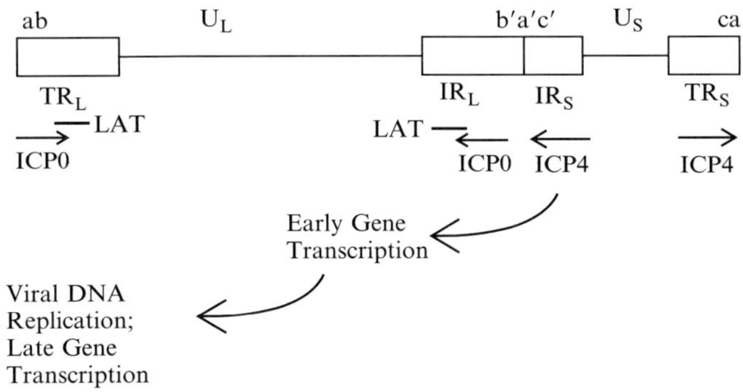


FIG. 1. The HSV-1 viral genome and an abbreviated outline of the lytic cycle. The unique long ($U_L$) and unique short ($U_S$) regions are flanked by terminal and inverted repeat sequences (TR and IR). Immediate-early genes ICP0 and ICP4 are transcribed from the long and short repeats, respectively. The one RNA made during latency is the latency-associated transcript (LAT), which is antisense to part of the ICP0 mRNA.

(DeLuca *et al.*, 1985; Dixon and Schaer, 1980; Cai and Schaffer, 1992; Chen and Silverstein, 1992; Campbell, M. E., Palfreyman and Preston, 1984). Upon replication of the viral DNA, a third set of viral genes, the late genes, is activated and expressed. These include cytotoxic proteins, structural proteins, and glycoproteins, among other functions. The progressive expression of immediate-early, early, and late viral genes is called the lytic cycle and it is the cycle used by the virus to grow productively in both tissue culture cells as well as in epithelial cells of an animal.

### C. Infection of Neurons Results in Either Lytic or Latent Infections

HSV replicates in a lytic fashion only upon infection of cells in culture. However, infection of epithelial cells of an animal results in spread of the virus to the peripheral neurons, where the virus replicates lytically and, alternatively, can establish a quiescent, latent infection.

To study HSV latency, there are several good animal models, including the mouse eye and the rabbit eye models (Hill *et al.*, 1975; Knotts *et al.*, 1974; Laycock *et al.*, 1991). In these models the virus is inoculated onto a scarified cornea of the eye. The virus replicates in the eye and subsequently enters nerve termini, where it follows one of two divergent pathways (Margolis *et al.*, 1992). In some neurons the virus replicates in a lytic fashion; this probably kills the neuron. Upon infection of other neurons, the virus proceeds down an alternative pathway in which the lytic cycle genes are not expressed and the virus establishes what is called a latent infection. During HSV latency the viral DNA is present in single or multiple copies in an episomal structure in the neuronal nucleus (Mellerick and Fraser,

1987). HSV latency is a relatively quiescent state in which transcription of the viral genome is limited to a single promoter (Dobson *et al.*, 1989). The viral RNAs that are expressed during latency have been dubbed the "latency-associated transcripts" or LAT (Stevens *et al.*, 1987; Croen *et al.*, 1987; Spivack and Fraser, 1987; Rock *et al.*, 1987). The exact role of LAT in latency is still unclear. Most of the primary LAT transcript is degraded, but a very stable intron is spliced out that partially overlaps with and is antisense to a part of the ICP0 mRNA (Farrell *et al.*, 1991; Krummenacher *et al.*, 1997; Rodahl and Haarr, 1997).

## III. Reducing Cytotoxicity of the Virus: The Gene Deletion Approach

The five immediate-early genes are ICP0, ICP4, ICP22, ICP27, and ICP47. Of these, only ICP4 and ICP27 are essential for viral replication. In trying to reduce cytotoxicity, the logical gene to mutate first is the ICP4 gene because the protein of this gene is absolutely required to activate the transcription of all early and late viral genes. Unfortunately, ICP4 protein also represses the transcription of the other immediate-early genes (Batchelor *et al.*, 1994). Thus, although a virus deleted for the ICP4 gene does not express early and late viral genes, it actually overexpresses the other four immediate-early genes. It turns out that these proteins are toxic to cells and infection with an ICP4 minus virus results in cell death, even though the early and late genes remain silent (Johnson *et al.*, 1994; Wu *et al.*, 1996; Samaniego *et al.*, 1998; Krisky *et al.*, 1998).

To resolve this problem, DeLuca and colleagues have produced a series of viruses with multiple deletions of the immediate-early genes. All of these viruses, including the original ICP4 minus virus, are grown as host range mutants. That is, they are grown on cell lines expressing the deleted viral gene. The phenotype of each deletion can be analyzed by infection of normal, control cells that do not express the immediate-early proteins. Infection of control cells with a virus that is deleted for both the ICP4 and the ICP22 gene results in less cytotoxicity compared to infection with a virus bearing the single deletion ICP4 (Krisky *et al.*, 1998). The same is true for infection with a virus doubly deleted for ICP4 and ICP27 (Krisky *et al.*, 1998). Even less cytotoxicity is observed with infection of the triple mutants, which is deleted for all three of these genes (Krisky *et al.*, 1998).

Quintuple mutants have been made in which all five immediate-early viral genes have been deleted (Samaniego *et al.*, 1998). This mutant is entirely nontoxic to cells. Although these viral genomes persist within the infected cell for prolonged periods, the expression of promoters from quintuple mutants is very low. This apparently is due to the absence of the ICP0 protein. ICP0 allows the ecient expression of genes from the viral genome. However, the ICP0 protein is apparently degraded in neurons, so that its inclusion within a vector may have dierent eects depending upon the target cell (Krisky *et al.*, 1998; Lilley *et al.*, 2001; Marshall *et al.*, 2000; Palmer *et al.*, 2000).

## IV. Reducing Cytotoxicity of the Virus: The Amplicon Approach

HSV is composed of 152 kb of DNA and it packages its DNA on the head-full hypothesis. The *cis* signals needed on the virus for replication and packaging are limited to an origin of replication and the packaging signals, called the "a" sequences. At high multiplicities of infection, the virus will form defective interfering particles, as first demonstrated by Niza Frenkel (Frenkel *et al.*, 1976; Spaete and Frenkel, 1982). Defective interfering particles contain multiple copies of a shorter fragment of the viral genome containing the *cis* signals. For example, the virus might contain 10 copies of a 15-kb region, making approximately 152 kb. These defective interfering particles are formed in the presence of wild-type virus and must be replicated by passing the defective viral stocks at a high multiplicity of infection, which ensures that each infected cell has sucient wild-type virus to complement the functions that are missing in the defective particles. The ratio of wild-type virus to defective interfering particle often changes with each passage of the defective viral stock.

Despite the fluctuation of titer in this system, the potential reduction in toxicity of a defective virus has attracted many investigators. The original idea was to start with a plasmid that contained the origin of replication and the "a" sequences. The transgene is cloned into this plasmid, creating what is called an amplicon, that is, a plasmid that is capable of being replicated and packaged by a wild-type helper virus. Some of the original amplicons were packaged using helper viruses that were either temperature sensitive for the ICP4 gene or were host range mutants in ICP4 (Geller and Breakefield, 1988). However, these virus stocks still contained the other four immediate-early genes, as well as the virion host shut-off protein, VHS, and they were cytotoxic.

Alfred Geller devised a clever scheme for improving amplicon stocks. It was known that a series of overlapping cosmid clones of HSV could be transfected into cells to generate virus by recombination with the overlapping, homologous sequences. Geller deleted the "a" sequences from this set of cosmids, but retained the sequences in the amplicon. Thus, cotransfection of an amplicon with the modified cosmid set of viral DNAs provided all the proteins required for complementation, but the only viral DNAs that could be packaged were the amplicons because only the amplicons contained the packaging signals (Geller *et al.*, 1990). This approach has allowed the generation of "helper-free" amplicons.

One of the disadvantages of the helper-free amplicons is their low titer. Typically stocks of these amplicons are only around $1 \times 10^4$ plaque-forming units

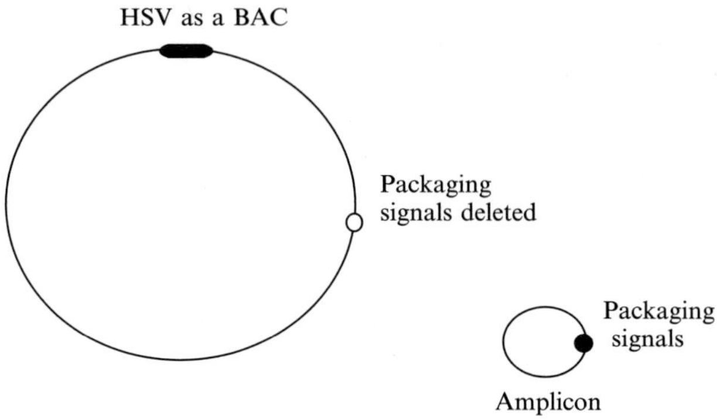


FIG. 2. The structure of an HSV DNA cloned into a BAC. The entire genome of HSV can be cloned into a BAC, as shown above. If this BAC is to be used as a helper virus, the packaging signals are deleted from the viral genome. This deleted form thus has all of the viral genes necessary for complementing the amplicon, but its DNA cannot be packaged into a capsid.

(pfu)/ml, in contrast to wild-type viral stocks that can exceed $1 \times 10^9$ pfu/ml. The likely explanation for the lower titer is the requirement for transfection and recombination of all of the cosmids to occur in the same cell as the amplicon in order to generate viral DNA that can replicate to amplify the viral DNA copy number. This problem was partially solved by using the modified viral DNA as cloned into a BAC vector, rather than as overlapping cosmids (Fig. 2). A BAC is a bacterial artificial chromosome. It replicates at one copy per bacterial cell and is capable of containing hundreds of kilobases of DNA. Herpesvirologists first used BACs to clone cytomegalovirus DNA, and later for other herpesvirus genomes, including HSV, Epstein–Barr virus, and pseudorabies virus (Messerle *et al.*, 1997; Borst *et al.*, 1999; Saeki *et al.*, 1998; Smith and Enquist, 1999; Delecluse *et al.*, 1998). A BAC used in an amplicon system would contain the entire viral DNA genome, absent the "a" sequences, and cloned into a BAC vector. In this way the viral DNA is prepared in the same manner as for any plasmid DNA, for example, by Qiagen preparation. The viral DNA in the BAC is cotransfected with the amplicon, and this system is much more ecient at replicating the helper-free amplicons. Titers approaching $1 \times 10^7$ pfu of helper-free amplicons have been produced using this technique.

## V. The Problem of Long-Term Expression during HSV Latency

Another aspect of the use of HSV as a vector is that of long-term promoter expression within the latent state. Two observations made HSV an attractive candidate for expression of transgenes in neurons. First, it was observed that during latency the lytic cycle genes are not expressed (Stevens *et al.*, 1987). Second, a viral promoter was identified that remained active during latency. This promoter is responsible for transcription of the latency-associated transcripts or LAT (Dobson *et al.*, 1989). It seemed a simple task to use the LAT promoter to drive a transgene of interest during the latent phase. A number of investigators made LAT promoter reporter viruses, in which the LAT promoter driving the *lac*Z gene was inserted into the viral genome in a recombinant virus. Neurons infected with these viruses could be stained with the chromophore x-gal, and the $\beta$-galactosidase protein, a product of the *lac*Z gene, would cleave the x-gal, turning the neurons blue. However, it turns out that the LAT promoter is more active at the beginning of the latent phase, and its activity declines by 15- to 20-fold after the first week of infection (Margolis *et al.*, 1993). Thus, infection with LAT reporter viruses gave rise to blue neurons when harvested at 1 week postinfection, but by 28 days postinfection almost every infected neuron was white, indicating a dramatic decline in LAT promoter activity.

A mystery in these experiments was why the LAT reporter constructs failed to make enough $\beta$-gal protein to be observed during latency, but the LAT promoter in the repeat region still made detectable LAT RNA. Many other promoters inserted into the virus also declined in activity. An exception to this was the pairing of the LTR from Moloney murine leukemia virus with the LAT promoter. LAT-LTR-lacZ reporter viruses maintained $\beta$-gal production throughout the latent period (Lokensgard *et al.*, 1994).

Another factor in the decline in LAT promoter during latency was the position of the LAT promoter within the viral genome. Reporter constructs placed within the long repeat region, at the natural site of the LAT promoter, continued to make RNA or protein during latency. An example of this is the LAT promotor-$\beta$-globin virus (Dobson *et al.*, 1989). However, LAT reporter viruses inserted outside of the long repeat region always failed to synthesize detectable RNA or protein during latency. This dierence in expression was traced to a region of the long repeat termed the LTE, for long-term expression. When the LTE is paired with the LAT promoter, detectable RNA is made during latency (Lokensgard, *et al.*, 1997). At first glance, the LTE and the LTR appear to be similar in function; however, the LTE is actually a strong enhancer of the LAT promoter and apparently functions by greatly raising the initial synthetic rate of the LAT promoter (Berthomme *et al.*, 2000). Exactly how the LTR functions to maintain latent expression is unknown, but it is not an enhancer of the LAT promoter (Fig. 3).

Both the LAT–LTE promoter and the LAT–LTR promoter provide the ability to express transgenes in latency in peripheral neurons. Lachman *et al.* have improved upon this system by using combinations of these two elements, as

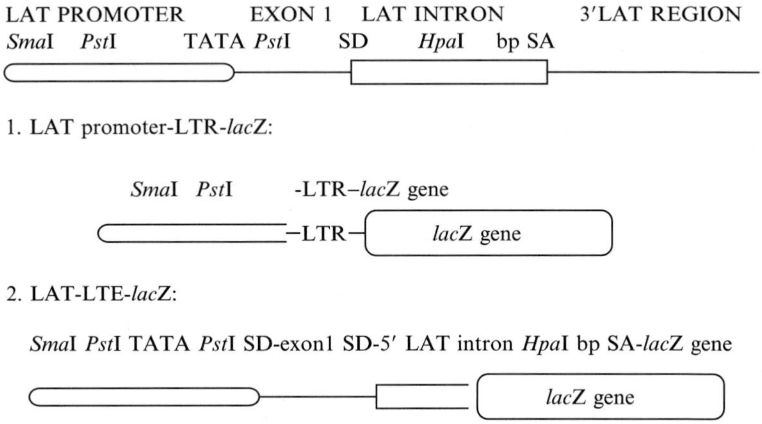


FIG. 3. The LAT promoter region and two reporter constructs for long-term expression. The top panel shows the LAT promoter region, extending from an upstream *Sma*I site through the TATA box and the second *Pst*I site. A leader RNA region, labeled here as exon 1, precedes the LAT intron, which is demarcated by splice donor (SD) and splice acceptor (SA) sites, and an internal *Hpa*I site. In the LAT–LTR–*lac*Z construct, the upstream LAT promoter region preceding the TATA box is paired with the LTR. This dual promoter, with initiation within the LTR region, is driving the *lac*Z gene. In the LAT–LTE–*lac*Z construct, the LTE is defined as exon 1 plus the 5′ part of the LAT intron until the *Hpa*I site. The entire LAT promoter and exon 1 region, extending into the LAT intron until the *Hpa*I site, is paired with the *lac*Z gene. However, there are splicing sites flanking the entire LTE region, and the wild-type LAT intron splice donor (mSD) has been mutated to be ineffective. This construct thus removes the entire LTE region as an intron, leaving the first ATG of the mRNA in the *lac*Z gene.

have others (Lachman and Efsthathiou, 1997; Palmer *et al.*, 2000; Marshall *et al.*, 2000).

In summary, substantial improvement in HSV as a vector has been obtained in reduction of the toxicity of the virus, either through deletion of genes or through the help-free amplicon approach. Expression of transgenes from the virus could be further improved, particularly in the central nervous system. Exactly how helpful HSV will be will depend on the system as well as on the ability to infect as many cells as possible without doing harm.

## References

Batchelor, A. H., Wilcox, K. W., and O'Hare, P. (1994). Binding and repression of the latency-associated promoter of herpes simplex virus by the immediate early 175 K protein. *J. Gen. Virol.* **75,** 753–767.

Batterson, W., and Roizman, B. (1983). Characterization of the herpes simplex virion-associated factor responsible for the induction of alpha genes. *J. Virol.* **46,** 371–377.

Berthomme, H., Lokensgard, J., Yang, L., Margolis, T., and Feldman, L. T. (2000). Evidence for a bidirectional element located downstream from the herpes simplex virus type 1 latency-associated promoter that increases its activity during latency. *J. Virol.* **74,** 3613–3622.

Borst, E. M., Hahn, G., Koszinowski, U. H., and Messerle, M. (1999). Cloning of the human cytomegalovirus (HCMV) genome as an infectious bacterial artificial chromosome in *Escherichia coli*: A new approach for construction of HCMV mutants. *J. Virol.* **73,** 8320–8329.

Burton, E. A., Wechuck, J. B., Wendell, S. K., Goins, W. F., Fink, D. J., and Glorioso, J. C. (2001). Multiple applications for replication-defective herpes simplex virus vectors. *Stem Cells* **19,** 358–377.

Cai, W., and Schaffer, P. A. (1992). Herpes simplex virus type 1 ICP0 regulates expression of immediate-early, early and late genes in productively infected cells. *J. Virol.* **66,** 2904–2915.

Campbell, M. E., Palfreyman, J. W., and Preston, C. M. (1984). Identification of herpes simplex virus DNA sequences which encode a trans-acting polypeptide responsible for stimulation of immediate early transcription. *J. Mol. Biol.* **180,** 1–19.

Chen, J., and Silverstein, S. (1992). Herpes simplex viruses with mutations in the gene encoding ICPO are defective in gene expression. *J. Virol.* **66,** 2916–2927.

Croen, K. D., Ostrove, J. M., Dragovic, L. J., *et al.* (1987). Latent herpes simplex virus in human trigeminal ganglia. Detection of an immediate early gene "anti-sense" transcript by in situ hybridization. *N. Engl. J. Med.* **317,** 1427–1432.

Delecluse, H. J., Hilsendegen, T., Pich, D., Zeidler, R., and Hammerschmidt, W. (1998). Propagation and recovery of intact, infectious Epstein-Barr virus from prokaryotic to human cells. *Proc. Natl. Acad. Sci. USA* **95,** 8245–8250.

DeLuca, N. A., McCarthy, A. M., and Schaffer, P. A. (1985). Isolation and characterization of deletion mutants of herpes simplex virus type 1 in the gene encoding immediate-early regulatory protein ICP4. *J. Virol.* **56,** 558–570.

Dixon, R. A., and Schaffer, P. A. (1980). Fine-structure mapping and functional analysis of temperature-sensitive mutants in the gene encoding the herpes simplex virus type 1 immediate early protein VP175. *J. Virol.* **36,** 189–203.

Dobson, A. T., Sedarati, F., Devi-Rao, G., Flanagan, W. M., Farrell, M., Wagner, E. K., Stevens, J. G., and Feldman, L. T. (1989). Identification of the latency-associated transcript promoter by expression of rabbit beta-globin mRNA in mouse sensory ganglia latently infected with a recombinant herpes simplex virus. *J. Virol.* **63,** 3844–3851.

Farrell, M., Dobson, A. T., and Feldman, L. T. (1991). Herpes simplex virus latency-associated transcript is a stable intron. *Proc. Natl. Acad. Sci. USA* **88,** 790–794.

Frenkel, N., Locker, H., Batterson, W., Hayward, G. S., and Roizman, B. (1976). Anatomy of herpes simplex virus DNA. VI. Defective DNA originates from the S component. *J. Virol.* **20,** 527–531.

Geller, A. I., and Breakefield, X. O. (1988). A defective HSV-1 vector expresses Escherichia coli $\beta$-galactosidase in cultured peripheral neurons. *Science* **241,** 1667–1669.

Geller, A. I., Yu, L., Wang, Y., and Fraefel, C. (1990). Helper virus-free herpes simplex virus-1 plasmid vectors for gene therapy of Parkinson's disease and other neurological disorders. *Proc. Natl. Acad. Sci. USA* **87,** 8950–8954.

He, X., Treacy, M. N., Simmons, D. M., Ingraham, H. A., Swanson, L. S., and Rosenfeld, M. G. (1989). Expression of a larger family of POU-domain regulatory genes in mammalian brain development. *Nature* **340,** 35–42.

Hill, T. J., Field, H. J., and Blyth, W. A. (1975). Acute and recurrent infection with herpes simplex virus in the mouse: A model for studying latency and recurrent disease. *J. Gen. Virol.* **28,** 341–353.

Honess, R. W., and Roizman, B. (1974). Regulation of herpesvirus macromolecular synthesis. I. Cascade regulation of the synthesis of three groups of viral proteins. *J. Virol.* **14,** 8–19.

Honess, R. W., and Roizman, B. (1975). Regulation of herpesvirus macromolecular synthesis: Sequential transition of polypeptide synthesis requires functional viral polypeptides. *Proc. Natl. Acad. Sci. USA* **72,** 1276–1280.

Johnson, P. A., Wang, M. J., and Friedmann, T. (1994). Improved cell survival by the reduction of immediate-early gene expression in replication-defective mutants of herpes simplex virus type 1 but not by mutation of the virion host shutoff function. *J. Virol.* **68,** 6347–6362.

Knotts, F. B., Cook, M. L., and Stevens, J. G. (1974). Pathogenesis of herpetic encephalitis in mice after ophthalmic inoculation. *J. Infect. Dis.* **130,** 16–27.

Krisky, D. M., Wolfe, D., Goins, W. F., *et al.* (1998). Deletion of multiple immediate-early genes from herpes simplex virus reduces cytotoxicity and permits long-term gene expression in neurons. *Gene Ther.* **5,** 1593–1603.

Krummenacher, C., Zabolotny, J. M., and Fraser, N. W. (1997). Selection of a nonconsensus branch point is influenced by an RNA stem-loop structure and is important to confer stability to the herpes simplex virus 2-kilobase latency-associated transcript. *J. Virol.* **71,** 5849–5860.

Lachman, R. H., and Efstathiou, S. (1997). Utilization of the herpes simplex virus type 1 latency-associated regulatory region to drive stable reporter gene expression in the nervous system. *J. Virol.* **71,** 3197–3207.

Laycock, K. A., Lee, S. F., Brady, R. H., and Pepose, J. S. (1991). Characterization of a murine model of recurrent herpes simplex viral keratitis induced by ultraviolet B radiation. *Invest. Ophthalmol. Vis. Sci.* **32,** 2741–2746.

Lilley, C. E., Groust, F., Han, Z., Palmer, J. A., Anderson, P. N., Latchman, D. S., and Coffin, R. S. (2001). Multiple immediate-early gene-deficient herpes simplex virus vectors allowing efficient gene delivery to neurons in culture and widespread gene delivery to the central nervous system in vivo. *J. Virol.* **75,** 4343–4356.

Lokensgard, J. R., Bloom, D. C., Dobson, A. T., and Feldman, L. T. (1994). Long-term promoter activity during herpes simplex virus latency. *J. Virol.* **68,** 7148–7158.

Lokensgard, J. R., Berthomme, H., and Feldman, L. T. (1997). The latency-associated promoter of herpes simplex virus type 1 requires a region downstream of the transcriptional start site for long-term expression during latency. *J. Virol.* **71,** 6714–6719.

Mackem, S., and Roizman, B. (1982). Structural features of the herpes simplex virus alpha gene 4, 0, and 27 promoter-regulatory sequences which confer alpha regulation on chimeric thymidine kinase genes. *J. Virol.* **44,** 939–949.

Margolis, T. P., Sedarati, F., Dobson, A. T., Feldman, L. T., and Stevens, J. G. (1992). Pathways of viral gene expression during acute neuronal infection with HSV-1. *Virology* **189,** 150–160.

Margolis, T. P., Bloom, D. C., Dobson, A. T., Feldman, L. T., and Stevens, J. G. (1993). Decreased reporter gene expression during latent infection with HSV LAT promoter constructs. *Virology* **197,** 585–592.

Marshall, K. R., Lachmann, R. H., Efstathiou, S., Rinaldi, A., and Preston, C. M. (2000). Long-term transgene expression in mice infected with a herpes simplex virus type 1 mutant severely impaired for immediate-early gene expression. *J. Virol.* **74,** 956–964.

Mellerick, D. M., and Fraser, N. W. (1987). Physical state of the latent herpes simplex virus genome in a mouse model system: Evidence suggesting an episomal state. *Virology* **158,** 265–275.

Messerle, M., Crnkovic, I., Hammerschmidt, W., Ziegler, H., and Koszinowski, U. H. (1997). Cloning and mutagenesis of a herpesvirus genome as an infectious bacterial artificial chromosome. *Proc. Natl. Acad. Sci. USA* **94,** 14759–14763.

O'Hare, P., and Goding, C. R. (1988). Herpes simplex regulatory elements and the immunoglobulin octomer domain bind a common factor and are both targets for virion transactivation. *Cell* **52,** 435–445.

Palmer, J. A., Branston, R. H., Lilley, C. E., Robinson, M. J., Groutsi, F., Smith, J., Latchman, D. S., and Coffin, R. S. (2000). Development and optimization of herpes simplex virus vectors for multiple long-term gene delivery to the peripheral nervous system. *J. Virol.* **74,** 5604–5618.

Preston, C. M., Frame, M. C., and Campbell, M. E. M. (1988). A complex formed between cell components and an HSV structural polypeptide binds to a viral immediate-early regulatory sequence. *Cell* **52,** 425–434.

Rock, D. L., Nesburn, A. B., Ghiasi, H., *et al.* (1987). Detection of latency-related viral RNAs in trigeminal ganglia of rabbits latently infected with herpes simplex virus type 1. *J. Virol.* **61,** 3820–3826.

Rodahl, E., and Haarr, L. (1997). Analysis of the 2-kilobase latency- associated transcript expressed in PC12 cells productively infected with herpes simplex virus type 1: Evidence for a stable, nonlinear structure. *J. Virol.* **71,** 1703–1707.

Sacks, W. R., Greene, C. C., Aschman, D. P., *et al.* (1985). Herpes simplex virus type 1 ICP27 is an essential regulatory protein. *J. Virol.* **55,** 796–805.

Saeki, Y., Ichikawa, T., Saeki, A., Chiocca, E. A., Tobler, K., Ackermann, M., Breakefield, X. O., and Fraefel, C. (1998). Herpes simplex virus type 1 DNA amplified as bacterial artificial chromosome in *Escherichia coli*: Rescue of replication-competent virus progeny and packaging of amplicon vectors. *Hum. Gene Ther.* **9,** 2787–2794.

Samaniego, L. A., Neiderhiser, L., and DeLuca, N. A. (1998). Persistence and expression of the herpes simplex virus genome in the absence of immediate-early proteins. *J. Virol.* **72,** 3307–3320.

Smith, G. A., and Enquist, L. W. (1999). Construction and transposon mutagenesis in *Escherichia coli* of a full-length infectious clone of pseudorabies virus, an alphaherpes virus. *J. Virol.* **73,** 6405–6414.

Spaete, R. R., and Frenkel, N. (1982). The herpes simplex amplicon: A new eucaryotic defective-virus cloning-amplfying vector. *Cell* **30,** 295–304.

Spivack, J. G., and Fraser, N. W. (1987). Detection of herpes simplex virus type 1 transcripts during latent infection in mice [published erratum appears in *J. Virol.* (1988) **62,** 663]. *J. Virol.* **61,** 3841–3847.

Stevens, J. G., Wagner, E. K., Devi-Rao, G. B., *et al.* (1987). RNA complementary to a herpesvirus alpha gene mRNA is prominent in latently infected neurons. *Science* **235,** 1056–1059.

Wu, N., Watkins, S. C., Schaffer, P. A., *et al.* (1996). Prolonged gene expression and cell survival after infection by a herpes simplex virus mutant defective in the "immediate-early genes encoding ICP4, ICP27, and ICP22. *J. Virol.* **70,** 6358".

# LENTIVIRAL VECTORS

Johan Jakobsson, Cecilia Ericson, Nina Rosenqvist, and Cecilia Lundberg

Wallenberg Neuroscience Center
Division of Neurobiology
BMC, A11
221 84 Lund, Sweden



## I. Introduction

The use of gene transfer to the brain has rapidly increased over the past years and the large numbers of scientific reports illustrate the enormous energy devoted to improving the technology. Lentiviruses are regarded as one of the most promising viral vectors in the field and the vectors are constantly being refined. Efforts include the development of second, third, and fourth generations of viral vectors, changing the tropism of the viral vectors, and finding ways to regulate transcription, posttranscription, and gene expression (Romano *et al.*, 2000; Kafri, 2001). Lentiviral transduction to the central nervous system (CNS) is now known to have many advantages, especially the ability to incorporate into the host genome, both into dividing and nondividing cells, and to maintain long-term transgene expression (Blomer *et al.*, 1997; Deglon *et al.*, 2000; Naldini *et al.*, 1996a; Trono, 2000). In this chapter we will describe different types of lentiviral vectors and discuss their use for gene transfer to the CNS.

## II. Lentiviral Vectors

The lentiviral family consists of several viruses, both primate and nonprimate. The most widely used lentiviral vectors for gene transfer to the CNS are based on human immunodeficiency virus type I (HIV-I), and the focus of this review will be on vectors derived from this virus. HIV-I has a complex genome consisting of a total of nine genes, including the *gag*, *pol*, and *env*, which are found in all




0074-7742/03 $35.00

retroviruses, as well as a further six accessory genes: *vif*, *vpr*, *nef*, *vpu*, *tat*, and *rev* (for a comprehensive review see Frankel and Young, 1998). Early vectors derived from HIV-I and used for CNS–gene transfer were replication-defective viral particles consisting of the enzymes and structural proteins of HIV-I and an envelope derived from another virus, usually the glycoprotein of vesicular stomatitis virus (VSV-G) (Naldini *et al.*, 1996b). The design of the system was influenced by other retroviral vectors such as murine leukemia virus-derived vectors (Ory *et al.*, 1996; Burns *et al.*, 1993) and typically consisted of three plasmids: (1) a packaging plasmid that expressed the viral proteins necessary for the capsid and integration, (2) a plasmid that expressed an envelope protein, and (3) a transfer plasmid that contained the transfer sequence, usually regulated by an internal promoter, as well as the *cis* sequences necessary for packaging and integration. These plasmids were then cotransfected into human cells and the medium was collected. When pseudotyped with VSV-G these vectors could be concentrated, using ultracentrifugation, to about $10^8$–$10^9$ transducing units/ml (Naldini *et al.*, 1996b).

The potential of HIV-I-based vectors was greatly enhanced by the appearance of so-called multiply attenuated vectors (Zufferey *et al.*, 1997). These vectors had deleted most of the accessory genes in the packaging construct leading to a drastic increase in the biosafety of the system. Recent further developments (Kim *et al.*, 1998; Dull *et al.*, 1998) have allowed all accessory proteins to be removed from the packaging construct, although the *rev* gene must be expressed from a separate fourth plasmid. This design excludes any possibility that a pathogenic replication-competent vector will arise through a recombination event as all pathogenic HIV-I proteins have been removed.

The introduction of a self-inactivating (SIN) deletion in the 3′ long terminal repeat (LTR) of the transfer construct has removed most of the residual promoter activity of the LTR (Miyoshi *et al.*, 1998; Zufferey *et al.*, 1998); this reduces the risk for replication-competent vectors to arise through a recombination event. It also decreases promoter competition between the LTR and an internal promoter. This quite extensive deletion (400 bp) does not affect the production or functionality of the vectors and studies in rat brain indicate at least the same level of expression when using SIN vectors (Zufferey *et al.*, 1998; Deglon *et al.*, 2000). Taken together these improvements suggest that the biosafety of HIV-I- based vectors is superior to that of other retroviral vectors that are used in clinical tests today (Trono, 2000; Vigna and Naldini, 2000).

The improvements in the design of the transfer construct have in recent years further enhanced efficiency of these vectors. The realization that HIV-I has an additional polypurine tract in the middle of the genome (cPPT) led to a reinsertion of this *cis*-acting sequence into the transfer construct (Zennou *et al.*, 2000; Follenzi *et al.*, 2000). Several studies have revealed an increased transduction and transgenic expression (3- to 10-fold) due to enhanced nuclear transport when this element is used (Sirven *et al.*, 2000, 2001; Zennou *et al.*, 2001). The inclusion of the woodchuck hepatitis virus posttranscriptional regulatory element has also enhanced expression levels (4- to 6-fold) in a number of different cell lines and tissues due to an increased stability of transgenic mRNA (Zufferey *et al.*, 1999). Most notably these alterations seem to affect expression in the CNS to an extent similar to the affect in other tissues (Deglon *et al.*, 2000; Zennou *et al.*, 2001). Park and Kay (2001) have also attempted to enhance transduction efficiencies by using double *rev* responsible elements in the transfer construct and detected a 2.5-fold increase in vector efficiency in hepatocytes.

A key issue to solve if clinical protocols will be approved is the development of regulatable vectors (Kafri, 2001; Trono, 2000). Kafri and co-workers (2000) as well as Reiser and collaborators (2000) introduced a tetracycline-dependent system into lentiviral vectors resulting in up to a 500-fold induction of green fluorescent protein (GFP) expression *in vitro* upon withdrawal of tetracycline. It is noteworthy that regulation was also demonstrated *in vivo* in the brain when tetracycline was administered in the drinking water (Kafri *et al.*, 2000). A problem encountered in both studies was the basal expression levels of GFP found even in the presence of tetracycline.

Site-specific recombination represents another way of regulating transgene expression. Verma and co-workers delivered the cre-recombinase with a lentiviral vector using a clever design that self-deleted the recombinase transgene directly after the first round of expression (Pfeifer *et al.*, 2001). This allowed efficient introduction of a transgene flanked by p-lox sites in the mouse brain without adverse immunological effects.

Lentiviral vectors are usually pseudotyped with the vesicular stomatitis virus G glycoprotein (VSV-G) (Zufferey *et al.*, 1997). This confers several advantages, including a wide tropism and improved stability that enables ultracentrifugation and allows the HIV-I Nef protein to be independent due to access to an endocytotic pathway (Kafri 2001). Lentiviruses pseudotyped with VSV-G have in several studies been shown to transduce cells of CNS origin *in vitro* as well as *in vivo* in the brain in several mammalian species (Bensadoun *et al.*, 2000; Blomer *et al.*, 1997; Deglon *et al.*, 2000; Kordower *et al.*, 2000).

However, using alternative envelopes may allow targeting of specific cell types and enable different routes of transduction. For instance, the envelope from rabies virus, in contrast to VSV- G, enabled retrograde transport of lentiviral particles. This study (Mazarakis *et al.*, 2001) was the first to show peripheral delivery of a transgene to the when using a lentiviral vector. Heard and co-workers showed that lentiviral particles pseudotyped with lyssavirus envelope were able to transduce cells in the rat brain, although at a lower rate than VSV-G pseudotyped vectors (Desmaris *et al.*, 2001).

Development of gene transfer vectors based on lentiviruses other than HIV-I is a growing field and has been successful in several cases. Several reports have described efficient transduction, including injections into the brain, when using vectors based on equine infectious anemia virus (EIVA) (Mitrophanous *et al.*,

1999; O'Rourke *et al.*, 2002; Yamada *et al.*, 2001). O'Rourke *et al.* (2002) found, however, that EIAV-based vectors were significantly less effective than HIV-derived vectors and that this was most pronounced when observing long-term expression. Johnston *et al.* (2001) have developed vectors derived from feline immunodeficiency virus (FIV) and recently Berkowitz *et al.* (2001) developed a vector derived from bovine immunodeficiency virus (BIV). Vectors based on caprine arthritis/encephalitis virus (CEAV), Jemberana disease virus, and simian immunodeficiency virus (SIV) have also been described (Vigna and Naldini, 2000). It has been argued that the use of nonprimate lentiviral vectors would confer a higher safety degree since potential replication-competent particles would be nonpathogenic in humans. Where this is actually the case is however, still under debate and it remains to be shown that non-primate-derived vectors actually represent a higher degree of biosafety.

It has been argued that stable producer cell lines must be developed before lentiviral vectors can be used in a clinical setting (Kafri, 2001; Trono, 2000). It has been hard establishing such cell lines, probably due to the adverse effects on the cells by the HIV-derived proteins and VSV-G (Kafri, 2001). Recently there have, however, been some successful reports that describe the development of producer cell lines. Several groups have attempted to develop cell lines using inducible systems (Farson *et al.*, 2001; Pacchia *et al.*, 2001; Klages *et al.*, 2000) and recently, Kafri and co-workers used tetracycline induction to develop a cell line capable of producing SIN- lentiviral vectors (Xu *et al.*, 2001). However, these cell lines still only represent a proof of principle and it remains to be seen whether production can be scaled up.

## III. Direct Injection of Lentiviral Vectors

Lentiviral vectors have been injected into the brain of mice, rats, and primates and transgenic expression was shown to be sustained at high levels for periods of up to a year. Efficient expression has been reported in several areas of the CNS including the striatum, the midbrain, the facial nucleus, and the retina (Bensadoun *et al.*, 2000; Deglon *et al.*, 2000; Kordower *et al.*, 2000; Miyoshi *et al.*, 1997; Rosenblad *et al.*, 2000). Transfer of therapeutical agents has been demonstrated to be beneficial in several disease models in both rodents and primates (see, e.g., Kordower *et al.*, 2000).

As discussed above, lentiviral vectors are regularly pseudotyped with VSV-G, which enables a wide variety of cells to be transduced. However, several reports indicate that when injecting these vectors into the striatum the majority of transduced cells are neuronal and that only a few transduced glial cells can be detected (Jakobsson *et al.*, 2002a; Blomer *et al.*, 1997; Deglon *et al.*, 2000; Kordower *et al.*, 2000; Naldini *et al.*,

1996a). There also seem to be regional differences in the expression pattern, for instance, Rosenblad *et al.* (2000) found that when injecting lentiviral vectors to the midbrain, mainly glial cells were transduced. There may be several reasons for this discrepancy in tropism in the CNS, including different promoter activity in the cell populations and different properties of the cells in regard to vector uptake.

The ability to direct transgenic expression to specific cell populations will be crucial for future clinical applications. Today many studies have been performed using vectors with strong ubiquitous promoters, such as the human cytomegalovirus (CMV) promoter and the mouse phosphoglycerate kinase (PGK) promoter. The ability to replace the ubiquitous promoters with cell- specific promoters such as the neuron specific enolase (NSE) promoter or glial fibrillary acidic protein (GFAP) promoter is a promising alternative. We have shown that an efficient and neuron-specific expression can be achieved in the rat striatum when using an NSE promoter (Fig. 1). A glial-specific lentiviral construct that is expressed efficiently *in vivo* remains to be developed.

## IV. *Ex Vivo* Gene Transfer Using Lentiviral Vectors

Viral vectors can be used to manipulate cells in culture that subsequently are transplanted into the CNS to deliver therapeutic proteins. For this purpose several types of vectors could be used, but lentiviral vectors have many

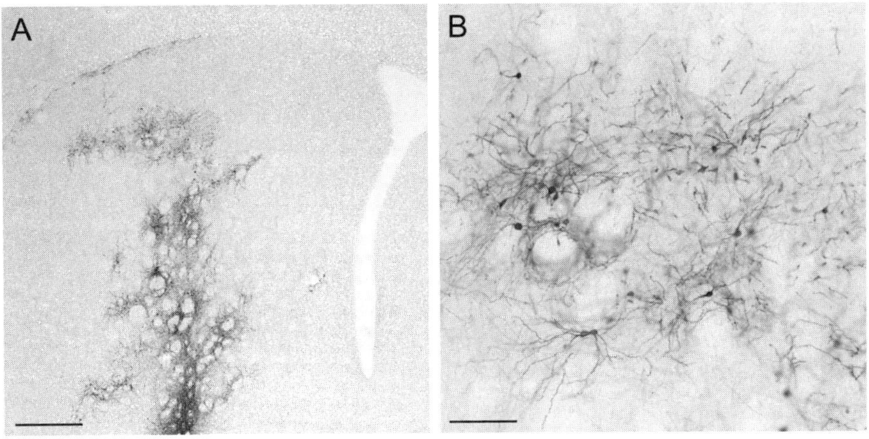


FIG. 1. Rat striatum transduced with a neuron-specific lentiviral vector (NSE promoter) expressing green fluorescent protein (GFP). Only neurons expressed the transgene using this promoter. Note the extensive GFP immunoreactivity in the dendritic tree of the transduced neurons. Animals were sacrified and perfused 6 weeks posttransduction. Brains were sectioned and immunhistochemically stained using a GFP antibody. Scale bars: (A) 500 $\mu$m; (B) 100 $\mu$m.

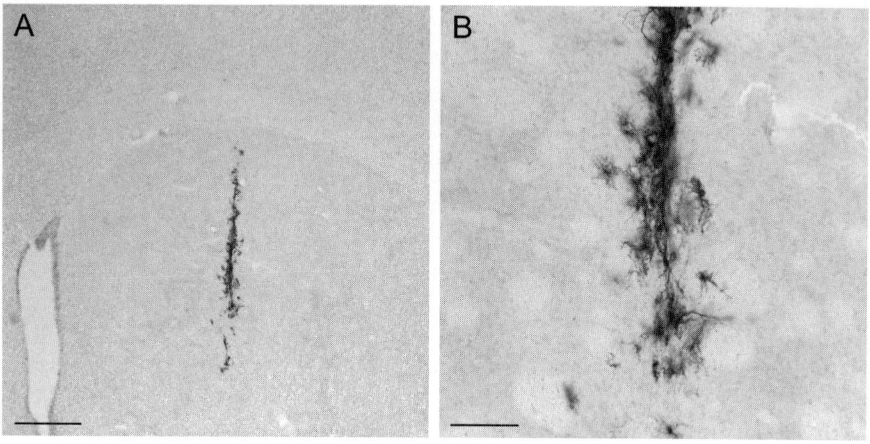


FIG. 2. Primary astrocytes transduced with a lentiviral vector encoding GFP were transplanted to rat striatum. Note that only cells adjacent to the injection track expressed GFP even though surviving, grafted astrocytes could be detected throughout the entire striatum (not shown here). Animals were sacrificed and perfused 6 weeks posttransduction. Brains were sectioned and immunhistochemically stained using a GFP antibody. Scale bars: (A) 500 $\mu$m; (B) 100 $\mu$m.

advantages, such as integration into the chromosomes, allowing for large transcriptional cassettes and the ability to transduce quiescent cells (Buchschacher *et al.*, 2000; Kafri, 2001; Lever, 2000; Naldini and Verma, 2000). *Ex vivo* gene therapy in the CNS may be preferable to direct *in vivo* vector injections in certain applications. For instance, in the case of neurodegenerative disorders the situation may arise in which there are no appropriate endogenous target cells left to infect and a more useful strategy would be to replace lost cells and try to repair damaged neural circuits. Another issue is the phenotype of the target cell. If local delivery of a therapeutic protein is desired, neurons projecting to different areas are not the best candidates for secreting the transgenic product. Instead the use of cultured, nonneuronal cells as transgene vehicles grafted to discrete locations of the CNS would result in local delivery. Furthermore, the ability to control the integration site and gene dose and not interfering with the endogenous cells are advantages presented by *ex vivo* gene transfer. The transplanted cells could even be encapsulated, increasing safety by providing the opportunity to easily withdraw treatment if necessary (Aebischer *et al.*, 1996). Another feature in favor of *ex vivo* gene therapy is the already existing transplantation protocols, making this approach closer to a clinical application, and also the development of stem cell research (Shihabuddin *et al.*, 1999; Flax *et al.*, 1998; Isacson *et al.*, 2001; Priller *et al.*, 2001).

Several reports have shown promising results when using lentiviral *ex vivo* approaches to targets outside the brain (see, e.g., Pawliuk *et al.*, 2001; Zahler *et al.*,

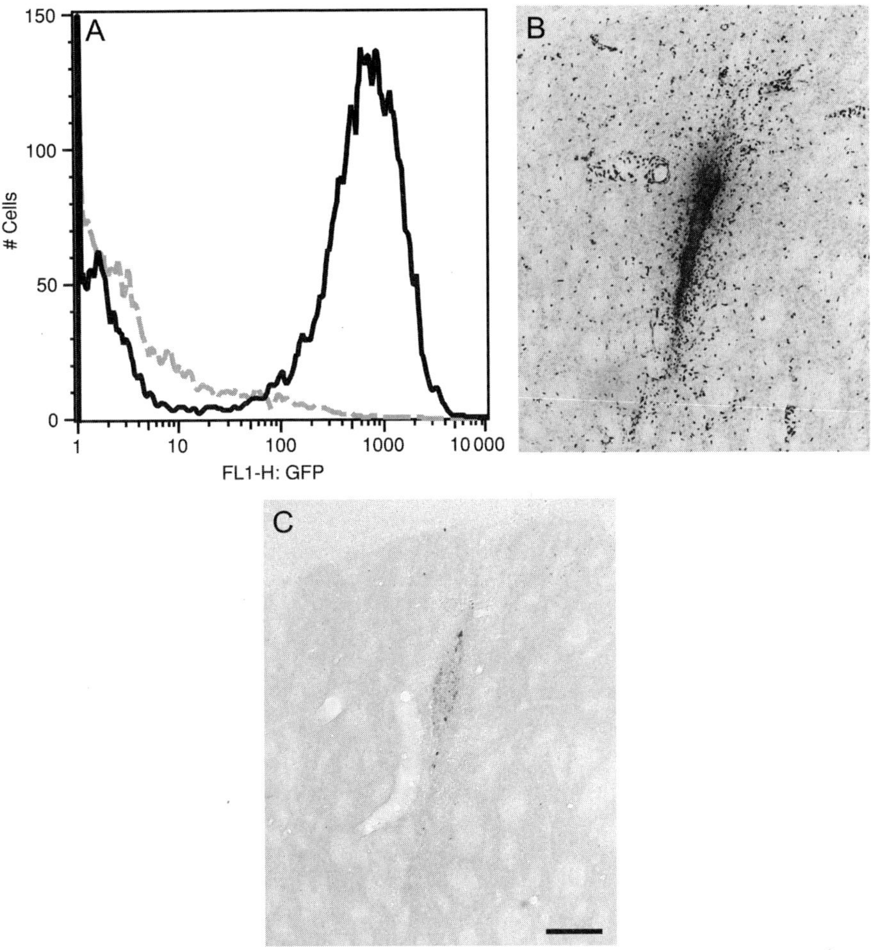


FIG. 3. (A) The black line demonstrates GFP expression in a lentivirally transduced clone of an immortalized neuronal progenitor cell line (RN33B) analyzed by flow cytometry. (B, C) The same clone grafted to the rat striatum. The cells were labelled using [$^3$H]thymidine prior to grafting and autoradiography showed that they engraft well after 7 days (B). The number of thymidine-labeled cells should be compared to the low number of cells expressing GFP at 7 days postgrafting (C). The clear loss of transgene expression *in vivo* could be induced *in vitro* as well by differentiating the cells at a nonpermissive temperature, 37°C, for 7 days (A, dashed line) suggesting that this could be used as an *in vitro* model of transgene silencing (Rosenqvist *et al.*, 2002). Scale bar represents 200 $\mu$m.

2000), demonstrating the potential of this strategy. Cell types that have been used as vehicles for lentiviral protein delivery to the CNS so far are human neural progenitor cells (Englund *et al.*, 2002), primary rat astrocytes (Fig. 2) (Ericson *et al.*, 2002), and immortalized progenitor cell lines (Englund *et al.*, 2000; Warfvinge *et al.*, 2001; Rosenqvist *et al.*, 2002). Promising results have been seen, although only a fraction of the transplanted cells seems to express detectable levels of the transgene (Fig. 3) (Englund *et al.*, 2000; Ericson *et al.*, 2002; Rosenqvist *et al.*, 2002). This downregulation of transgene expression is not specific for lentivirally modified cells. Rather it seems to be dependent on the cell type, differentiation, and also the actual transgene, rather then the vectors used (Leff *et al.*, 1998; Martinez-Serrano and Bjorklund, 1995; Ridoux *et al.*, 1994; Lundberg *et al.*, 1996). The problematic silencing of transgenes in transplanted neural cells seems to be due to changes in chromatin structure when the cells differentiate, mediated by processes such as histone deacetylation (Rosenqvist *et al.*, 2002). It might therefore be possible to overcome this phenomenon by making constructs with decreased tendencies to become condensed by, for example, inserting insulator elements (Jakobsson *et al.*, 2002). Another option is to pharmacologically manipulate the cells by agents leaving the chromatin less condensed, thereby achieving higher and more long-term transgene expression (Rosenqvist *et al.*, 2002). The growing knowledge of genome regulation will undoubtedly reveal ways to modify or manipulate the transgenic cells and might result in novel ways to prevent transgene silencing.

## V. Concluding Remarks

Lentiviral vectors possess many of the hallmarks of a suitable vector for gene transfer to the brain. The development of safer vectors and producer cell lines will make the vectors a valid alternative for clinical trial in the not so distant future. There are many disorders of the central nervous system that cannot be treated now and there is a need for new alternative strategies, such as gene therapy. Therefore the lentiviral vectors and their use in the brain are very important and valuable venues for future research.

## Acknowledgments

This work was supported by the Swedish MRC (13479), the Swedish Gene Therapy Program (13285), and the European Commission (NEUROGET, contract GLK3-1999-0702).

## References

Aebischer, P., Schluep, M., Deglon, N., Joseph, J. M., Hirt, L., Heyd, B., Goddard, M., Hammang, J. P., Zurn, A. D., Kato, A. C., Regli, F., and Baetge, E. E. (1996). Intrathecal delivery of CNTF using encapsulated genetically modified xenogeneic cells in amyotrophic lateral sclerosis patients. *Nat. Med.* **2,** 696–699.

Bensadoun, J. C., Deglon, N., Tseng, J. L., Ridet, J. L., Zurn, A. D., and Aebischer, P. (2000). Lentiviral vectors as a gene delivery system in the mouse midbrain: Cellular and behavioral improvements in a 6-OHDA model of Parkinson's disease using GDNF. *Exp. Neurol.* **164,** 15–24.

Berkowitz, R., Ilves, H., Lin, W. Y., Eckert, K., Coward, A., Tamaki, S., Veres, G., and Plavec, I. (2001). Construction and molecular analysis of gene transfer systems derived from bovine immunodeficiency virus. *J. Virol.* **75,** 3371–3382.

Blomer, U., Naldini, L., Kafri, T., Trono, D., Verma, I. M., and Gage, F. H. (1997). Highly efficient and sustained gene transfer in adult neurons with a lentivirus vector. *J. Virol.* **71,** 6641–6649.

Buchschacher, G. L., Jr., and Wong-Staal, F. (2000). Development of lentiviral vectors for gene therapy for human diseases. *Blood* **95,** 2499–2504.

Burns, J. C., Friedmann, T., Driever, W., Burrascano, M., and Yee, J. K. (1993). Vesicular stomatitis virus G glycoprotein pseudotyped retroviral vectors: Concentration to very high titer and efficient gene transfer into mammalian and nonmammalian cells. *Proc. Natl. Acad. Sci. USA* **90,** 8033–8037.

Deglon, N., Tseng, J. L., Bensadoun, J. C., Zurn, A. D., Arsenijevic, Y., Pereira de Almeida, L., Zufferey, R., Trono, D., and Aebischer, P. (2000). Self-inactivating lentiviral vectors with enhanced transgene expression as potential gene transfer system in Parkinson's disease. *Hum. Gene Ther.* **11,** 179–190.

Desmaris, N., Bosch, A., Salaun, C., Petit, C., Prevost, M. C., Tordo, N., Perrin, P., Schwartz, O., de Rocquigny, H., and Heard, J. M. (2001). Production and neurotropism of lentivirus vectors pseudotyped with lyssavirus envelope glycoproteins. *Mol. Ther.* **4,** 149–156.

Dull, T., Zufferey, R., Kelly, M., Mandel, R. J., Nguyen, M., Trono, D., and Naldini, L. (1998). A third-generation lentivirus vector with a conditional packaging system. *J. Virol.* **72,** 8463–8471.

Englund, U., Ericson, C., Rosenblad, C., Mandel, R. J., Trono, D., Wictorin, K., and Lundberg, C. (2000). The use of a recombinant lentiviral vector for ex vivo gene transfer into the rat CNS. *Neuroreport* **11,** 3973–3977.

Englund, U., Fricker-Gates, R. A., Lundberg, C., Bjorklund, A., and Wictorin, K. (2002). Transplantation of human neural progenitor cells into the neonatal rat brain: Extensive migration and differentiation with long-distance axonal projections. *Exp. Neurol.* **173,** 1–21.

Ericson, C., Wictorin, K., and Lundberg, C. (2002). Ex vivo and in vitro studies of transgene expression in rat astrocytes transduced with lentiviral vectors. *Exp. Neurol.* **173,** 22–30.

Farson, D., Witt, R., McGuinness, R., Dull, T., Kelly, M., Song, J., Radeke, R., Bukovsky, A., Consiglio, A., and Naldini, L. (2001). A new-generation stable inducible packaging cell line for lentiviral vectors. *Hum. Gene Ther.* **12,** 981–997.

Flax, J. D., Aurora, S., Yang, C., Simonin, C., Wills, A. M., Billinghurst, L. L., Jendoubi, M., Sidman, R. L., Wolfe, J. H., Kim, S. U., and Snyder, E. Y. (1998). Engraftable human neural stem cells respond to developmental cues, replace neurons, and express foreign genes. *Nat. Biotechnol.* **16,** 1033–1039.

Follenzi, A., Ailles, L. E., Bakovic, S., Geuna, M., and Naldini, L. (2000). Gene transfer by lentiviral vectors is limited by nuclear translocation and rescued by HIV-1 pol sequences. *Nat. Genet.* **25,** 217–222.

Frankel, A. D., and Young, J. A. (1998). HIV-1: Fifteen proteins and an RNA. *Annu. Rev. Biochem.* **67,** 1–25.

Isacson, O., van Horne, C., Schumacher, J. M., and Brownell, A. L. (2001). Improved surgical cell therapy in Parkinson's disease. Physiological basis and new transplantation methodology. *Adv. Neurol.* **86,** 447–454.

Jakobsson, J., Ericson, C., Jansson, M., Bjork, E., and Lundberg, C. (2003a). Targeted transgene expression in the rat brain using lentrival vectors. Submitted.

Jakobsson, J., Rosenqvist, N., and Lundberg, C. (2003b). A chromatin insulator protects lentiviral transgene expression from silencing in neural cell lines. Submitted.

Johnston, J. B., Olson, M. E., Rud, E. W., and Power, C. (2001). Xenoinfection of nonhuman primates by feline immunodeficiency virus. *Curr. Biol.* **11,** 1109–1113.

Kafri, T. (2001). Lentivirus vectors: Difficulties and hopes before clinical trials. *Curr. Opin. Mol. Ther.* **3,** 316–326.

Kafri, T., van Praag, H., Gage, F. H., and Verma, I. M. (2000). Lentiviral vectors: Regulated gene expression. *Mol. Ther.* **1,** 516–521.

Kim, V. N., Mitrophanous, K., Kingsman, S. M., and Kingsman, A. J. (1998). Minimal requirement for a lentivirus vector based on human immunodeficiency virus type 1. *J. Virol.* **72,** 811–816.

Klages, N., Zufferey, R., and Trono, D. (2000). A stable system for the high-titer production of multiply attenuated lentiviral vectors. *Mol. Ther.* **2,** 170–176.

Kordower, J. H., Emborg, M. E., Bloch, J., Ma, S. Y., Chu, Y., Leventhal, L., McBride, J., Chen, E. Y., Palfi, S., Roitberg, B. Z., Brown, W. D., Holden, J. E., Pyzalski, R., Taylor, M. D., Carvey, P., Ling, Z., Trono, D., Hantraye, P., Deglon, N., and Aebischer, P. (2000). Neurodegeneration prevented by lentiviral vector delivery of GDNF in primate models of Parkinson's disease. *Science* **290,** 767–773.

Leff, S. E., Rendahl, K. G., Spratt, S. K., Kang, U. J., and Mandel, R. J. (1998). In vivo L-DOPA production by genetically modified primary rat fibroblast or 9L gliosarcoma cell grafts via coexpression of GTPcyclohydrolase I with tyrosine hydroxylase. *Exp. Neurol.* **151,** 249–264.

Lever, A. M. (2000). Lentiviral vectors: Progress and potential. *Curr. Opin. Mol. Ther.* **2,** 488–496.

Lundberg, C., Horellou, P., Mallet, J., and Bjorklund, A. (1996). Generation of DOPA-producing astrocytes by retroviral transduction of the human tyrosine hydroxylase gene: In vitro characterization and in vivo effects in the rat Parkinson model. *Exp. Neurol.* **139,** 39–53.

Martinez-Serrano, A., and Bjorklund, A. (1995). Gene transfer to the mammalian brain using neural stem cells: A focus on trophic factors, neuroregeneration, and cholinergic neuron systems. *Clin. Neurosci.* **3,** 301–309.

Mazarakis, N. D., Azzouz, M., Rohll, J. B., Ellard, F. M., Wilkes, F. J., Olsen, A. L., Carter, E. E., Barber, R. D., Baban, D. F., Kingsman, S. M., Kingsman, A. J., O'Malley, K., and Mitrophanous, K. A. (2001). Rabies virus glycoprotein pseudotyping of lentiviral vectors enables retrograde axonal transport and access to the nervous system after peripheral delivery. *Hum. Mol. Genet.* **10,** 2109–2121.

Mitrophanous, K., Yoon, S., Rohll, J., Patil, D., Wilkes, F., Kim, V., Kingsman, S., Kingsman, A., and Mazarakis, N. (1999). Stable gene transfer to the nervous system using a non-primate lentiviral vector. *Gene Ther.* **6,** 1808–1818.

Miyoshi, H., Takahashi, M., Gage, F. H., and Verma, I. M. (1997). Stable and efficient gene transfer into the retina using an HIV-based lentiviral vector. *Proc. Natl. Acad. Sci. USA* **94,** 10319–10323.

Miyoshi, H., Blomer, U., Takahashi, M., Gage, F. H., and Verma, I. M. (1998). Development of a self-inactivating lentivirus vector. *J. Virol.* **72,** 8150–8157.

Naldini, L., and Verma, I. M. (2000). Lentiviral vectors. *Adv. Virus Res.* **55,** 599–609.

Naldini, L., Blomer, U., Gage, F. H., Trono, D., and Verma, I. M. (1996a). Efficient transfer, integration, and sustained long-term expression of the transgene in adult rat brains injected with a lentiviral vector. *Proc. Natl. Acad. Sci. USA* **93,** 11382–11388.

Naldini, L., Blomer, U., Gallay, P., Ory, D., Mulligan, R., Gage, F. H., Verma, I. M., and Trono, D. (1996b). In vivo gene delivery and stable transduction of nondividing cells by a lentiviral vector. *Science* **272,** 263–267.

O'Rourke, J. P., Newbound, G. C., Kohn, D. B., Olsen, J. C., and Bunnell, B. A. (2002). Comparison of gene transfer efficiencies and gene expression levels achieved with equine infectious anemia virus- and human immunodeficiency virus type 1-derived lentivirus vectors. *J. Virol.* **76,** 1510–1515.

Ory, D. S., Neugeboren, B. A., and Mulligan, R. C. (1996). A stable human-derived packaging cell line for production of high titer retrovirus/vesicular stomatitis virus G pseudotypes. *Proc. Natl. Acad. Sci. USA* **93,** 11400–11406.

Pacchia, A. L., Adelson, M. E., Kaul, M., Ron, Y., and Dougherty, J. P. (2001). An inducible packaging cell system for safe, efficient lentiviral vector production in the absence of HIV-1 accessory proteins. *Virology* **282,** 77–86.

Park, F., and Kay, M. A. (2001). Modified HIV-1 based lentiviral vectors have an effect on viral transduction efficiency and gene expression in vitro and in vivo. *Mol. Ther.* **4,** 164–173.

Pawliuk, R., Westerman, K. A., Fabry, M. E., Payen, E., Tighe, R., Bouhassira, E. E., Acharya, S. A., Ellis, J., London, I. M., Eaves, C. J., Humphries, R. K., Beuzard, Y., Nagel, R. L., and Leboulch, P. (2001). Correction of sickle cell disease in transgenic mouse models by gene therapy. *Science* **294,** 2368–2371.

Pfeifer, A., Brandon, E. P., Kootstra, N., Gage, F. H., and Verma, I. M. (2001). Delivery of the Cre recombinase by a self-deleting lentiviral vector: Efficient gene targeting in vivo. *Proc. Natl. Acad. Sci. USA* **98,** 11450–11455.

Priller, J., Flugel, A., Wehner, T., Boentert, M., Haas, C. A., Prinz, M., Fernandez-Klett, F., Prass, K., Bechmann, I., de Boer, B. A., Frotscher, M., Kreutzberg, G. W., Persons, D. A., and Dirnagl, U. (2001). Targeting gene-modified hematopoietic cells to the central nervous system: Use of green fluorescent protein uncovers microglial engraftment. *Nat. Med.* **7,** 1356–1361.

Reiser, J., Lai, Z., Zhang, X. Y., and Brady, R. O. (2000). Development of multigene and regulated lentivirus vectors. *J. Virol.* **74,** 10589–10599.

Ridoux, V., Robert, J. J., Zhang, X., Perricaudet, M., Mallet, J., and Le Gal La Salle, G. (1994). The use of adenovirus vectors for intracerebral grafting of transfected nervous cells. *Neuroreport* **5,** 801–804.

Romano, G., Michell, P., Pacilio, C., and Giordano, A. (2000). Latest developments in gene transfer technology: Achievements, perspectives, and controversies over therapeutic applications. *Stem Cells* **18,** 19–39.

Rosenblad, C., Gronborg, M., Hansen, C., Blom, N., Meyer, M., Johansen, J., Dago, L., Kirik, D., Patel, U. A., Lundberg, C., Trono, D., Bjorklund, A., and Johansen, T. E. (2000). In vivo protection of nigral dopamine neurons by lentiviral gene transfer of the novel GDNF-family member neublastin/artemin. *Mol. Cell. Neurosci.* **15,** 199–214.

Rosenqvist, N., Hard af Segerstad, C., Samuelsson, C., Johansen, J., and Lundberg, C. (2002). Activation of silenced transgene expression in neural precursor cell lines by inhibitors of histone deacetylation. *J. Gene Med.* **4,** 248–257.

Shihabuddin, L. S., Palmer, T. D., and Gage, F. H. (1999). The search for neural progenitor cells: Prospects for the therapy of neurodegenerative disease. *Mol. Med. Today* **5,** 474–480.

Sirven, A., Pflumio, F., Zennou, V., Titeux, M., Vainchenker, W., Coulombel, L., Dubart-Kupperschmitt, A., and Charneau, P. (2000). The human immunodeficiency virus type-1 central DNA flap is a crucial determinant for lentiviral vector nuclear import and gene transduction of human hematopoietic stem cells. *Blood* **96,** 4103–4110.

Sirven, A., Ravet, E., Charneau, P., Zennou, V., Coulombel, L., Guetard, D., Pflumio, F., and Dubart-Kupperschmitt, A. (2001). Enhanced transgene expression in cord blood CD34(+)-derived hematopoietic cells, including developing T cells and NOD/SCID mouse repopulating cells, following transduction with modified trip lentiviral vectors. *Mol. Ther.* **3,** 438–448.

Trono, D. (2000). Lentiviral vectors: Turning a deadly foe into a therapeutic agent. *Gene Ther.* **7,** 20–23.

Vigna, E., and Naldini, L. (2000). Lentiviral vectors: Excellent tools for experimental gene transfer and promising candidates for gene therapy. *J. Gene Med.* **2,** 308–316.

Warfvinge, K., Kamme, C., Englund, U., and Wictorin, K. (2001). Retinal integration of grafts of brain-derived precursor cell lines implanted subretinally into adult, normal rats. *Exp. Neurol.* **169,** 1–12.

Xu, K., Ma, H., McCown, T. J., Verma, I. M., and Kafri, T. (2001). Generation of a stable cell line producing high-titer self-inactivating lentiviral vectors. *Mol. Ther.* **3,** 97–104.

Yamada, K., Olsen, J. C., Patel, M., Rao, K. W., and Walsh, C. E. (2001). Functional correction of fanconi anemia group C hematopoietic cells by the use of a novel lentiviral vector. *Mol. Ther.* **3,** 485–490.

Zahler, M. H., Irani, A., Malhi, H., Reutens, A. T., Albanese, C., Bouzahzah, B., Joyce, D., Gupta, S., and Pestell, R. G. (2000). The application of a lentiviral vector for gene transfer in fetal human hepatocytes. *J. Gene Med.* **2,** 186–193.

Zennou, V., Petit, C., Guetard, D., Nerhbass, U., Montagnier, L., and Charneau, P. (2000). HIV-1 genome nuclear import is mediated by a central DNA flap. *Cell* **101,** 173–185.

Zennou, V., Serguera, C., Sarkis, C., Colin, P., Perret, E., Mallet, J., and Charneau, P. (2001). The HIV-1 DNA flap stimulates HIV vector-mediated cell transduction in the brain. *Nat. Biotechnol.* **19,** 446–450.

Zufferey, R., Nagy, D., Mandel, R. J., Naldini, L., and Trono, D. (1997). Multiply attenuated lentiviral vector achieves efficient gene delivery in vivo. *Nat. Biotechnol.* **15,** 871–875.

Zufferey, R., Dull, T., Mandel, R. J., Bukovsky, A., Quiroz, D., Naldini, L., and Trono, D. (1998). Self-inactivating lentivirus vector for safe and efficient in vivo gene delivery. *J. Virol.* **72,** 9873–9880.

Zufferey, R., Donello, J. E., Trono, D., and Hope, T. J. (1999). Woodchuck hepatitis virus posttranscriptional regulatory element enhances expression of transgenes delivered by retroviral vectors. *J. Virol.* **73,** 2886–2892.

# RETROVIRAL VECTORS FOR GENE DELIVERY TO NEURAL PRECURSOR CELLS

Ryoichiro Kageyama, Hiromi Hirata, and Jun Hatakeyama

Institute for Virus Research
Kyoto University
Kyoto 606-8507, Japan



## I. Introduction

A retroviral vector is an infectious vehicle that efficiently transfers exogenous genes into mitotic cells. After infection, a DNA copy (a provirus form) of the RNA viral genome is stably integrated into the host cell genome and transmitted precisely to the progeny of the infected cells. If the introduced genes are under the control of constitutive promoters, they will be permanently expressed in all the progeny of the infected cells unless those promoters are inactivated. Therefore, this vector system is suitable to stably change the phenotypes of dividing cells such as neural precursor (stem) cells and tumor cells. Recent studies revealed that neuronal and glial differentiation is controlled by transcription factors. By application of retroviruses that carry such transcription factor genes, it is now possible to instruct neural precursor cells to adopt specific cell fates, which is useful for regeneration of the nervous system. The retroviral system is also applicable to treatment of tumors, such






0074-7742/03 $35.00

as introduction of tumor suppressor genes or apoptosis genes to tumor cells. Thus, this system offers an important tool for gene therapy for various diseases. The feature that retrovirus is not infectious to nondividing cells is advantageous to regeneration from neural precursor cells and treatment of tumors because the retrovirus system does not affect mature neurons. Although retrovirus cannot be used for introduction of a large size (>8 kb) of DNA, this size restriction is not a problem in most cases because most transcription factor genes are much smaller in size.

In this chapter, we will describe the overview of retrovirus, the features of the replication-incompetent retroviral system, and the application of retroviral vectors to the nervous system. For more detailed information and original references about retrovirus, other reviews should be consulted (Coffin *et al.*, 1997; Goff, 2001).

## II. Overview of Retrovirus

### A. Structure of Retrovirus

#### 1. *General Structure*

The retrovirus particle is spherical and about 100 nm in diameter. It consists of the inner core and the outer envelope. The core contains two copies of a plus strand of the viral RNA genome and viral proteins, such as capsid (encoded by *gag*) and reverse transcriptase (encoded by *pol*), whereas the envelope consists of the viral coat proteins (encoded by *env*) and the lipid bilayer derived from the host cells. Retroviruses are classified into two types, simple and complex, based on the RNA genome structure: the RNA genome of simple retrovirus contains only three essential genes, *gag*, *pol*, and *env* whereas that of complex retrovirus contains additional gene(s).

#### 2. *Structure of Viral Genome*

The RNA genome of simple retrovirus such as Moloney murine leukemia virus (MoMLV) contains the following structures in this order: R, U5, primer-binding site, packaging signal ($\Psi$), splicing donor site, *gag*, *pol*, splicing acceptor site, *env*, polypurine tract, U3, and R (Fig. 1). In addition, it has the cap structure at the 5′-terminus and poly(A) tail at the 3′-terminus, like the cellular mRNA (Fig. 1). The features of each structure are as follows.

**R:** The R is a short sequence repeated twice, at the 5′- and 3′-ends. This region usually contains AAUAAA, which serves as a polyadenylation signal in the 3′-R.

**U5:** U5 is the unique 5′ sequence and contains one of the *att* sites, which are required for proviral integration into the host genome.

**Primer-binding site (PB):** This site is complementary to a host tRNA, which is used as a primer for minus-strand DNA synthesis.

**Packaging signal ($\psi$):** This region is required for encapsidation of the viral RNA genome into the virus particles. The signal for efficient packaging is extended into the *gag* region, and this extended signal is often designated as $\psi^+$.

**Splicing donor site (SD):** This site is utilized for formation of the spliced RNA, which is necessary for translation of *env*. This site is located inside the packaging signal.

***gag*:** The products of *gag* are essential for virion assembly and release. In the case of MoMLV, there are two translation initiation sites: the upstream CUG codon and downstream AUG codon.

***pol*:** *pol* encodes aspartyl protease (Pro), reverse transcriptase (RT), and integrase (IN). Pro is required for processing of viral proteins. RT is used for generation of double-strand DNA (the proviral DNA) from the viral RNA genome whereas IN is required for integration of the proviral DNA into the host genome. In some viruses, Pro is encoded within *gag* or as a separate gene.

**Splicing acceptor site (SA):** This site is utilized for formation of the spliced RNA, which is necessary for translation of *env*.

**env:** *env* encodes the envelope proteins. This gene determines the ecotropic/amphotropic features. Ecotropic retrovirus only infects rodents whereas amphotropic retrovirus infects a wider range of species, including humans.

**Polypurine tract (PP):** This tract, located just upstream of U3, is the initiation site for the plus-strand DNA synthesis.

**U3:** U3 is the unique 3′ sequence, which serves as a transcriptional enhancer/promoter in the proviral DNA form. This region also contains one of the *att* sites, which are required for DNA integration.

The provirus form (the DNA copy of the viral RNA genome) has additional U3 and U5 sequences at the 5′ and 3′ termini, respectively. As a result, there are two U3–R–U5 structures, called long terminal repeats (LTRs), at both ends of the provirus form (Fig. 1). This proviral form is used for retroviral vectors.

The RNA genome of complex retrovirus contains additional genes. For example, the human T cell leukemia virus (HTL V)-1 genome contains a gene for Tax, which positively regulates transcription from the LTR promoter.

## B. Life Cycle of Retrovirus

Retrovirus infection involves the following sequential steps: (1) entry into host cells, (2) reverse transcription, (3) integration, (4) expression of viral genes, and (5)

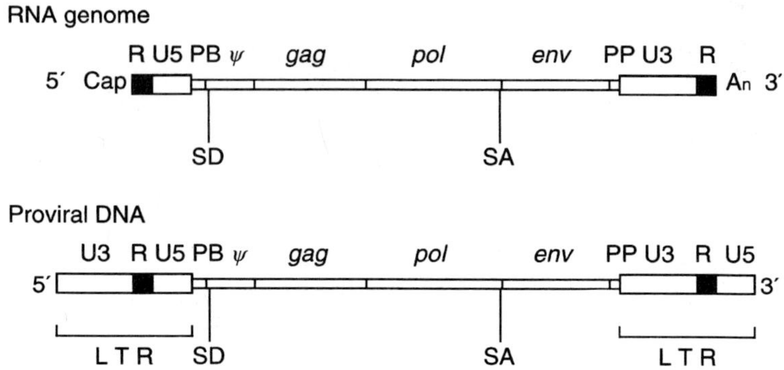


FIG. 1. Structure of retroviral genome and proviral DNA. The viral RNA genome (plus-strand RNA) contains the cap structure at the 5′-terminus and poly(A) tail (An) at the 3′-terminus, like the cellular mRNA. In addition, it has the following structures in this order: R, U5, primer-binding site (PB), packaging signal (ψ), *gag*, *pol*, *env*, polypurine tract (PP), U3, and R. The proviral form (double-strand DNA) has additional U3 and U5 at the upstream end of the 5′-R and the downstream end of the 3′-R, respectively. The U3–R–U5 sequences at both ends are called long terminal repeats (LTRs). SA, splicing acceptor site; SD, splicing donor site.

production of viral particles (Fig. 2). At each step, retroviruses skillfully utilize the host machinery for their propagation.

1. *Entry into Host Cells*

First, retroviruses interact with surface molecules of host cells. The products of *env*, present at the virion surface, contact with the host membrane proteins (virus receptors). The ecotropic virus, which is infectious only to rodents, uses a transporter for basic amino acids, whereas amphotropic virus, which is infectious to a wide range of species, uses a sodium-dependent phosphate symporter as viral receptors. After interaction of retroviruses and viral receptors, the virus and host cell membranes fuse together, and the virion core is delivered into the cytoplasm of the host cells (Fig. 2).

2. *Reverse Transcription*

As soon as the virion core enters the cytoplasm, reverse transcriptase included in the core initiates reverse transcription and produces the minus- strand DNA. For this process, tRNA annealed to the primer-binding site of the viral RNA genome is used as a primer for reverse transcription (Fig. 3A). This transcription stops at the 5′-end of the viral RNA. However, because the RNaseH activity of reverse transcriptase degrades the RNA strand of the DNA:RNA hybrid (Fig. 3B), the resultant single- strand DNA containing the sequence complementary to R can hybridize to the 3′-R of the RNA genome (Fig. 3C). This jump to the

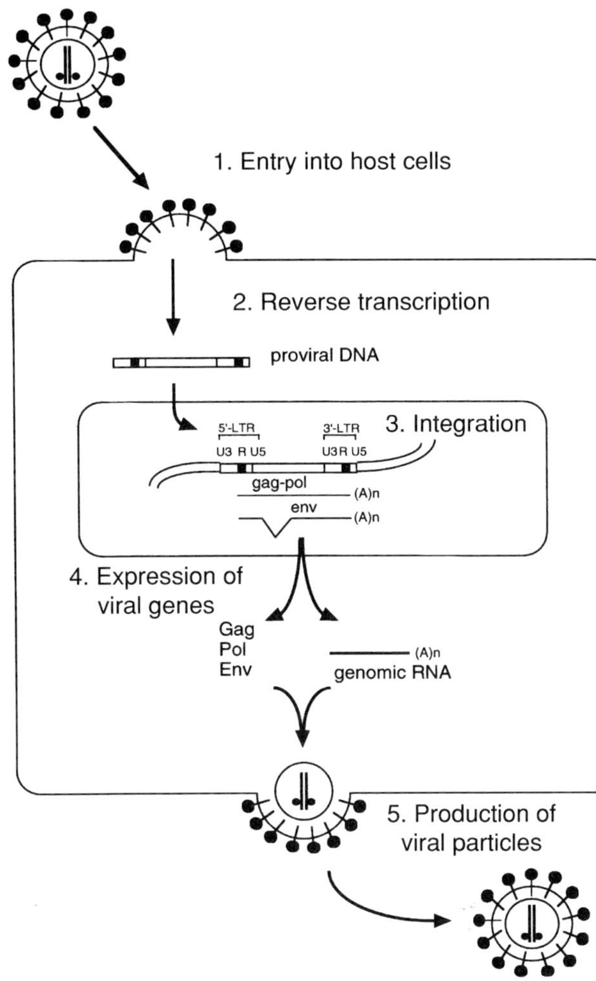


FIG. 2. Life cycle of retrovirus. Retrovirus infection involves the following sequential steps: (1) Entry into host cells, (2) Reverse transcription of the RNA genome and generation of the double-strand proviral DNA, (3) Integration of the proviral DNA into the host genome, (4) Transcription of the viral genes from the 5′-LTR, splicing, and translation, and (5) Production of viral particles. Integration of proviral DNA occurs only in dividing cells, and the integration site (only one per cell) is essentially random. See text for details.

3′-end then enables further reverse transcription to the 5′-end (Fig. 3D). Next, the plus strand of DNA is produced, using the minus-strand DNA as a template by reverse transcriptase. In this process, the polypurine tract, which is resistant to RNaseH, is used as a primer (Fig. 3E and F). The plus-strand DNA synthesis

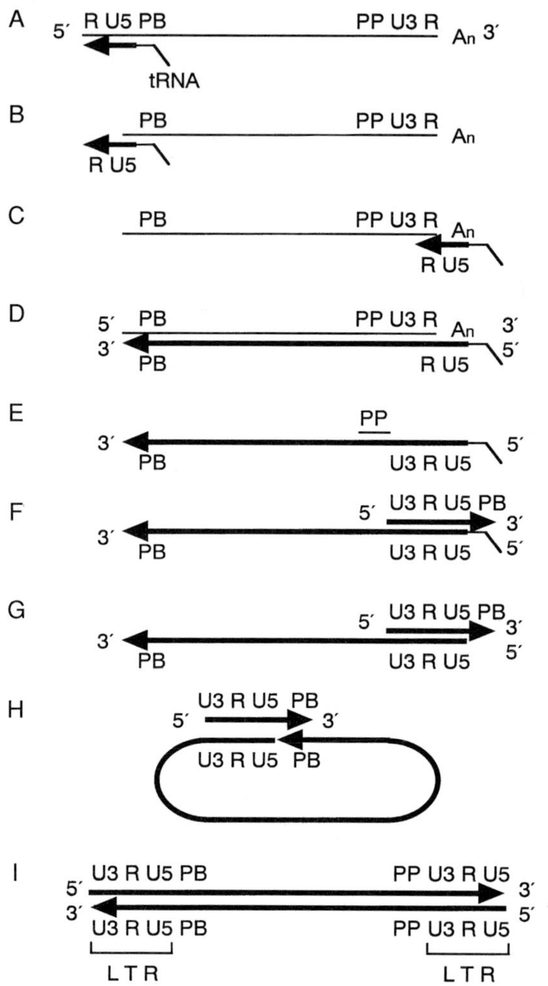


FIG. 3. Formation of the double-strand proviral DNA from RNA genome. (A) Reverse transcriptase included in the core initiates reverse transcription and produces the minus-strand DNA using tRNA as a primer, which is annealed to the primer-binding site (PB) of the viral RNA genome. This transcription stops at the 5′-end of the viral RNA. (B) RNaseH activity of reverse transcriptase degrades the RNA strand of the DNA:RNA hybrid. (C) The resultant single-strand DNA containing the sequence complementary to R can hybridize to the 3′-R of the RNA genome. (D) The jump to the 3′-end enables further reverse transcription to the 5′-end of the RNA genome. (E) RNaseH activity of reverse transcriptase degrades the RNA strand of the DNA:RNA hybrid. The polypurine tract (PP), located just upstream of U3, is resistant to RNaseH and serves as a primer for the plus-strand DNA synthesis. (F) The plus-strand DNA is produced, using the minus-strand DNA as a template by reverse transcriptase. The plus-strand DNA synthesis proceeds to the annealed tRNA, which has the complementary sequence to the primer-binding site (PB). (G) The tRNA that was used as a primer is degraded by RNaseH. (H) The resultant single-strand region containing the

proceeds to the annealed tRNA, which has the complementary sequence to the primer-binding site (Fig. 3F). After degradation of tRNA by RNaseH (Fig. 3G), the resultant single-strand region containing the primer-binding site hybridizes to the 3′-end of the minus-strand DNA, which contains the sequence complementary to the primer-binding site (Fig. 3H). This jump enables the synthesis of the complete double-strand proviral DNA (Fig. 3I). Because of the two jumps, the U3 and U5 are duplicated at the 5′- and 3′-ends, respectively.

3. *Integration*

The double-strand proviral DNA next enters the nucleus (Fig. 2). This process requires the breakdown of the nuclear membrane and therefore occurs only in dividing cells. After entry into the nucleus, the proviral DNA is integrated into the host chromosomal DNA through the *att* sites of the U5 and U3 regions. This process is controlled by integrase. Only a single copy of the proviral DNA is integrated. The integration sites of the proviral DNA are essentially random.

4. *Expression of Viral Genes*

The U3 of the 5′-LTR functions as a promoter, which directs transcription from the U3–R border. Transcription proceeds to the 3′-LTR, and the RNA is cleaved and polyadenylated at the R–U5 border of the 3′-LTR (Fig. 2). For MoMLV, two distinct RNAs are generated from the viral genome (Fig. 2). One is the full- length form containing the region from the 5′-R to the 3′-R. This RNA serves both as the viral genome to be packaged into virion particles and as the template for translation of *gag* and *pol*. The other one is the spliced form and serves as the template for translation of *env*. In the latter form, the *gag* and *pol* regions are removed as an intron.

From the full-length form of RNA, both Gag and Gag–Pol fusion proteins are produced. Because the open reading frame of *gag* ends at the stop codon, translation of the full-length RNA mostly generates only Gag protein. However, occasionally termination at the stop codon is suppressed, and translation continues to the open reading frame of *pol*, thereby producing Gag–Pol fusion protein. This fusion protein is later processed by a protease encoded by *pol*.

---

primer-binding site (PB) hybridizes to the 3′-end of the minus-strand DNA, which contains the sequence complementary to the primer-binding site (PB). This hybridization transiently makes a circular form since the U3–R–U5 portion of the plus strand remains hybridized to the U3–R–U5 portion of the minus strand. (I) The jump to the 3′-end of the minus-strand DNA enables the synthesis of the complete double-strand proviral DNA. Because of the two jumps, the U3 and U5 are duplicated at the 5′- and 3′-ends, respectively, generating two copies of a long terminal repeat (LTR). RNA and DNA are shown by thin and thick lines, respectively.

5. *Production of Viral Particles*

The full-length RNA form is now packaged into virion particles (Fig. 2). The $\psi$ sequence of the RNA genome is recognized as a packaging signal by Gag, and two copies of the viral RNA are packaged into one virus particle. In addition, reverse transcriptase and host tRNA, which are used for proviral DNA synthesis, are incorporated into virus particles. Finally, the virus particles are budded from the cell surface.

### C. Pseudotype Formation

Mixed infection of two different types of viruses sometimes results in production of virions containing the genome of one virus and the envelope proteins of the other, called pseudotyped viruses. For example, mixed infection of vesicular stomatitis virus (VSV) and retrovirus generates the pseudotyped retrovirus containing the retroviral genome encapsidated by the VSV envelope proteins. Because the infectious host range is determined by the envelope proteins, this pseudotyped retrovirus has the same host range as VSV, which is infectious to a wider range of species than amphotropic retrovirus. This feature is utilized to produce a pseudotyped recombinant retrovirus vector (see below).

## III. Generation of Recombinant Retroviral Vectors

### A. Structure of Recombinant Retrovirus

The proviral form (double-strand DNA) is used as a retroviral vector. For generation of recombinant retrovirus, the 5′- and 3′-LTR sequences, the primer binding site and the $\psi$ sequence are required in *cis*. In contrast, *gag*, *pol*, and *env* genes are required only *trans* and therefore these regions are deleted and replaced with other genes in most recombinant retrovirus vectors. These vectors generate infectious virus only from cells that express *gag*, *pol*, and *env* genes (packaging cells). The resultant recombinant retroviruses are infectious but cannot expand from cell to cell (replication incompetent). However, in early versions of retroviral vectors, viral genes are only partially removed, and recombination between the viral genes present in both retroviral vectors and packaging cells often produces replication-competent retrovirus (helper virus). Because helper virus may cause diseases such as leukemia in host animals, the viral genes are deleted as much as possible in later versions of retroviral vectors.

Many retroviral vectors are produced from MoMLV. These vectors direct exogenous gene expression from the 5′-LTR or an inserted promoter. Most vectors also carry markers, such as drug- resistance genes (*neo, gpt, dhfr*) and color-marking genes [*green fluorescent protein (GFP), β-galactosidase, alkaline phosphatase*], to identify the virus-infected cells. Here are some of the vectors that have been commonly used.

**pLN:** This retroviral vector is derived from MoMLV (Fig. 4A) (Miller and Rosman, 1989). This vector contains the packaging signal extending into the *gag* region, which is required for efficient packaging (Bender *et al.*, 1987). However, because translation of *gag* is not necessary for packaging, two changes are introduced into this vector. First, the region containing the upstream translation initiation codon (CTG) of *gag* is replaced with the homologous region from Moloney murine sarcoma virus, which does not make Gag product from the CTG codon. Second, the downstream translation initiation codon (ATG) of *gag* from MoMLV is mutated to the stop codon (TAG) (Fig. 4A). Thus, this vector does not produce Gag proteins. Another feature of this vector is that the *pol* and *env* regions are completely removed. To obtain a high titer of virus [$>10^6$ colony-forming units (cfu)/ml] from this vector, it is necessary to establish permanent virus-producing lines by selection with a drug resistance gene (*neo*).

From pLN, several vectors have been developed, such as pLXSN and pLXSG. In pLXSN, the simian virus 40 (SV40) early promoter is inserted upstream of *neo* so that the *neo* is expressed from the SV40 early promoter (Fig. 4B) (Miller and Rosman, 1989). In addition, the polylinker sites are introduced just upstream of the SV40 promoter. Any cDNA inserted into the polylinker site can be expressed from the 5′-LTR, thereby enabling expression of two genes. In pLXSG, the fusion of *β-galactosidase* and *neo* genes (*geo*) is inserted instead of *neo* (Fig. 4C) (Ishibashi *et al.*, 1994). This vector enables histological identification of virus-infected cells by X-gal staining.

**pBabe:** Like pLN, this vector has the extended packaging signal and nonsense mutation at the ATG of *gag*. In addition, because the splicing donor site could lead to an aberrant splicing event using cryptic splicing acceptor sites within inserted genes, a mutation is introduced into the splicing donor (GT to GC) in pBabe to avoid this artifact (Fig. 4D) (Morgenstern and Land, 1990). From pBabe, pLIA has been developed. This vector carries the internal ribosomal entry site (IRES) and human placental alkaline phosphatase (PLAP) gene instead of SV40 promoter-*neo* (Bao and Cepko, 1997). Thus, pLIA enables bicistronic expression of the cDNA inserted at the cloning site and PLAP.

**pCL:** One of the limiting steps for production of a high titer retrovirus is a low expression of the full-length retroviral RNA in packaging cells. To overcome this problem, pCL vector has been developed (Naviaux *et al.*, 1996). The cytomegalovirus (CMV) enhancer is introduced into the U3 region of

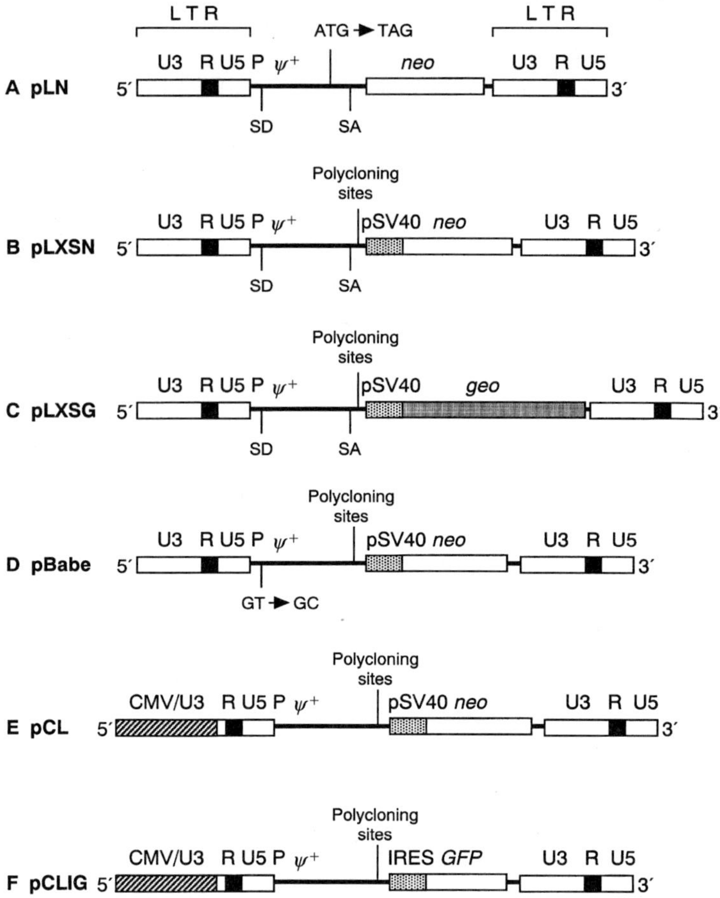


FIG. 4. Retroviral vectors. Retroviral vectors are generated from proviral DNAs. The essential regions are 5′-LTR, primer-binding site (P), packaging signal, and 3′-LTR. The packaging signal is extended into the *gag* region ($\psi^+$). Other viral regions are deleted from the vectors. (A) pLN. Because *gag* translation is not required, the translation initiation codon of *gag* is mutated (ATG to TAG). (B) pLXSN. The polycloning sites and the SV40 early promoter are inserted upstream of *neo* of pLN. The gene inserted into the polycloning sites is expressed from the 5′-LTR whereas *neo* is expressed from the SV40 early promoter. (C) pLXSG. This vector is the same as pLXSN except that it has *β-galactosidase–neo* fusion gene (*geo*) instead of *neo*. This vector allows histological identification of retrovirus-infected cells by X-gal staining. (D) pBabe. This vector is generated from pLN. The splicing donor site (SD) is also mutated (GT to GC) to prevent aberrant splicing using cryptic splicing acceptor sites within inserted genes. (E) pCL. The majority of the U3 region in the 5′-LTR is replaced with CMV enhancer in order to increase the viral titer. This enhancer region is not included in the RNA genome because the RNA to be encapsidated into virions is synthesized from the U3–R border. (F) pCLIG. The 5′-LTR of pCL and the rest of pBabe are fused. In addition, the SV40 promoter–*neo* region is exchanged with IRES-GFP. This vector directs bicistronic expression from the 5′-LTR.

the 5′-LTR, which significantly increases production of the viral RNA in packaging cells (Fig. 4E). This vector exhibits an almost 100-fold increase of the viral titer, and transient transfection of pCL into packaging cell lines produces the titer of $>10^6$ cfu/ml. Thus, this vector is suitable for genes that are toxic to packaging cells. It should be noted that the retrovirus produced from pCL does not carry the CMV enhancer because the U3 region of the 5′-LTR is not included in the viral genome. pCLIG is one of the derivatives of pCL and pBabe. pCLIG directs expression of the inserted cDNA and GFP from the 5′-LTR (Fig. 4F) (Hojo *et al.*, 2000). GFP is expressed through the IRES, which enables bicistronic expression.

**pHIT110/111 and pkat3:** These vectors have not only the CMV enhancer/promoter instead of U3 in the 5′-LTR but also the SV40 origin at the downstream of the 3′-LTR (Finer *et al.*, 1994; Soneoka *et al.*, 1995). The SV40 origin region enables replication of the vectors in packaging cells expressing SV40 large T antigen. These vectors can produce a high titer of retrovirus by transient transfection into packaging cells.

## B. Packaging Cells

Packaging cells express the viral genes (*gag, pol, env*) but do not produce infectious retroviral particles because the $\psi$ sequence is deleted from the viral genes. These cells make infectious retrovirus if retroviral vectors, which have the $\psi$ sequence, are introduced (Fig. 5A). Early versions of packaging cells make helper virus by recombination between the homologous regions of the viral genes present in both packaging cells and the retroviral vectors. Helper virus exhibits horizontal expansion of virus infection, causing diseases such as leukemia. Thus, it is important to develop helper-free packaging cells to prevent such problems. In later versions of packaging cells, homologous regions are deleted from the viral genes.

There are two types of packaging cells, ecotropic and amphotropic virus-producer lines, which have different *env* genes. In addition, the pseudotyped retroviral vector system has been also developed. Here are some of the packaging cells that have been commonly used.

### 1. *Ecotropic*

**$\psi$-2:** This cell is produced by introducing the MoMLV vector lacking the $\psi$ sequence into NIH3T3 cells (Mann *et al.*, 1983). $\psi$-2 generates a high titer of ecotropic retrovirus. However, because this cell contains most of the viral sequence, it frequently generates helper virus.

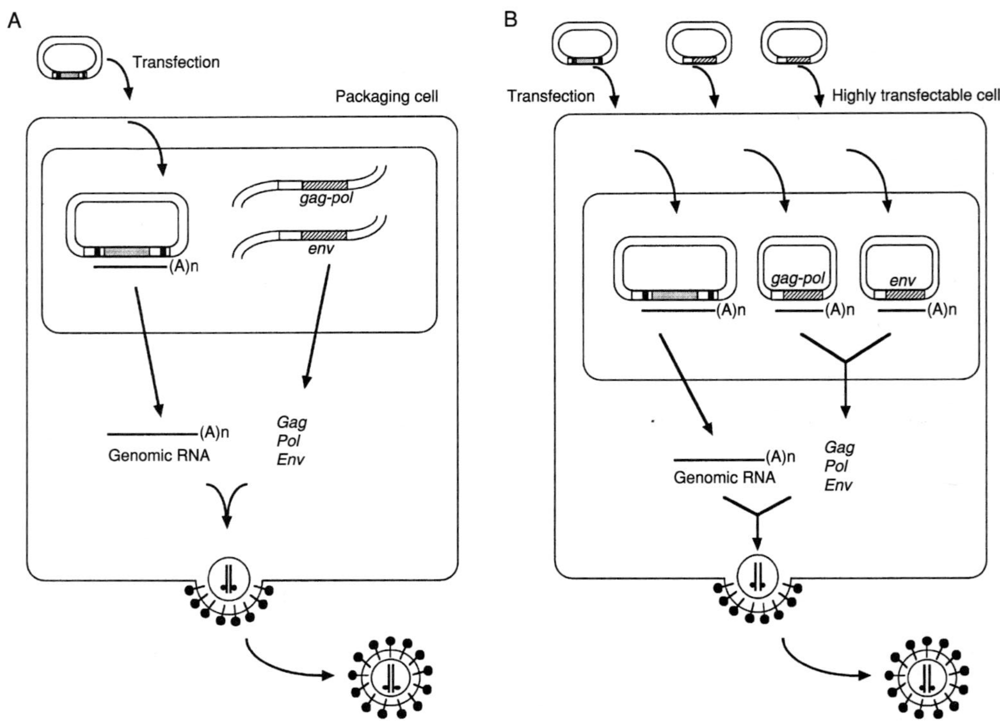


Fig. 5. Strategy for generation of retroviruses. (A) Strategy using packaging cells. Packaging cells express *gag*, *pol*, and *env*, and thus can produce retrovirus particles when the retroviral vectors are introduced. Different packaging cell lines are required to produce different types of retroviruses (ecotropic, amphotropic, and pseudotyped). (B) Strategy using nonpackaging cells. Retrovirus can be generated by introducing the *gag–pol* expression vector and the *env* expression vector together with a retroviral vector into nonpackaging cells. Ecotropic, amphotropic, and pseudotyped retroviruses can be produced by changing the *env* expression vector.

**$\psi$CRE:** This cell is produced by introducing the *gag–pol* expression vector pCRIPenv- and the *env* expression vector pCRIPgag-2 separately into NIH3T3 cells (Danos and Mulligan, 1988). pCRIPenv- is made from MoMLV but lacks the $\psi$ sequence, the 3′-region of *env*, and the 3′-LTR. pCRIPgag-2 also lacks the $\psi$ sequence and the 3′-LTR. Because both pCRIPenv- and pCRIPgag-2 have less homologous regions than the viral genes used in $\psi$-2 and are introduced separately, the chance to produce helper virus is very low.

**BOSC23:** This cell is generated by introducing pCRIPenv- and pCRIPgag-2 separately into 293T cells (Pear *et al.*, 1993). 293T cells are Ad5-transformed human embryonic kidney cells and express SV40 large T antigen. 293T cells are highly transfectable and enables replication of vectors containing the SV40 origin. Thus, transient transfection of retroviral vectors carrying the SV40 origin produces infectious virus with the titer of $>10^6$ cfu/ml.

2. *Amphotropic*

**$\psi$-AM:** This cell is generated by introducing pMAV-$\psi^-$ into NIH3T3 cells (Cone and Mulligan, 1984). pMAV-$\psi^-$ is a hybrid vector in which *env* of MoMLV is replaced with that of amphotropic virus 4070A. In addition, pMAV-$\psi^-$ has a deletion of the $\psi$ sequence. Because this cell contains most of the viral sequence, it frequently generates helper virus by recombination with homologous regions in retroviral vectors.

**PA317:** This cell is generated by introducing pPAM3 into NIH3T3 cells (Miller and Buttimore, 1986). pPAM3 is a hybrid vector in which *env* of MoMLV is replaced with that of amphotropic virus 4070A. In addition, pPAM3 has deletions of the 5′-region of the 5′-LTR, the $\psi$ sequence, and the 3′-LTR. This cell has a very low chance to make helper virus.

**$\psi$CRIP:** This cell is produced by introducing pCRIPenv- and pCRIPAMgag-separately into NIH3T3 cells (Danos and Mulligan, 1988). pCRIPenv- is a *gag–pol* expression vector, the same as used for $\psi$CRE cells. pCRIPAMgag-directs amphotropic *env* expression and lacks the $\psi$ sequence and the 3′-LTR. Like $\psi$CRE ecotropic packaging cells, the chance to produce helper virus from $\psi$CRIP is very low.

3. *Pseudotype*

VSV G glycoprotein (VSV-G) gene is commonly used with MoMLV-derived vector for pseudotyped retrovirus (Yee *et al.*, 1994). This pseudotyped virus has two advantages: (1) Because the virions are stable, they can be concentrated more than 100-fold without losing activity by ultracentrifugation; (2) VSV-G displays a wider infectious host range than amphotropic retrovirus. However, because persistent expression of VSV-G is toxic to mammalian cells, VSV-G expression vector (such as pHCMV-G, which directs VSV-G expression from the CMV enhancer/promoter) should be transiently cotransfected with retroviral vector into the packaging cells expressing only *gag* and *pol* (such as 293GP cells). The virus should be harvested within 3 days after transfection (Yee *et al.*, 1994). 293GP cells are generated from 293 cells by introducing MoMLV *gag–pol* genes under the control of the CMV promoter (Burns *et al.*, 1993).

#### 4. *Another Method without Using Packaging Cells*

Many passages of packaging cells often result in reduction of efficiency for virus production. In that case, subcloning and rescreening of packaging cells are required to obtain a high titer virus. To avoid this problem, *gag*, *pol*, and *env* expression vectors are transiently cotransfected with the retrovirus vector into highly transfectable cells such as 293 cells (Fig. 5B). For example, cotransfection of the *gag–pol* expression vector, *env* expression vector, and retroviral vector all containing the SV40 origin (such as pkat and pHIT vectors) into 293T cells can produce retrovirus with the titer of $\sim 10^7$ cfu/ml after 48h (Soneoka *et al.*, 1995). This transient transfection system is particularly suitable for virus carrying genes that are toxic to cells because packaging cells producing such virus become sick and easily lose virus-producing ability. In addition, with this system, ecotropic, amphotropic, and pseudotyped viruses can be easily generated by simply changing *env* expression vectors without changing packaging cells. Another advantage of this system is that it does not produce helper virus.

### C. Generation of Recombinant Retrovirus

To generate infectious virus, retroviral vectors should be introduced into packaging cells by transfection (Cepko, 1992). For generating the virus with the titer of $\sim 10^6$ cfu/ml, transient transfection is enough. For pseudotyped retroviral vectors, transient transfection can make the titer of $10^8$–$10^9$ cfu/ml after concentration of the virus. However, for ecotropic retrovirus, permanent virus-producer lines should be established to get such a high titer because ecotropic viruses are fragile and cannot be concentrated >100-fold. For efficient expression, the inserted cDNAs should be as short as possible. Particularly, if AATAAA or similar sequences are present in the 3′-noncoding region, it would cause premature termination of transcription of the viral RNA genome, thereby reducing the virus titer. Therefore, preferably only the coding region should be inserted into the retroviral vectors. In addition, the 5′-LTR promoter does not work well in some cell types and, in such a case, a strong promoter is necessary. For example, it has been reported that the EF1$\alpha$ promoter works better in the brain than the LTR promoter and is successfully used in retroviral vectors (Gaiano *et al.*, 1999).

Here is one of our protocols for generation and concentration of retrovirus. All procedures require special care to prevent biological hazard. Particularly, extreme precaution is necessary to handle amphotropic and pseudotyped viruses.

## D. Protocol for Generation of Retrovirus by Transient Transfection

**Step 1**: (Day 1) Plate $3 \times 10^5$ packaging cells (or nonpackaging cells) in a 100-mm dish.
**Step 2**: (Day 2) The cells should be within 30–40% confluence. Prepare the following transfection solution.
Retroviral vector, 10 $\mu$g
(In the case of using nonpackaging cells such as 293 cells, also mix 5 $\mu$g each of the *gag–pol* and *env* expression vectors)
Culture medium (such as DMEM) without serum, 1 ml
Lipofectamine (Invitrogen), 25 $\mu$l
**Step 3**: Mix the solution and keep it at room temperature for 45 min.
**Step 4**: Rinse the cells with serum-free culture medium once. Then, add 4 ml of serum-free culture medium.
**Step 5**: Add the transfection solution at Step 2 to the cells and incubate for 6 h.
**Step 6**: Discard the medium, and add 5 ml of fresh medium containing serum.
**Step 7**: Incubate the cells for 2 days.
**Step 8**: (Day 4) Harvest the supernatant and filtrate through a 0.45-$\mu$m filter.
**Step 9**: Concentrate with Centriprep 100 (Amicon) at 4°C.
**Step 10**: Divide virus solution into small aliquots and store them at −80°C. To minimize the loss of the virus titer, freeze-thaw should be avoided.
**Step 11**: Check the virus titer as follows.
**a.** Plate cells (such as NIH3T3) in six-well plates on the previous day.
**b.** Next day, remove the medium and add the following:
Virus solution + culture medium with serum, 500 $\mu$l
4 mg/ml of Polybrene, 1 $\mu$l
**c.** Incubate for 1–3 h at 37°C in 5% $CO_2$.
**d.** Add 1.5 ml of medium with serum to dilute Polybrene.
**e.** Incubate for 1–2 days at 37°C in 5% $CO_2$.
**f.** If the virus carries color-marking genes such as *β-galactosidase*, stain cells with an appropriate staining procedure and count the stained cells (because pairs of stained cells present next to each other are usually derived from single infected cells, count them as one). Calculate the virus titer

(cfu/ml) = the number of stained cells/virus volume (ml). If the virus carries drug resistance genes, split the cells 1:10, culture them in the medium containing an appropriate drug for about 1 week, and then count the drug-resistant colonies. Calculate the virus titer (cfu/ml) = the number of colonies/[virus volume (ml) × 0.1 × 2 (in case the cells are doubled before changed to drug-containing medium)].

### E. Helper Virus

If there are homologous regions between the viral genes in packaging cells and in the retroviral vectors, there is a chance of recombination to generate *gag–pol–env*-carrying replication-competent retrovirus, called helper virus. Such virus could result in repeated infection and expansion in host animals, causing leukemia. In addition, it could carry other genes derived from host cells, which might cause unrelated effects. To avoid these problems, it is important to generate helper-free retrovirus. Most packaging cells and retroviral vectors developed recently have a minimum of homologous regions of the viral genes, and thus the chance to produce helper virus is very low. However, it is still necessary to confirm that there is no helper virus to avoid any defects by repeated infection. For detection of helper virus, the supernatant is prepared from the cells, which are infected with a test virus solution and then cultured for several days. If the test virus contains helper virus, this supernatant also contains virus, which can be detected by the ability of marker gene transduction to other cells or by $S^+L^-$ assay (Eckner and Hettrick, 1977) or reverse transcriptase assay (Goff *et al.*, 1981).

## IV. Gene Delivery to the Nervous System by Retroviral Vectors

Because retrovirus is infectious only to mitotic cells, the major targets in the nervous system are neural precursor (stem) cells and tumor cells (Sanes *et al.*, 1986; Turner and Cepko, 1987; Culver *et al.*, 1992). Retrovirus-mediated gene introduction can stably change the phenotypes of dividing cells and will be useful for neuronal and glial regeneration from neural precursor cells as well as for treatment of tumors. Here, application of retrovirus to neural precursor cells is described.

Neural precursor cells, which give rise to both neurons and glial cells, are present in the ventricular zone of the developing brain. It has been shown that these cells remain as astrocyte-like cells in the subventricular zone of the adult brain (Temple, 2001). Although their neurogenesis ability is limited, adult neural precursor cells have a potential to become a wide variety of neurons if they are properly instructed by environmental cues such as growth factors and by intrinsic cues such as transcription factors. Similarly, embryonal neural precursor cells require proper intrinsic and extrinsic cues to differentiate into diverse types of neurons. Many attempts have been made to instruct neural precursor cells to adopt the desired cell fates by infection of retroviruses directing expression of transcription factors. It has now been increasingly clear that basic helix–loop–helix (bHLH) factors regulate maintenance of neural precursor cells and fate determination of neurons and glial cells (Kageyama and Nakanishi, 1997; Lee, 1997). In addition, combinations of bHLH and homeobox factors are important for specification of neuronal subtypes (Tanabe and Jessell, 1996). Thus, the information about the transcription factor code will be useful for regeneration of specific neurons and glial cells. Here, we describe some of the recent studies about retrovirus-mediated introduction of transcription factor genes, which regulate maintenance of neural precursor cells and differentiation into neurons and glial cells.

### A. Maintenance of Neural Precursor Cells

The bHLH gene *Hes1* is expressed by neural precursor cells in the ventricular zone of the embryonal brain. *Hes1* expression disappears when precursor cells differentiate into neurons. In *Hes1*-deficient mice, neural precursor cells are not properly maintained and prematurely differentiate into neurons (Ishibashi *et al.*, 1995; Tomita *et al.*, 1996). These results suggest that *Hes1* is important for maintenance of neural precursor cells. This notion is confirmed by misexpression study with *Hes1*-transducing retrovirus, which was applied to the developing nervous system. SG-Hes1 virus directs *Hes1* expression from the 5′-LTR and *geo* (fusion of *β-galactosidase* and *neo*) expression from the internal SV40 promoter whereas SG virus (made from pLXSG) directs only *geo* expression (Fig. 6A) (Ishibashi *et al.*, 1994). When SG-Hes1 virus is injected into embryonal brain, all the virus-infected cells remain in the ventricular zone whereas cells infected with SG virus migrate out of the ventricular zone and differentiate into both neurons and glial cells (Fig. 6A) (Ishibashi *et al.*, 1994). Similar results are obtained with CLIG-Hes1 virus, which directs *Hes1* and *GFP* expression: Neural precursor cell cultures infected with CLIG-Hes1 virus are inhibited from neuronal differentiation and remain as nestin-positive undifferentiated cells whereas cells infected with CLIG virus differentiate into neurons ($MAP2^+$, Fig. 6B) (Hirata *et al.*, 2000). Interestingly, undifferentiated cells maintained by *Hes1* undergo both neuronal and glial differentiation when *Hes1* expression is downregulated (Ohtsuka *et al.*, 2001). These results together indicate that misexpression of *Hes1* can inhibit

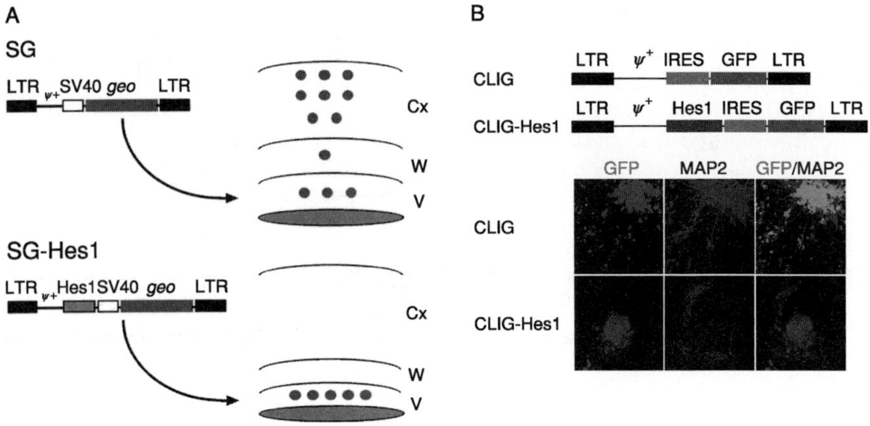


FIG. 6. Maintenance of neural precursor cells by *Hes1*-transducing retrovirus. (A) SG virus (produced from pLXSG) directs *geo* (fusion of *β-galactosidase* and *neo*) expression from the internal SV40 promoter whereas SG–Hes1 virus directs *Hes1* expression from the 5′-LTR promoter and *geo* expression from the internal SV40 promoter. One microliter of SG or SG–Hes1 retrovirus ($10^6$ cfu/ml) in 0.2 mg/ml of Polybrene was injected into lateral ventricles of mouse embryos at Days 12–13, and the fates of the virus-infected cells were examined by X-gal staining at postnatal Days 0–4. The majority of the cells infected with SG virus migrated out of the ventricular zone and differentiated into neurons and glial cells whereas those infected with SG–Hes1 virus remained in the ventricular zone. Cx, cortex; V, ventricular zone; W, white matter. (B) CLIG directs only GFP expression whereas CLIG–Hes1 directs Hes1 and GFP expression. Neural precursor cells prepared from mouse embryos were infected with CLIG or CLIG–Hes1 virus. Cells infected with CLIG virus differentiated into neurons with multiple neurites ($MAP2^+$) whereas those infected with CLIG–Hes1 did not. (See Color Insert.)

neuronal and glial differentiation and maintain neural precursor cells in the embryonal brain. It has been reported that the related bHLH genes *Hes3* and *Hes5* also have the same activities in the embryonal brain (Ohtsuka *et al.*, 1999, 2001; Hirata *et al.*, 2000, 2001).

Another useful gene to maintain neural precursor cells is the v-*myc* oncogene, which encodes a bHLH-leucine zipper-type factor. Neural precursor cells from adult rat hippocampus are infected with the retroviral vector carrying v-*myc* under the control of a tetracycline-suppressable transactivator and promoter system (Hoshimaru *et al.*, 1996). These cells are immortalized and proliferate as nestin-positive neural precursor cells in the absence of tetracycline. Strikingly, addition of tetracycline downregulates v-*myc* expression and converts the nestin-positive cells into mature neurons. Thus, retroviruses carrying *Hes*/v-*myc* genes under the control of regulatable promoters can be used for expansion of neural precursor cells.

## B. Neuronal Differentiation

### 1. *The neuronal bHLH Genes Promote Neurogenesis*

The neuronal bHLH genes *Mash1*, *Math3*, and *Neurogenin* are expressed in the developing nervous system. Disruption of these genes leads to the loss of neurons and concomitant acceleration of gliogenesis, indicating that these bHLH genes promote neuronal versus glial fate determination (Tomita *et al.*, 2000; Nieto *et al.*, 2001). In accordance with these results, cells infected with retroviruses carrying *Mash1*, *Math3*, *NeuroD*, or *Neurogenin* efficiently differentiate into neurons in the retina, neural crest, and brain (Lo *et al.*, 1998; Morrow *et al.*, 1999; Bae *et al.*, 2000; Cai *et al.*, 2000; Sun *et al.*, 2001). Thus, these retroviruses can be used to instruct neural precursor cells to adopt the neuronal fate in various regions of the nervous system. Interestingly, in neural crest, *Mash1* has been shown to induce both pan-neuronal features and autonomic neuronal identity (Lo *et al.*, 1998). Thus, it is likely that bHLH genes not only regulate neuronal fate determination but also specify the neuronal subtype identity.

Although these bHLH genes have neuronal fate determination activity, they cause neuronal degeneration at later stages in the brain (Cai *et al.*, 2000). These results suggest that persistent expression of some bHLH genes is toxic to cortical neurons. Thus, for regeneration of neurons in the brain, transient expression of the bHLH genes may be better.

### 2. *Homeobox Genes Are Important for Specification of Neuronal Subtypes*

Homeobox genes have been intensively analyzed and shown to play an essential role in neuronal subtype specification (Tanabe and Jessell, 1996). Although retroviral misexpression study has been mostly done with chick, some work has also been done with mammals. The homeobox gene *Crx* is essential for photoreceptor cell differentiation (Furukawa *et al.*, 1999). Infection of *Crx*-transducing retrovirus in the developing retina promotes photoreceptor development (Furukawa *et al.*, 1997). Significantly, *Crx* also exhibits a photoreceptor genesis activity in neural precursor cells of a nonretinal origin. Neural precursor cells prepared from iris proliferate in the presence of bFGF and differentiate into neurons in the presence of serum. However, none of the iris-derived neurons expresses retina-specific markers such as rhodopsin, indicating that iris-derived cells do not differentiate into retinal neurons by themselves. However, treatment of the iris cells with *Crx*-transducing retrovirus efficiently generates rhodopsin-expressing photoreceptors (Haruta *et al.*, 2001). Thus, this retrovirus system is very effective for regeneration of photoreceptors from neural precursor cells of a different origin. Because iris is relatively easy to take from patients, this method is promising for treatment of photoreceptor degeneration.

The homeobox gene *Phox2a* is essential for differentiation of subsets of autonomic neurons (Morin *et al.*, 1997). Retroviral misexpression of *Phox2a* induces autonomic features in neural precursor cells, although it does not make neurons (Lo *et al.*, 1998). Thus, it is likely that homeobox genes are primarily involved in neuronal subtype specification rather than neuronal fate determination.

3. *Combinations of bHLH and Homeobox Genes Are Important for Specification of Neuronal Subtypes*

Although bHLH and homeobox genes regulate neuronal differentiation, each gene alone is not sufficient for specification of correct neuronal subtypes in many cases. Recent data indicate that combinations of bHLH and homeobox genes are important for specification of correct neuronal subtypes in the neural retina. Neural retina consists of six types of neurons and one type of glia, and these seven types of cells differentiate from common precursor cells. The bHLH gene *Math3* and the homeobox gene *Chx10* are both expressed by bipolar cells, retinal interneurons. Although cells infected with CLIG–Chx10 or CLIG–Math3 do not differentiate into bipolar cells, cells infected with CLIG–Chx10–Math3, which carries both *Chx10* and *Math3* genes, efficiently differentiate into bipolar cells (Fig. 7) (Hatakeyama *et al.*, 2001). Similarly, cells infected with CLIG–Pax6–NeuroD, which carries the homeobox gene *Pax6* and the bHLH gene *NeuroD*, efficiently differentiate into amacrine cells, other retinal interneurons, whereas cells infected with CLIG-Pax6 or CLIG-NeuroD do not (Inoue *et al.*, 2002). Thus, combinations of bHLH and homeobox genes are important for generation of specific neurons, and the retrovirus system offers a good tool for such dual gene expression. It has been demonstrated that homeobox genes provide region-specific identity whereas bHLH genes regulate neuronal versus glial fate determination. bHLH genes are also involved in neuronal subtype specification when proper homeobox genes are coexpressed.

4. *Other Transcription Factors Can Also Influence Neuronal Subtype Specification*

Transcription factors other than bHLH and homeobox factors are also known to influence neuronal subtype specification. For example, Nurr1, an orphan nuclear receptor, is essential for midbrain dopaminergic neurons (Zetterstrom *et al.*, 1997). Misexpression of Nurr1 with retroviral vector in neural precursor cells upregulates expression of tyrosine hydroxylase, which is essential for dopamine synthesis, although Nurr1 is not able to promote neurogenesis (Sakurada *et al.*, 1999). Thus, Nurr1 is not involved in neuronal fate determination but controls specification of dopaminergic features.

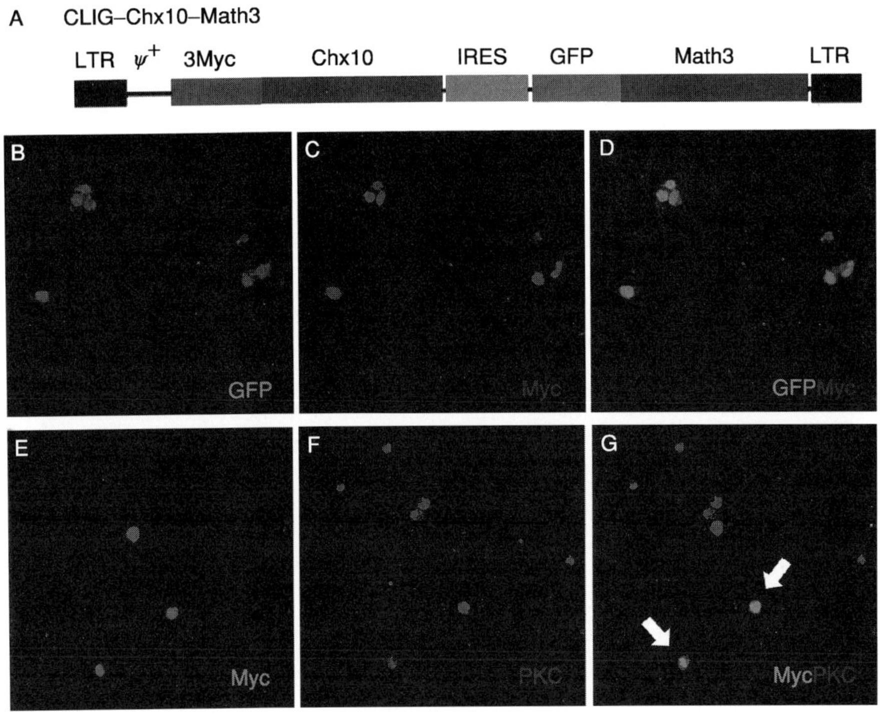


FIG. 7. Combinations of bHLH and homeobox genes promote specification of interneurons. (A) Structure of CLIG–Chx10–Math3. CLIG–Chx10–Math3 carries the homeobox gene *Chx10* and the bHLH gene *Math3*, which are under the control of the 5′-LTR. In this vector, Chx10 has three repeats of Myc tags whereas Math3 is fused with GFP. Dual gene expression can be monitored by Myc and GFP expression. (B–D) Neural precursor cells prepared from retina of mouse embryos were infected with CLIG–Chx10–Math3. Almost all cells infected with CLIG–Chx10–Math3 virus coexpressed GFP and Myc, indicating that this virus efficiently directs dual gene expression. (E–G) Many of the cells infected with CLIG–Chx10–Math3 ($Myc^{+}$) differentiated into bipolar cells ($PKC^{+}$, arrows in G). This retroviral system is useful for generation of specific neuronal subtypes. (See Color Insert.)

## C. Glial Differentiation

### 1. *Müller Glial Differentiation in the Retina*

The bHLH genes *Hes5* and *Hesr2* are specifically expressed by Müller glial cells in the developing postnatal retina. Cells infected with CLIG–Hes5 or CLIG–Hesr2 efficiently differentiate into Müller glial cells in the postnatal retina (Hojo *et al.*, 2000; Satow *et al.*, 2001). Similar activity has been reported for *Hes1* and the homeobox gene *Rax* (Furukawa *et al.*, 2000). Because *Hes1* and *Hes5* promote maintenance of neural precursor cells in the embryonal brain and retina, it is likely that postnatal cells respond differently from embryonal cells to the same factors. It remains to be determined what genes are responsible for the stage-specific different effects.

2. *Oligodendrocyte Differentiation*

The bHLH genes *Olig1* and *Olig2* are expressed by precursors for oligodendrocytes, which myelinate axon fascicles. Cells infected with LIA-Olig1 virus, which directs alkaline phosphatase and *Olig1* expression, efficiently differentiate into oligodendrocytes in the brain (Lu *et al.*, 2001). A similar activity has been reported for *Olig2* (Zhou *et al.*, 2001). Thus, *Olig1* and *Olig2* are important regulators for olligodendrocyte formation, and this retrovirus system may be useful for treatment of demyelinated diseases.

The studies described above indicate that specific cell types can be induced from neural precursor cells by retrovirus-mediated misexpression of transcription factors. Particularly, combinations of transcription factors are important for specification of many neuronal subtypes. Retroviral vectors successfully transduce combinations of bHLH and homeobox genes and induce specific cell types. These vectors can be used for direct injection into patients or for infection to cell cultures, which are then transplanted into patients. In both cases, the retrovirus-mediated gene delivery system offers a useful method for regeneration of specific neurons from neural precursor cells, which cannot efficiently become neurons by themselves in the adult brain. However, there are still many issues to be solved for clinical use of the retroviral system. Particularly, the safety of the retrovirus system should always be considered to avoid retrovirus-induced diseases such as leukemia. In addition, it is not yet known to what extent the cells regenerated by recombinant retrovirus functionally rescue the brain defects. Thus, the long-term behavior of such regenerated cells should be monitored carefully. Furthermore, because the LTR promoter is often inactivated, it is necessary to use stable promoters. Regulatable promoters will be also required in some cases, such as cortical neurogenesis, because persistent expression of some bHLH genes is toxic to some neurons. Although there are many such problems to be solved, the retroviral system is suitable for efficient and stable gene introduction to neural precursor cells and will be applicable to regeneration therapy for various neurological diseases.

## Acknowledgments

We thank Ms. Kana Yamamoto for help in preparing the manuscript. This work was supported by research grants from the Ministry of Education, Culture, Sports, Science and Technology of Japan and Japan Society for the Promotion of Science.

## References


Bae, S. -K., Bessho, Y., Hojo, M., and Kageyama, R. (2000). The bHLH gene *Hes6*, an inhibitor of *Hes1*, promotes neuronal differentiation. *Development* **127,** 2933–2943.

Bao, Z.-Z., and Cepko, C. L. (1997). The expression and function of *Notch* pathway genes in the developing rat eye. *J. Neurosci.* **17,** 1425–1434.

Bender, M. A., Palmer, T. D., Gelinas, R. E., and Miller, A. D. (1987). Evidence that the packaging signal of Moloney murine leukemia virus extends into the *gag* region. *J. Virol.* **61,** 1639–1646.

Burns, J. C., Friedmann, T., Driever, W., Burrascano, M., and Yee, J.-K. (1993). Vesicular stomatitis virus G glycoprotein pseudotyped retroviral vectors: Concentration to very high titer and efficient gene transfer into mammalian and nonmammalian cells. *Proc. Natl. Acad. Sci. USA* **90,** 8033–8037.

Cai, L., Morrow, E. M., and Cepko, C. L. (2000). Misexpression of basic helix-loop-helix genes in the murine cerebral cortex affects cell fate choices and neuronal survival. *Development* **127,** 3021–3030.

Cepko, C. (1992). Transduction of genes using retrovirus vectors. *In* "Current Protocols in Molecular Biology" (F. M. Ausubel, R. Brent, R. E. Kingston, D. D. Moore, J. G. Seidman, J. A. Smith, and K. Struhl, Eds.), Suppl. 17, pp. 9.10.1–9.14.3. John Wiley & Sons, New York.

Coffin, J. M., Hughes, S. H., and Varmus, H. E. (1997). "Retroviruses." Cold Spring Harbor Laboratory Press, Cold Spring Harbor, NY.

Cone, R. D., and Mulligan, R. C. (1984). High-efficiency gene transfer into mammalian cells: Generation of helper-free recombinant retrovirus with broad mammalian host range. *Proc. Natl. Acad. Sci. USA* **81,** 6349–6353.

Culver, K. W., Ram, Z., Wallbridge, S., Ishii, H., Oldfield, E. H., and Blaese, R. M. (1992). In vivo gene transfer with retroviral vector-producer cells for treatment of experimental brain tumors. *Science* **256,** 1550–1552.

Danos, O., and Mulligan, R. C. (1988). Safe and efficient generation of recombinant retroviruses with amphotropic and ecotropic host ranges. *Proc. Natl. Acad. Sci. USA* **85,** 6460–6464.

Eckner, R. J., and Hettrick, K. L. (1977). Defective Friend spleen focus-forming virus: Interfering properties and isolation free from standard leukemia-inducing helper virus. *J. Virol.* **24,** 383–396.

Finer, M. H., Dull, T. J., Qin, L., Farson, D., and Roberts, M. R. (1994). *kat*: A high-efficiency retroviral transduction system for primary human T lymphocytes. *Blood* **83,** 43–50.

Furukawa, T., Morrow, E. M., and Cepko, C. L. (1997). *Crx*, a novel *otx*-like homeobox gene, shows photoreceptor-specific expression and regulates photoreceptor differentiation. *Cell* **91,** 531–541.

Furukawa, T., Morrow, E. M., Li, T., Davis, F. C., and Cepko, C. L. (1999). Retinopathy and attenuated circadian entrainment in *Crx*-deficient mice. *Nat. Genet.* **23,** 466–470.

Furukawa, T., Mukherjee, S., Bao, Z.-Z., Morrow, E. M., and Cepko, C. L. (2000). *rax*, *Hesl*, and *notch1* promote the formation of Müller glia by postnatal retinal progenitor cells. *Neuron* **26,** 383–394.

Gaiano, N., Kohtz, J. D., Turnbull, D. H., and Fishell, G. (1999). A method for rapid gain-of-function studies in the mouse embryonic nervous system. *Nat. Neurosci.* **2,** 812–819.

Goff, S. P. (2001). Retroviridae: The retroviruses and their replication. *In* "Fields' Virology" (D. M. Knipe and P. M. Howley, Eds.), Vol. 2, pp. 1871–1939. Lippincott, Williams & Wilkins, Philadelphia.

Goff, S., Traktman, P., and Baltimore, D. (1981). Isolation and properties of Moloney murine leukemia virus mutants: Use of a rapid assay for release of virion reverse transcriptase. *J. Virol.* **38,** 239–248.

Haruta, M., Kosaka, M., Kanegae, Y., Saito, I., Inoue, T., Kageyama, R., Nishida, A., Honda, Y., and Takahashi, M. (2001). Induction of photoreceptor-specific phenotypes in adult mammalian iris tissue. *Nat. Neurosci.* **4,** 1163–1164.

Hatakeyama, J., Tomita, K., Inoue, T., and Kageyama, R. (2001). Roles of homeobox and bHLH genes in specification of a retinal cell type. *Development* **128,** 1313–1322.

Hirata, H., Ohtsuka, T., Bessho, Y., and Kageyama, R. (2000). Generation of structurally and functionally distinct factors from the basic helix-loop-helix gene *Hes3* by alternative first exons. *J. Biol. Chem.* **275,** 19083–19089.

Hirata, H., Tomita, K., Bessho, Y., and Kageyama, R. (2001). *Hes1* and *Hes3* regulate maintenance of the isthmic organizer and development of the mid/hindbrain. *EMBO J.* **20,** 4454–4466.

Hojo, M., Ohtsuka, T., Hashimoto, N., Gradwohl, G., Guillemot, F., and Kageyama, R. (2000). Glial cell fate specification modulated by the bHLH gene *Hes5* in mouse retina. *Development* **127,** 2515–2522.

Hoshimaru, M., Ray, J., Sah, D. W. Y., and Gage, F. H. (1996). Differentiation of the immortalized adult neuronal progenitor cells line HC2S2 into neurons by regulatable suppression of the v-*myc* oncogene. *Proc. Natl. Acad. Sci. USA* **93,** 1518–1523.

Inoue, T., Hojo, M., Bessho, Y., Tano, Y., Lee, J. E., and Kageyama, R. (2002). *Math3* and *NeuroD* regulate amacrine cell fate specification in the retina. *Development* **129,** 831–842.

Ishibashi, M., Moriyoshi, K., Sasai, Y., Shiota, K., Nakanishi, S., and Kageyama, R. (1994). Persistent expression of helix-loop-helix factor HES-1 prevents mammalian neural differentiation in the central nervous system. *EMBO J.* **13,** 1799–1805.

Ishibashi, M., Ang, S.-L., Shiota, K., Nakanishi, S., Kageyama, R., and Guillemot, F. (1995). Targeted disruption of mammalian *hairy* and *Enhancer of split* homolog-1 (*HES-1*) leads to up-regulation of neural helix-loop-helix factors, premature neurogenesis and severe neural tube defects. *Genes Dev.* **9,** 3136–3148.

Kageyama, R., and Nakanishi, S. (1997). Helix-loop-helix factors in growth and differentiation of the vertebrate nervous system. *Curr. Opin. Genet. Dev.* **7,** 659–665.

Lee, J. E. (1997). Basic helix-loop-helix genes in neural development. *Curr. Opin. Neurobiol.* **7,** 13–20.

Lo, L., Tiveron, M.-C., and Anderson, D. J. (1998). MASH1 activates expression of the paired homeodomain transcription factor Phox2a, and couples pan-neuronal and subtype-specific components of autonomic neuronal identity. *Development* **125,** 609–620.

Lu, Q. R., Cai, L., Rowitch, D., Cepko, C. L., and Stiles, C. D. (2001). Ectopic expression of *Olig1* promotes oligodendrocyte formation and reduces neuronal survival in developing mouse cortex. *Nat. Neurosci.* **4,** 973–974.

Mann, R., Mulligan, R. C., and Baltimore, D. (1983). Construction of a retrovirus packaging mutant and its use to produce helper-free defective retrovirus. *Cell* **33,** 153–159.

Miller, A. D., and Buttimore, C. (1986). Redesign of retrovirus packaging cell lines to avoid recombination leading to helper virus production. *Mol. Cell. Biol.* **6,** 2895–2902.

Miller, A. D., and Rosman, G. J. (1989). Improved retroviral vectors for gene transfer and expression. *BioTechniques.* **7,** 980–990.

Morgenstern, J. P., and Land, H. (1990). Advanced mammalian gene transfer: High titer retroviral vectors with multiple drug selection markers and a complementary helper-free packaging cell line. *Nuceic Acids Res.* **18,** 3587–3596.

Morin, X., Cremer, H., Hirsch, M.-R., Kapur, R. P., Goridis, C., and Brunet, J.-F. (1997). Defects in sensory and autonomic ganglia and absence of locus coeruleus in mice deficient for the homeobox gene *Phox2a*. *Neuron* **18,** 411–423.

Morrow, E. M., Furukawa, T., Lee, J. E., and Cepko, C. L. (1999). NeuroD regulates multiple functions in the developing neural retina in rodent. *Development* **126,** 23–36.

Naviaux, R. K., Costanzi, E., Haas, M., and Verma, I. M. (1996). The pCL vector system: Rapid production of helper-free, high-titer, recombinant retroviruses. *J. Virol.* **70,** 5701–5705.

Nieto, M., Schuurmans, C., Britz, O., and Guillemot, F. (2001). Neural bHLH genes control the neuronal versus glial fate decision in cortical progenitors. *Neuron* **29,** 401–413.

Ohtsuka, T., Ishibashi, M., Gradwohl, G., Nakanishi, S., Guillemot, F., and Kageyama, R. (1999). *Hes1* and *Hes5* as Notch effectors in mammalian neuronal differentiation. *EMBO J.* **18,** 2196–2207.

Ohtsuka, T., Sakamoto, M., Guillemot, F., and Kageyama, R. (2001). Roles of the basic helix-loop-helix genes *Hes1* and *Hes5* in expansion of neural stem cells of the developing brain. *J. Biol. Chem.* **276,** 30467–30474.

Pear, W. S., Nolan, G. P., Scott, M. L., and Baltimore, D. (1993). Production of high-titer helper-free retroviruses by transient transfection. *Proc. Natl. Acad. Sci. USA* **90,** 8392–8396.

Sakurada, K., Ohshima-Sakurada, M., Palmer, T. D., and Gage, F. H. (1999). Nurr1, an orphan nuclear receptor, is a transcriptional activator of endogenous tyrosine hydroxylase in neural progenitor cells derived from the adult brain. *Development* **126,** 4017–4026.

Sanes, J. R., Rubenstein, J. L. R., and Nicolas, J.-F. (1986). Use of a recombinant retrovirus to study post-implantation cell lineage in mouse embryos. *EMBO J.* **5,** 3133–3142.

Satow, T., Bae, S.-K., Inoue, T., Inoue, C., Bessho, Y., Hashimoto, N., and Kageyama, R. (2001). The bHLH gene *hesr2* promotes gliogenesis in mouse retina. *J. Neurosci.* **21,** 1265–1273.

Soneoka, Y., Cannon, P. M., Ramsdale, E. E., Griffiths, J. C., Romano, G., Kingsman, S. M., and Kingsman, A. J. (1995). A transient three-plasmid expression system for the production of high titer retroviral vectors. *Nuceic Acids Res.* **23,** 628–633.

Sun, Y., Nadal-Vicens, M., Misono, S., Lin, M. Z., Zubiaga, A., Hua, X., Fan, G., and Greenberg, M. E. (2001). Neurogenin promotes neurogenesis and inhibits glial differentiation by independent mechanisms. *Cell* **104,** 365–376.

Tanabe, Y., and Jessell, T. M. (1996). Diversity and pattern in the developing spinal cord. *Science* **274,** 1115–1123.

Temple, S. (2001). The development of neural stem cells. *Nature* **414,** 112–117.

Tomita, K., Ishibashi, M., Nakahara, K., Ang, S.-L., Nakanishi, S., Guillemot, F., and Kageyama, R. (1996). Mammalian *hairy* and *Enhancer of split* homolog 1 regulates differentiation of retinal neurons and is essential for eye morphogenesis. *Neuron* **16,** 723–734.

Tomita, K., Moriyoshi, K., Nakanishi, S., Guillemot, F., and Kageyama, R. (2000). Mammalian *achaete-scute* and *atonal* homologs regulate neuronal versus glial fate determination in the central nervous system. *EMBO J.* **19,** 5460–5472.

Turner, D. L., and Cepko, C. L. (1987). A common progenitor for neurons and glia persists in rat retina late in development. *Nature* **328,** 131–136.

Yee, J.-K., Miyanohara, A., LaPorte, P., Bouic, K., Burns, J. C., and Friedmann, T. (1994). A general method for the generation of high-titer, pantropic retroviral vectors: Highly efficient infection of primary hepatocytes. *Proc. Natl. Acad. Sci. USA* **91,** 9564–9568.

Zetterstrom, R. H., Solomin, L., Jansson, L., Hoffer, B. J., Olson, L., and Perlmann, T. (1997). Dopamine neuron agenesis in Nurr1-deficient mice. *Science* **276,** 248–250.

Zhou, Q., Choi, G., and Anderson, D. J. (2001). The bHLH transcription factor Olig2 promotes oligodendrocyte differentiation in collaboration with Nkx2.2. *Neuron* **31,** 791–807.

# SECTION II

# GENE THERAPY WITH VIRUS VECTORS FOR SPECIFIC DISEASES OF THE NERVOUS SYSTEM

# THE PRINCIPLES OF MOLECULAR THERAPIES FOR GLIOBLASTOMA

George Karpati and Josephine Nalbantoglu

Montreal Neurological Institute
Montreal QC H3A 2B4, Canada



## I. Introduction

The therapy of glioblastoma (GBS) continues to be a major challenge for neurosurgeons, oncologists, radiotherapists, and other health professionals involved in the care of patients (Viola and Martuza, 1996). Despite a constant flow of new knowledge pertaining to the relevant basic biology of GBS and the intensive preclinical studies of various therapeutic modalities and strategies, effective therapeutic results in human trials have not yet emerged (Ampil *et al.*, 1998; Andratschke *et al.*, 2001; Viola and Martuza, 1996). In the subsequent parts of this review we shall touch upon the possible reasons for the delay in succeeding and suggest promising therapeutic paradigms (Tunici *et al.*, 2001).

## II. Review

### A. Key Factors That Determine Long-Lasting and Safe Therapeutic Efficiency for GBS

To approach this problem productively, it is necessary to analyze the basic factors and circumstances the optimization of which will be required for the eradication (cure) of GBS in the vast majority of patients (Barker and Israel, 1999; Coons and Ashby, 1999; Van de Kelft, 1997; von Deimling *et al.*, 1995).




0074-7742/03 $35.00

1. *Choice of the Most Optimal Molecular Tumoricidal Agent(s) and Their Vectors*

The key points here are efficiency, lack of pathogenicity for the normal elements of the neuropil, and cost-effective production in GMP mode. In consideration of efficiency, tumoricidal agents must access and enter the GBS cells and there they must be lethal.

For an efficient transport of the tumoricidal agent into GBS cells, viral vectors have been the most efficient (Boviatsis *et al.*, 1994). We favor the use of a first-generation adenovirus vector in preference to a retrovirus (Culver *et al.*, 1992; Ram *et al.*, 1993, 1997; Takamiya *et al.*, 1992), herpes simplex virus, or adeno-associated virus (Okada *et al.*, 1996), because it can transduce both dividing and nondividing cell populations and by creating a destructive inflammatory reaction, it can augment the tumoricidal effect, without it spilling over to the perineoplastic edematous or normal brain (Chen *et al.*, 1994). However, if this proves to be a problem, the helper-dependent ("gutted") adenovirus vector can be used, which does not evoke significant inflammation (Dong *et al.*, 1996; Thomas *et al.*, 2000). Another new generation adenovirus vector has been engineered that has a limited replication capacity and can thus extend the exposure of the GBS cells to the tumoricidal agent. Lentivirus vector may also be useful as it transduces both dividing and nondividing cells and it integrates into the host cell genomes (Naldini *et al.*, 1996). It has not yet been used for molecular therapy research for GBS. More recently stem cells with neural differentiating properties have been suggested as tumoricidal agents (Noble, 2000). Such cells may be embedded in a medium that is not neurotoxic but sustains their survival.

Tumoricidal agents can be chosen by exploiting the known aberrations in the basic biology of GBS cells versus normal astrocytes. These items include abnormalities in cell cycle control and signal transduction (Mercer *et al.*, 1990), alteration of growth factor receptor profile (Asai *et al.*, 1994), as well as apoptotic pathways and death receptor characteristics (Ambar *et al.*, 1999; Wu *et al.*, 1997). Nonspecific inhibition of glioma cell growth can also be achieved by specifically targeted agents, such as double-stranded RNA-dependent protein kinase PKP (Shir and Levitzki, 2002). An important fact is that the GBS cell population is biologically and clonally heterogeneous and thus the susceptibility to a particular tumoricidal agent or to a vector carrying a tumoricidal agent could vary. To cope with this situation, combinatorial approaches with multiple vectors and multiple successive applications are likely to be necessary. For maximum lethality, a good "bystander" effect is also a major asset (Hamel *et al.*, 1996).

In consideration of the lack of pathogenicity for the normal neuropil, the same principles can be exploited for efficiency as cited above, but here the route of administration can also be an important factor. In addition, the use of a GBS-specific promoter controlling the transcription of the therapeutic transgene expression could spare the normal cells of the neuropil and vascular endothelium of the brain (McKie *et al.*, 1998; Parr *et al.*, 1997).

The issue of cost effectiveness of the tumoricidal agents produced by GMP technology can ultimately be a major constraint in the translation of the preclinical experience into human trials.

2. *Choice of the Route of Administration*

For central nervous system (CNS) malignancies there are two major routes of administration that had been used in preclinical experiments and clinical trials: (1) direct injection into the principal tumor mass with or without prior debulking (in the former case the exposed tumor bed is easily accessible, whereas in the latter case, stereotaxic targeting is necessary) and (2) intraarterial injection using the carotid arteries or major arterial tumor feeders as ports of entry. (Intrathecal injection is not applicable to GBS.) Although intraarterially injected vectors carrying tumoricidal genes or agents can relatively easily pass the impaired vascular barrier to GBS, further reducing the blood–brain barrier may provide additional advantage (Elliott *et al.*, 1996).

It is of critical importance to appreciate that the route of administration of the tumoricidal agents has an all important determining effect upon the ultimate efficiency of anti-GBS therapy. This is related to two distinct features of GBS (Coons and Ashby, 1999). First, in addition to the main GBS tumor mass visible on imaging of the brain, there appears to be microscopic tumor pockets present not only in the immediate perineoplastic mantle but strewn throughout the hemisphere. Furthermore, the corpus callosum is invaded by GBS cells early on, and therefore even early radical hemispherectomy cannot cure GBS.

To overcome these problems, based on the study of Li *et al.* (1998) we advocate multiple successive ipsilateral intracarotid injections of the tumoricidal agents followed by the same on the contralateral side. Although this appears to be a burden in terms of cost effectiveness, there seems to be no alternative to access all GBS cells that otherwise could serve as the source of tumor recurrence and a hindrance to cure. The described intraarterial route of administration of a cytotoxic agent could also be useful in minimizing the neovascularity of GBS, which is believed to be an important accessory to the eradication of GBS.

3. *The Tumor Size Issue*

In addition to the presence of disseminated micropockets of GBS cells, the size of the principal tumor mass has a major influence upon the therapeutic efficiency of different forms of molecular therapy (Izquierdo *et al.*, 1997). In our experiments with C6 rat gliomas installed in rat cerebrum, we ascertained that a single intratumoral or intraarterial administration of an adenoviral vector containing a suicide gene and/or a tumor-suppresser gene could eradicate a C6 glioma, as long as it had a relatively small mass. For eradication of larger tumors, multiple administrations were necessary. Unfortunately, human GBS is usually diagnosed when the principal tumor mass has already attained a relatively large size. This will predict that large amounts of therapeutic material in multiple successive administrations will be necessary, particularly for inoperable tumors.

## B. Major Possible Strategies for Molecular Therapies of GBS

As indicated earlier, it is most likely that for the complete eradication of GBS, more than one molecular strategy will be necessary in a given patient. However, it appears that for an optimal outcome, molecular and nonmolecular modalities should be combined. In the latter category, the use of corticosteroids, radiotherapy, and drugs for inhibition of neovascularization is likely to be the most important. In addition, debulking of operable tumors by advanced techniques (i.e., gamma knife, etc.) is also a valuable adjunct.

Molecular therapy in the context of this review is defined as the introduction of genetic material into GBS cells either for direct killing of the cells or to facilitate the ability of the host to eradicate them.

### 1. *Prodrug Activating Systems (Suicide Gene Therapy)*

Prodrug activating systems (Galipeau *et al.*, 1999; Moolten, 1986) are based on the intracellular conversion of a systemically administered nontoxic prodrug into a highly cytotoxic molecule by an enzyme encoded by a gene introduced independently into the GBS cells by a viral vector (Manome *et al.*, 1996; Rainov *et al.*, 1998). The two most widely used systems are (1) *Escherichia* coli cytosine deaminase (CDA) as the converting enzyme and 5-fluorocytosine (5-FC) as the prodrug with 5-fluorouracil (5-FU) as the cytotoxic molecule (Huber *et al.*, 1994) and (2) herpes simplex virus thymidine kinase (HSV-tk) as the converting enzyme and gancyclovir (GCV) as the prodrug transformed to a cytotoxic agent, phosphogancyclovir (P-GCV) (Izquierdo *et al.*, 1996; Klatzmann *et al.*, 1998; Stockhammer *et al.*, 1997). The genes encoding the converting enzymes may be delivered by viral vectors, of which we favor first-generation adenovirus. The CDA/5-FC system seems to have a better bystander effect than the HSV-tk/GCV system (Fick *et al.*, 1995). Although the prodrug activating systems are generally highly efficient tumoricidal methods, there may be appreciable variation of their lethal effects in different clonal GBS cell populations. The combined use of the two prodrug activating systems might be advantageous (Aghi *et al.*, 1998).

### 2. *Application of Tumor Suppresser Genes*

Of tumor suppresser genes, p53 enjoys the widest use (Harris, 1996; Lang *et al.*, 1999). Overexpression of p53 in GBS cells induces apoptosis (Gomez-Manzano *et al.*, 1996; Kock *et al.*, 1996) that probably involves downregulation of the antiapoptotic protein Bcl-2 and upregulation of the proapoptotic Bax and the death receptor Fas and DR5 (Miyashita *et al.*, 1994; Miyashita and Reed, 1995; Shinoura *et al.*, 1998). Although loss of p53 is a widespread change in GBS cells, the proapoptotic effect of p53 overexpression is not predicated on a priori p53 loss (Li *et al.*, 1997). p53 also seems to sensitize GBS cells to the tumoricidal effect of radiotherapy (Badie *et al.*, 1998; Geng *et al.*, 1998). Other tumor suppresser molecules [$Ca^{2+}$-permeable 2-(aminomethyl) phenylacetic acid (AMPA) receptors MMAC1/PTEN] also impair glioma cell proliferation (Cheney *et al.*, 1998; Ishiuchi *et al.*, 2002).

3. *Use of Genes Encoding Molecules That Influence Cell Cycle Events*

The nuclearly localized retinoblastoma protein is an inhibitor of the cell cycle and its overexpression results in inhibition of glioma cell proliferation *in vitro* and *in vivo* (Fueyo *et al.*, 1998).

4. *Inhibition of Growth Factor Effects on Neoplastic Cells*

The excessive activity of epithelial growth factor (EGFR) promotes GBS growth due to mutation of its receptors (Kuan *et al.*, 2000; Yamazaki *et al.*, 1998). The growth-promoting effect of EGFR can be mitigated by the introduction of a mutant *erb*B gene (O'Rourke *et al.*, 1997). Inhibition of the growth-promoting effect of insulin-like growth factor (IGF-1) through the use of antisense mRNA was shown to reduce the growth rate of malignant gliomas *in vitro* and *in vivo* (Resnicoff *et al.*, 1996). The third growth factor whose inhibition was shown to negatively affect growth of malignant gliomas is platelet-derived growth factor (PDGF).

5. *Immunomodulatory Measures*

The rationale for using immunomodulatory measures (Beutler *et al.*, 1999) rests on the observations that the host's immune system is handicapped in its capacity to destroy tumor cells or inhibit neovascularization (Asano *et al.*, 1995; Cheng *et al.*, 1996).

The cellular/molecular targets in malignant gliomas that come under this category include downregulation of class 1 and 2 MHC expression on tumor cells (Beutler *et al.*, 1999), the production of immunosuppressive cytokines such as tumor growth factor-$\beta$ and interleukin 10, and lymphocidal agents such as FasL. Counteracting these features could enhance antitumoral immune activity of the host. Such effects may be obtained by the introduction of genes into glioma cells that encode cytokines (Benedetti *et al.*, 1999; Husain *et al.*, 1998; Kanno *et al.*, 1999) (i.e., interleukin 2 or 4 or 12, interferon-$\gamma$) or costimulatory factors (i.e., B7-1 or B7-2) or granulocyte–macrophage colony-stimulating factor (Wallenfriedman *et al.*, 1999). Antitumor vaccines have shown early promise in human trials for GBS (Plate *et al.*, 1992; Schneider *et al.*, 2001; Shweiki *et al.*, 1992).

### 6. *Inhibition or Reversal of Neovascularization*

The rationale for the use of inhibition or reversal of neovascularization rests on the well-established fact that rapid growth of GBS is accompanied by the generation of abundant new intratumoral blood vessels without which the tumor could not grow (Brem *et al.*, 1972). In fact, not infrequently the tumor growth outstrips the rate of the minimum necessary neovascularization resulting in massive tumor necrosis. Another deleterious consequence of GBS neovascularization is that the nascent immature blood vessels are prone to rupture causing massive intratumoral hemorrhages. Several basic findings provide useful information for designing therapeutic strategies for inhibition or reversal of neovascularization. These include high levels in GBS of vascular endothelial growth factor (VEGEF) and its receptors (Im *et al.*, 1999; Plate *et al.*, 1993) as well as angiopoietins and its receptors (Tie 1 and 2) (Lin *et al.*, 1997). There are several strategies by which these factors can be mitigated with reduction of hypervascularity. In addition, the following agents have been shown to be useful *in vivo* models of GBS to control hypervascularization: angiostatin (Tanaka *et al.*, 1998), thrombospondin (Hsu *et al.*, 1996), pigment epithelium-derived factor (PEDF) (Dawson *et al.*, 1999), thalidomide (D'Amato *et al.*, 1994), and TNP-70 (Taki *et al.*, 1994). The cytotoxic effects of the tumoricidal agents designed to kill neoplastic cells could also exert lethal effects on the nascent endothelial cells after intravascular injection. It is anticipated that inhibition or reversal of hypervascularity of GBS by itself cannot eradicate GBS. Whether it is even indispensable in the combinatorial paradigm for a curative effect remains to be seen.

### 7. *Control of Intracerebral Spread of GBS Cells*

Although GBS does not metastazise outside of the CNS, it shows aggressive spread within the brain (Chintala *et al.*, 1997). This appears to be responsible for the early appearance of the GBS cells in commissural pathways and at sites removed from the main tumor mass. Thus, inhibiting the migration of GBS cells could therapeutically be useful but only in conjunction with other tumoricidal agents or methods. Metalloproteinases have been shown to be upregulated in GBS, which may be a major factor in not only spread of the tumor but the associated avid neovascularization (Price *et al.*, 1999). Agents that have been shown to inhibit GBS cell migration and invasiveness *in vivo* and *in vitro* include metalloproteinase inhibitors (AG3340) and p16/CDKN2 (Price *et al.*, 1999).

### 8. *Oncolytic Viruses*

The use of oncolytic viruses (Heise *et al.*, 1997; Ikeda *et al.*, 1999) is also classified as "molecular," as genetic material is introduced for therapeutic purposes. The most developed agent is a human herpes simplex virus, which has been "detoxified" for the normal brain by genetic manipulation but retained some proliferate capacity in GBS with tumoricidal effect (Martuza *et al.*, 1991; Mineta *et al.*, 1995). Its long-term effectiveness and safety have not yet been proven. The Newcastle virus, which is pathogenic for chickens and GBS cells but not for normal human cells, has some promise (Csatary and Bakacs, 1999). Results of controlled preclinical studies, however, have not been published. Selective oncolytic effects of reovirus by RAS pathway activation has been reported (Norman and Lee, 2000).

### 9. *Targeted Toxins*

Potent cytotoxins selectively targeted to GBS cells can be a very effective tumoricidal strategy (Laske *et al.*, 1997).

## C. Preclinical Studies and Clinical Trials

To ascertain the feasibility, efficacy, synergy, and safety of the cited modalities for GBS therapy, carefully designed preclinical experiments must be conducted in appropriate animal models (Barth, 1998; Beutler *et al.*, 1997; Shapiro, 1997; Weissenberger *et al.*, 1997). Satisfactory results from such experiments are a prerequisite for regulatory agencies to consider approval of human trials.

Although none of the currently used animal models is a perfect facsimile of human GBS, there are several models that are very useful. These include intracerebral tumors that arise after stereotaxic injection of genetically stably marked (Laz Z of GFP) human glioma cell lines into the head of the caudate nucleus of immunodeficient mice (Li *et al.*, 1999). Such tumours grow rapidly and prove fatal by 6 weeks postinjection. Tumoricidal agents can be injected stereotaxically into the tumor mass or into the carotid artery. The therapeutic effect can be monitored by imaging and microscopic parameters.

Another model is a hemispheric malignant glioma that arises in the immunocompetent rat brain after the same injection technique of C6 rat glioma cell line as described for mice (Albright *et al.*, 1977). Monitoring the therapeutic effect is similar to the one described for mice. The drawback of the rodent models is that the small size of the carotid arteries precludes repeated intracarotid injections. Thus, the curability of large tumors is difficult to ascertain. For that purpose, a canine model of malignant glioma appears to be promising.

Human trials are predicated upon satisfactory preclinical data showing at least safety and some efficiency as well as the approval by local and state regulatory agencies (Andratschke *et al.*, 2001).

With favorable preclinical results, the all-important risk/benefit ratio for GBS is such that approval for phase 1 clinical trials has not proven difficult. However, the availability of a sufficient amount of GMP-grade therapeutic material may prove to be a prohibitive financial burden without the participation of the commercial sector in the study. The clinical trials should be structured so that they could answer several critical questions: Is the procedure feasible? Is the procedure safe? What are the cost projections? What is the optimal combinatorial paradigm? Compare the results with and without prior debulking of the main tumor mass.

## References


Aghi, M., Kramm, C. M., Chou, T. C., Breakefield, X. O., and Chiocca, E. A. (1998). Synergistic anticancer effects of ganciclovir/thymidine kinase and 5-fluorocytosine/cytosine deaminase gene therapies. *J. Natl. Cancer Inst.* **90,** 370–380.

Albright, A. L., Gill, T. J., III, and Geyer, S. J. (1977). Immunogenetic control of brain tumor growth in rats. *Cancer Res.* **37,** 2512–2521.

Ambar, B. B., Frei, K., Malipiero, U., Morelli, A. E., Castro, M. G., Lowenstein, P. R., and Fontana, A. (1999). Treatment of experimental glioma by administration of adenoviral vectors expressing Fas ligand. *Hum. Gene Ther.* **10,** 1641–1648.

Ampil, F. L., Nanda, A., Willis, B. K., and Apple, S. (1998). Treatment in patients with glioblastoma multiforme and poor performance status: Is it worthwhile? *Radiat. Med.* **16,** 109–112.

Andratschke, N., Grosu, A. L., Molls, M., and Nieder, C. (2001). Perspectives in the treatment of malignant gliomas in adults. *Anticancer Res.* **21,** 3541–3550.

Asai, A., Miyagi, Y., Sugiyama, A., Gamanuma, M., Hong, S. H., Takamoto, S., Nomura, K., Matsutani, M., Takakura, K., and Kuchino, Y. (1994). Negative effects of wild-type p53 and s-Myc on cellular growth and tumorigenicity of glioma cells. Implication of the tumor suppressor genes for gene therapy. *J. Neurooncol.* **19,** 259–268.

Asano, M., Yukita, A., Matsumoto, T., Kondo, S., and Suzuki, H. (1995). Inhibition of tumor growth and metastasis by an immunoneutralizing monoclonal antibody to human vascular endothelial growth factor/vascular permeability factor 121. *Cancer Res.* **55,** 5296–5301.

Badie, B., Kramar, M. H., Lau, R., Boothman, D. A., Economou, J. S., and Black, K. L. (1998). Adenovirus-mediated p53 gene delivery potentiates the radiation-induced growth inhibition of experimental brain tumors. *J. Neurooncol.* **37,** 217–222.

Barker, F. G., and Israel, M. A. (1999). Molecular genetics. *In* "The Gliomas" (M. S. Bergers and C. B. Wilson, Eds.), pp. 39–51. W. B. Saunders Company, Philadelphia.

Barth, R. F. (1998). Rat brain tumor models in experimental neuro-oncology: The 9L, C6, T9, F98, RG2 (D74), RT-2 and CNS-1 gliomas. *J. Neurooncol.* **36,** 91–102.

Benedetti, S., Bruzzone, M. G., Pollo, B., DiMeco, F., Magrassi, L., Pirola, B., Cirenei, N., Colombo, M. P., and Finocchiaro, G. (1999). Eradication of rat malignant gliomas by retroviral-mediated, in vivo delivery of the interleukin 4 gene. *Cancer Res.* **59,** 645–652.

Beutler, A. S., Banck, M. S., and Aguzzi, A. (1997). Curing rat glioblastoma: Immunotherapy or graft rejection? *Science* **276,** 20–21.

Beutler, A. S., Banck, M. S., Wedekind, D., and Hedrich, H. J. (1999). Tumor gene therapy made easy: Allogeneic major histocompatibility complex in the C6 rat glioma model. *Hum. Gene Ther.* **10,** 95–101.

Boviatsis, E. J., Chase, M., Wei, M. X., Tamiya, T., Hurford, R. K., Jr., Kowall, N. W., Tepper, R. I., Breakefield, X. O., and Chiocca, E. A. (1994). Gene transfer into experimental brain tumors mediated by adenovirus, herpes simplex virus, and retrovirus vectors. *Hum. Gene Ther.* **5,** 183–191.

Brem, S., Cotran, R., and Folkman, J. (1972). Tumor angiogenesis: A quantitative method for histologic grading. *J. Natl. Cancer Inst.* **48,** 347–356.

Chen, S. H., Shine, H. D., Goodman, J. C., Grossman, R. G., and Woo, S. L. (1994). Gene therapy for brain tumors: Regression of experimental gliomas by adenovirus-mediated gene transfer in vivo. *Proc. Natl. Acad. Sci. USA* **91,** 3054–3057.

Cheney, I. W., Johnson, D. E., Vaillancourt, M. T., Avanzini, J., Morimoto, A., Demers, G. W., Wills, K. N., Shabram, P. W., Bolen, J. B., Tavtigian, S. V., and Bookstein, R. (1998). Suppression of tumorigenicity of glioblastoma cells by adenovirus-mediated MMAC1/PTEN gene transfer. *Cancer Res.* **58,** 2331–2334.

Cheng, S. Y., Huang, H. J., Nagane, M., Ji, X. D., Wang, D., Shih, C. C., Arap, W., Huang, C. M., and Cavenee, W. K. (1996). Suppression of glioblastoma angiogenicity and tumorigenicity by inhibition of endogenous expression of vascular endothelial growth factor. *Proc. Natl. Acad. Sci. USA* **93,** 8502–8507.

Chintala, S. K., Fueyo, J., Gomez-Manzano, C., Venkaiah, B., Bjerkvig, R., Yung, W. K., Sawaya, R., Kyritsis, A. P., and Rao, J. S. (1997). Adenovirus-mediated p16/CDKN2 gene transfer suppresses glioma invasion in vitro. *Oncogene* **15,** 2049–2057.

Coons, S. W., and Ashby, L. S. (1999). Pathology of intracranial neoplasms. *Neuroimaging Clin. N. Am.* **9,** 615–649.

Csatary, L. K., and Bakacs, T. (1999). Use of Newcastle disease virus vaccine (MTH-68/H) in a patient with high-grade glioblastoma. *JAMA* **281,** 1588–1589.

Culver, K. W., Ram, Z., Wallbridge, S., Ishii, H., Oldfield, E. H., and Blaese, R. M. (1992). In vivo gene transfer with retroviral vector-producer cells for treatment of experimental brain tumors. *Science* **256,** 1550–1552.

D'Amato, R. J., Loughnan, M. S., Flynn, E., and Folkman, J. (1994). Thalidomide is an inhibitor of angiogenesis. *Proc. Natl. Acad. Sci. USA* **91,** 4082–4085.

Dawson, D. W., Volpert, O. V., Gillis, P., Crawford, S. E., Xu, H., Benedict, W., and Bouck, N. P. (1999). Pigment epithelium-derived factor: A potent inhibitor of angiogenesis. *Science* **285,** 245–248.

Dong, Y., Wen, P., Manome, Y., Parr, M., Hirshowitz, A., Chen, L., Hirschowitz, E. A., Crystal, R., Weichselbaum, R., Kufe, D. W., and Fine, H. A. (1996). In vivo replication-deficient adenovirus vector-mediated transduction of the cytosine deaminase gene sensitizes glioma cells to 5-fluorocytosine. *Hum. Gene Ther.* **7,** 713–720.

Elliott, P. J., Hayward, N. J., Huff, M. R., Nagle, T. L., Black, K. L., and Bartus, R. T. (1996). Unlocking the blood-brain barrier: A role for RMP-7 in brain tumor therapy. *Exp. Neurol.* **141,** 214–224.

Fick, J., Barker, F. G., Dazin, P., Westphale, E. M., Beyer, E. C., and Israel, M. A. (1995). The extent of heterocellular communication mediated by gap junctions is predictive of bystander tumor cytotoxicity in vitro. *Proc. Natl. Acad. Sci. USA* **92,** 11071–11075.

Fueyo, J., Gomez-Manzano, C., Yung, W. K., Liu, T. J., Alemany, R., Bruner, J. M., Chintala, S. K., Rao, J. S., Levin, V. A., and Kyritsis, A. P. (1998). Suppression of human glioma growth by adenovirus-mediated Rb gene transfer. *Neurology* **50,** 1307–1315.

Galipeau, J., Li, H., Paquin, A., Sicilia, F., Karpati, G., and Nalbantoglu, J. (1999). Vesicular stomatitis virus G pseudotyped retrovector mediates effective in vivo suicide gene delivery in experimental brain cancer. *Cancer Res.* **59,** 2384–2394.

Geng, L., Walter, S., Melian, E., and Vaughan, A. T. (1998). Transfection of a vector expressing wild-type p53 into cells of two human glioma cell lines enhances radiation toxicity. *Radiat. Res.* **150,** 31–37.

Gomez-Manzano, C., Fueyo, J., Kyritsis, A. P., Steck, P. A., Roth, J. A., McDonnell, T. J., Steck, K. D., Levin, V. A., and Yung, W. K. (1996). Adenovirus-mediated transfer of the p53 gene produces rapid and generalized death of human glioma cells via apoptosis. *Cancer Res.* **56,** 694–699.

Hamel, W., Magnelli, L., Chiarugi, V. P., and Israel, M. A. (1996). Herpes simplex virus thymidine kinase/ganciclovir-mediated apoptotic death of bystander cells. *Cancer Res.* **56,** 2697–2702.

Harris, C. C. (1996). Structure and function of the p53 tumor suppressor gene: Clues for rational cancer therapeutic strategies. *J. Natl. Cancer Inst.* **88,** 1442–1455.

Heise, C., Sampson-Johannes, A., Williams, A., McCormick, F., Von Hoff, D. D., and Kirn, D. H. (1997). ONYX-015, an E1B gene-attenuated adenovirus, causes tumor-specific cytolysis and antitumoral efficacy that can be augmented by standard chemotherapeutic agents. *Nat. Med.* **3,** 639–645.

Hsu, S. C., Volpert, O. V., Steck, P. A., Mikkelsen, T., Polverini, P. J., Rao, S., Chou, P., and Bouck, N. P. (1996). Inhibition of angiogenesis in human glioblastomas by chromosome 10 induction of thrombospondin-1. *Cancer Res.* **56,** 5684–5691.

Huber, B. E., Austin, E. A., Richards, C. A., Davis, S. T., and Good, S. S. (1994). Metabolism of 5-fluorocytosine to 5-fluorouracil in human colorectal tumor cells transduced with the cytosine deaminase gene: Significant antitumor effects when only a small percentage of tumor cells express cytosine deaminase. *Proc. Natl. Acad. Sci. USA* **91,** 8302–8306.

Husain, S. R., Behari, N., Kreitman, R. J., Pastan, I., and Puri, R. K. (1998). Complete regression of established human glioblastoma tumor xenograft by interleukin-4 toxin therapy. *Cancer Res.* **58,** 3649–3653.

Ikeda, K., Ichikawa, T., Wakimoto, H., Silver, J. S., Deisboeck, T. S., Finkelstein, D., Harsh, G. R., Louis, D. N., Bartus, R. T., Hochberg, F. H., and Chiocca, E. A. (1999). Oncolytic virus therapy of multiple tumors in the brain requires suppression of innate and elicited antiviral responses. *Nat. Med.* **5,** 881–887.

Im, S. A., Gomez-Manzano, C., Fueyo, J., Liu, T. J., Ke, L. D., Kim, J. S., Lee, H. Y., Steck, P. A., Kyritsis, A. P., and Yung, W. K. (1999). Antiangiogenesis treatment for gliomas: Transfer of antisense-vascular endothelial growth factor inhibits tumor growth in vivo. *Cancer Res.* **59,** 895–900.

Ishiuchi, S., Tsuzuki, K., Yoshida, Y., Yamada, N., Hagimura, N., Okado, H., Miwa, A., Kurihara, H., Nakazato, Y., Tamura, M., Sasaki, T., and Ozawa, S. (2002). Blockage of Ca(2+)-permeable AMPA receptors suppresses migration and induces apoptosis in human glioblastoma cells. *Nat. Med.* **8,** 971–978.

Izquierdo, M., Martin, V., de Felipe, P., Izquierdo, J. M., Perez-Higueras, A., Cortes, M. L., Paz, J. F., Isla, A., and Blazquez, M. G. (1996). Human malignant brain tumor response to herpes simplex thymidine kinase (HSVtk)/ganciclovir gene therapy. *Gene Ther.* **3,** 491–495.

Izquierdo, M., Cortes, M. L., Martin, V., de Felipe, P., Izquierdo, J. M., Perez-Higueras, A., Paz, J. F., Isla, A., and Blazquez, M. G. (1997). Gene therapy in brain tumours: Implications of the size of glioblastoma on its curability. *Acta. Neurochir. Suppl. (Wien.)* **68,** 111–117.

Kanno, H., Hattori, S., Sato, H., Murata, H., Huang, F. H., Hayashi, A., Suzuki, N., Yamamoto, I., Kawamoto, S., Minami, M., Miyatake, S., Shuin, T., and Kaplitt, M. G. (1999). Experimental gene therapy against subcutaneously implanted glioma with a herpes simplex virus-defective vector expressing interferon-gamma. *Cancer Gene Ther.* **6,** 147–154.

Klatzmann, D., Valery, C. A., Bensimon, G., Marro, B., Boyer, O., Mokhtari, K., Diquet, B., Salzmann, J. L., and Philippon, J. (1998). A phase I/II study of herpes simplex virus type 1 thymidine kinase "suicide" gene therapy for recurrent glioblastoma. Study Group on Gene Therapy for Glioblastoma. *Hum. Gene Ther.* **9,** 2595–2604.

Kock, H., Harris, M. P., Anderson, S. C., Machemer, T., Hancock, W., Sutjipto, S., Wills, K. N., Gregory, R. J., Shepard, H. M., Westphal, M., and Maneval, D. C. (1996). Adenovirus-mediated p53 gene transfer suppresses growth of human glioblastoma cells in vitro and in vivo. *Int. J. Cancer* **67,** 808–815.

Kuan, C. T., Wikstrand, C. J., and Bigner, D. D. (2000). EGFRvIII as a promising target for antibody-based brain tumor therapy. *Brain Tumor Pathol.* **17,** 71–78.

Lang, F. F., Yung, W. K., Sawaya, R., and Tofilon, P. J. (1999). Adenovirus-mediated p53 gene therapy for human gliomas. *Neurosurgery* **45,** 1093–1104.

Laske, D. W., Youle, R. J., and Oldfield, E. H. (1997). Tumor regression with regional distribution of the targeted toxin TF-CRM107 in patients with malignant brain tumors. *Nat. Med.* **3,** 1362–1368.

Li, H., Lochmuller, H., Yong, V. W., Karpati, G., and Nalbantoglu, J. (1997). Adenovirus-mediated wild-type p53 gene transfer and overexpression induces apoptosis of human glioma cells independent of endogenous p53 status. *J. Neuropathol. Exp. Neurol.* **56,** 872–878.

Li, H., Alonso-Vanegas, M., Colicos, M. A., Jung, S. S., Lochmuller, H., Sadikot, A. F., Snipes, G. J., Seth, P., Karpati, G., and Nalbantoglu, J. (1999). Intracerebral adenovirus-mediated p53 tumor suppressor gene therapy for experimental human glioma. *Clin. Cancer Res.* **5,** 637–642.

Li, Z., Rakkar, A., Katayose, Y., Kim, M., Shanmugam, N., Srivastava, S., Moul, J. W., McLeod, D. G., Cowan, K. H., and Seth, P. (1998). Efficacy of multiple administrations of a recombinant adenovirus expressing wild-type p53 in an immune-competent mouse tumor model. *Gene Ther.* **5,** 605–613.

Lin, P., Polverini, P., Dewhirst, M., Shan, S., Rao, P. S., and Peters, K. (1997). Inhibition of tumor angiogenesis using a soluble receptor establishes a role for Tie2 in pathologic vascular growth. *J. Clin. Invest.* **100,** 2072–2078.

Manome, Y., Wen, P. Y., Dong, Y., Tanaka, T., Mitchell, B. S., Kufe, D. W., and Fine, H. A. (1996). Viral vector transduction of the human deoxycytidine kinase cDNA sensitizes glioma cells to the cytotoxic effects of cytosine arabinoside in vitro and in vivo. *Nat. Med.* **2,** 567–573.

Martuza, R. L., Malick, A., Markert, J. M., Ruffner, K. L., and Coen, D. M. (1991). Experimental therapy of human glioma by means of a genetically engineered virus mutant. *Science* **252,** 854–856.

McKie, E. A., Graham, D. I., and Brown, S. M. (1998). Selective astrocytic transgene expression in vitro and in vivo from the GFAP promoter in a HSV RL1 null mutant vector—potential glioblastoma targeting. *Gene Ther.* **5,** 440–450.

Mercer, W. E., Shields, M. T., Amin, M., Sauve, G. J., Appella, E., Romano, J. W., and Ullrich, S. J. (1990). Negative growth regulation in a glioblastoma tumor cell line that conditionally expresses human wild-type p53. *Proc. Natl. Acad. Sci. USA* **87,** 6166–6170.

Mineta, T., Rabkin, S. D., Yazaki, T., Hunter, W. D., and Martuza, R. L. (1995). Attenuated multi-mutated herpes simplex virus-1 for the treatment of malignant gliomas. *Nat. Med.* **1,** 938–943.

Miyashita, T., and Reed, J. C. (1995). Tumor suppressor p53 is a direct transcriptional activator of the human bax gene. *Cell* **80,** 293–299.

Miyashita, T., Krajewski, S., Krajewska, M., Wang, H. G., Lin, H. K., Liebermann, D. A., Hoffman, B., and Reed, J. C. (1994). Tumor suppressor p53 is a regulator of bcl-2 and bax gene expression in vitro and in vivo. *Oncogene* **9,** 1799–1805.

Moolten, F. L. (1986). Tumor chemosensitivity conferred by inserted herpes thymidine kinase genes: Paradigm for a prospective cancer control strategy. *Cancer Res.* **46,** 5276–5281.

Naldini, L., Blomer, U., Gage, F. H., Trono, D., and Verma, I. M. (1996). Efficient transfer, integration, and sustained long-term expression of the transgene in adult rat brains injected with a lentiviral vector. *Proc. Natl. Acad. Sci. USA* **93,** 11382–11388.

Noble, M. (2000). Can neural stem cells be used to track down and destroy migratory brain tumor cells while also providing a means of repairing tumor-associated damage? *Proc. Natl. Acad. Sci. USA* **97,** 12393–12395.

Norman, K. L., and Lee, P. W. (2000). Reovirus as a novel oncolytic agent. *J. Clin. Invest.* **105,** 1035–1038.

Okada, H., Miyamura, K., Itoh, T., Hagiwara, M., Wakabayashi, T., Mizuno, M., Colosi, P., Kurtzman, G., and Yoshida, J. (1996). Gene therapy against an experimental glioma using adeno-associated virus vectors. *Gene Ther.* **3,** 957–964.

O'Rourke, D. M., Qian, X., Zhang, H. T., Davis, J. G., Nute, E., Meinkoth, J., and Greene, M. I. (1997). Trans receptor inhibition of human glioblastoma cells by erbB family ectodomains. *Proc. Natl. Acad. Sci. USA* **94,** 3250–3255.

Parr, M. J., Manome, Y., Tanaka, T., Wen, P., Kufe, D. W., Kaelin, W. G., Jr., and Fine, H. A. (1997). Tumor-selective transgene expression in vivo mediated by an E2F-responsive adenoviral vector. *Nat. Med.* **3,** 1145–1149.

Plate, K. H., Breier, G., Weich, H. A., and Risau, W. (1992). Vascular endothelial growth factor is a potential tumour angiogenesis factor in human gliomas in vivo. *Nature* **359,** 845–848.

Plate, K. H., Breier, G., Millauer, B., Ullrich, A., and Risau, W. (1993). Up-regulation of vascular endothelial growth factor and its cognate receptors in a rat glioma model of tumor angiogenesis. *Cancer Res.* **53,** 5822–5827.

Price, A., Shi, Q., Morris, D., Wilcox, M. E., Brasher, P. M., Rewcastle, N. B., Shalinsky, D., Zou, H., Appelt, K., Johnston, R. N., Yong, V. W., Edwards, D., and Forsyth, P. (1999). Marked inhibition of tumor growth in a malignant glioma tumor model by a novel synthetic matrix metalloproteinase inhibitor AG3340. *Clin. Cancer Res.* **5,** 845–854.

Rainov, N. G., Dobberstein, K. U., Sena-Esteves, M., Herrlinger, U., Kramm, C. M., Philpot, R. M., Hilton, J., Chiocca, E. A., and Breakefield, X. O. (1998). New prodrug activation gene therapy for cancer using cytochrome P450 4B1 and 2-aminoanthracene/4-ipomeanol. *Hum. Gene Ther.* **9,** 1261–1273.

Ram, Z., Culver, K. W., Walbridge, S., Blaese, R. M., and Oldfield, E. H. (1993). In situ retroviral-mediated gene transfer for the treatment of brain tumors in rats. *Cancer Res.* **53,** 83–88.

Ram, Z., Culver, K. W., Oshiro, E. M., Viola, J. J., DeVroom, H. L., Otto, E., Long, Z., Chiang, Y., McGarrity, G. J., Muul, L. M., Katz, D., Blaese, R. M., and Oldfield, E. H. (1997). Therapy of malignant brain tumors by intratumoral implantation of retroviral vector-producing cells. *Nat. Med.* **3,** 1354–1361.

Resnicoff, M., Li, W., Basak, S., Herlyn, D., Baserga, R., and Rubin, R. (1996). Inhibition of rat C6 glioblastoma tumor growth by expression of insulin-like growth factor I receptor antisense mRNA. *Cancer Immunol. Immunother.* **42,** 64–68.

Schneider, T., Gerhards, R., Kirches, E., and Firsching, R. (2001). Preliminary results of active specific immunization with modified tumor cell vaccine in glioblastoma multiforme. *J. Neurooncol.* **53,** 39–46.

Shapiro, W. R. (1997). Bias in uncontrolled brain tumor trials. *Can. J. Neurol. Sci.* **24,** 269–270.

Shinoura, N., Yoshida, Y., Sadata, A., Hanada, K. I., Yamamoto, S., Kirino, T., Asai, A., and Hamada, H. (1998). Apoptosis by retrovirus- and adenovirus-mediated gene transfer for Fas ligand to glioma cells: implications for gene therapy. *Hum. Gene Ther.* **9,** 1983–1993.

Shir, A., and Levitzki, A. (2002). Inhibition of glioma growth by tumor-specific activation of double-stranded RNA-dependent protein kinase PKR. *Nat. Biotechnol.* **20,** 895–900.

Shweiki, D., Itin, A., Soffer, D., and Keshet, E. (1992). Vascular endothelial growth factor induced by hypoxia may mediate hypoxia-initiated angiogenesis. *Nature* **359,** 843–845.

Stockhammer, G., Brotchi, J., Leblanc, R., Bernstein, M., Schackert, G., Weber, F., Ostertag, C., Mulder, N. H., Mellstedt, H., Seiler, R., Yonekawa, Y., Twerdy, K., Kostron, H., De Witte, O., Lambermont, M., Velu, T., Laneuville, P., Villemure, J. G., Rutka, J. T., Warnke, P., Laseur, M., Mooij, J. J., Boethius, J., Mariani, L., and Gianella-Borradori, A. (1997). Gene therapy for glioblastoma (correction of gliobestome) multiform: In vivo tumor transduction with the herpes simplex thymidine kinase gene followed by ganciclovir. *J. Mol. Med.* **75,** 300–304.

Takamiya, Y., Short, M. P., Ezzeddine, Z. D., Moolten, F. L., Breakefield, X. O., and Martuza, R. L. (1992). Gene therapy of malignant brain tumors: A rat glioma line bearing the herpes simplex virus type 1-thymidine kinase gene and wild type retrovirus kills other tumor cells. *J. Neurosci. Res.* **33,** 493–503.

Taki, T., Ohnishi, T., Arita, N., Hiraga, S., Saitoh, Y., Izumoto, S., Mori, K., and Hayakawa, T. (1994). Anti-proliferative effects of TNP-470 on human malignant glioma in vivo: Potent inhibition of tumor angiogenesis. *J. Neurooncol.* **19,** 251–258.

Tanaka, T., Cao, Y., Folkman, J., and Fine, H. A. (1998). Viral vector-targeted antiangiogenic gene therapy utilizing an angiostatin complementary DNA. *Cancer Res.* **58,** 3362–3369.

Thomas, C. E., Schiedner, G., Kochanek, S., Castro, M. G., and Lowenstein, P. R. (2000). Peripheral infection with adenovirus causes unexpected long-term brain inflammation in animals injected intracranially with first-generation, but not with high-capacity, adenovirus vectors: Toward realistic long-term neurological gene therapy for chronic diseases. *Proc. Natl. Acad. Sci. USA* **97,** 7482–7487.

Tunici, P., Gianni, D., and Finocchiaro, G. (2001). Gene therapy of glioblastomas: From suicide to homicide. *Prog. Brain. Res.* **132,** 711–719.

Van de Kelft, E. (1997). Molecular pathogenesis of astrocytoma and glioblastoma multiforme. *Acta Neurochir. (Wien.)* **139,** 589–599.

Viola, J. J., and Martuza, R. L. (1996). Gene therapies for glioblastomas. *Baillieres. Clin. Neurol.* **5,** 413–424.

von Deimling, A., Louis, D. N., and Wiestler, O. D. (1995). Molecular pathways in the formation of gliomas. *Glia* **15,** 328–338.

Wallenfriedman, M. A., Conrad, J. A., DelaBarre, L., Graupman, P. C., Lee, G., Garwood, M., Gregerson, D. S., Jean, W .C., Hall, W. A., and Low, W. C. (1999). Effects of continuous localized infusion of granulocyte-macrophage colony-stimulating factor and inoculations of irradiated glioma cells on tumor regression. *J. Neurosurg.* **90,** 1064–1071.

Weissenberger, J., Steinbach, J. P., Malin, G., Spada, S., Rulicke, T., and Aguzzi, A. (1997). Development and malignant progression of astrocytomas in GFAP-v-src transgenic mice. *Oncogene* **14,** 2005–2013.

Wu, G. S., Burns, T. F., McDonald, E. R., III, Jiang, W., Meng, R., Krantz, I. D., Kao, G., Gan, D. D., Zhou, J. Y., Muschel, R., Hamilton, S. R., Spinner, N. B., Markowitz, S., Wu, G., and el Deiry, W. S. (1997). KILLER/DR5 is a DNA damage-inducible p53-regulated death receptor gene. *Nat. Genet.* **17,** 141–143.

Yamazaki, H., Kijima, H., Ohnishi, Y., Abe, Y., Oshika, Y., Tsuchida, T., Tokunaga, T., Tsugu, A., Ueyama, Y., Tamaoki, N., and Nakamura, M. (1998). Inhibition of tumor growth by ribozyme-mediated suppression of aberrant epidermal growth factor receptor gene expression. *J. Natl. Cancer Inst.* **90,** 581–587.

# ONCOLYTIC HERPES SIMPLEX VIRUS FOR TUMOR THERAPY

Jennifer C. C. Hu

Cancer Cell Biology
Hammersmith Hospital Campus
Imperial College School of Medicine
London W12 0NN, United Kingdom

Robert S. Coffin

Bio Vex Ltd
Abingdon OX14 4RX, United Kingdom



## I. Introduction

Oncolytic viruses selectively replicate in and kill tumor cells and as such have shown considerable promise for the treatment of cancer (reviewed by Hawkins *et al.*, 2002; Ring, 2002). Oncolytic virus therapy has a number of potential advantages over other gene-based approaches to cancer treatment in that virus




0074-7742/03 $35.00

replication will not only directly kill tumor cells, but will also disseminate the therapeutic agent further through the tumor tissue than nonreplicating therapies. Thus, if the oncolytic virus were also to carry a therapeutic gene, the efficiency of gene delivery to the tumor should be enhanced as compared to gene vectors that do not include a replicative component.

To date, although gene therapy for cancer has often given promising results in animal models, the results of clinical trials have generally been disappointing. This is probably due to the fact that gene delivery is accomplished only in a small portion of the tumor tissue and accompanying problems with the large vector doses then required to give significant gene transfer. Nonreplicating vectors based on adenoviruses and retroviruses have been tested in the clinic for direct intratumoral delivery of potentially therapeutic genes such as p53 (reviewed by Fujiwara *et al.*, 2000), various prodrug activators such as thymidine kinase and nitroreductase (reviewed by Xu and McLeod, 2001), or immune stimulatory molecules such as interleukin 2 (IL-2) (e.g., Belldegrun *et al.*, 2001; Stewart *et al.*, 1999) or various interferons (reviewed by Ferrantini and Belardelli, 2000). Significant efficacy has not, however, so far been demonstrated. Intratumoral administration of naked DNA or DNA in liposome formulations to deliver similar genes has also been attempted, but without clinical success, probably due to the very low gene transfer efficiency achieved (e.g., see Kaushik, 2001).

Taken together, these results indicate that if gene therapy for cancer via direct intratumoral injection is to be clinically effective, greater gene transfer efficiencies are required. Such an improvement may be provided through the use of replicative vectors.

Various oncolytic viruses have been developed including adenoviruses, herpesviruses (HSV), reoviruses, retroviruses, and more recently vesicular stomatitis virus (VSV) and measles virus (oncolytic virus therapy reviewed by Hawkins *et al.*, 2002; Ring, 2002). These generally exploit the fact that for a virus to replicate in a target cell cellular antiviral response pathways must usually be disabled. Viruses such as herpesviruses and adenoviruses encode genes to block the antiviral response and allow replication, whereas other viruses such as reoviruses and VSV cannot generally replicate in human cells (although they can infect), as they do not encode antiviral response-blocking genes. As tumor cells often have a disabled antiviral response pathway, viruses such as reoviruses and VSV can replicate in tumor cells but not normal cells where replication is blocked. Likewise, the removal of antiviral response avoidance genes from viruses such as adenoviruses and herpesviruses provides the property of tumor-selective replication as replication is then blocked in normal cells (see below). Alternatively, tumor-selective replicative viruses have been developed through the use of tumor-specific promoters to drive the expression of an essential virus gene only in the tumor. Clinical testing using this approach includes the use of adenoviruses where the E1A region of the virus is driven by a prostate-specific promoter (Chen *et al.*, 2001).

Clinical testing with viruses in which pathways blocking the antiviral response have been disabled has used adenoviruses in which the E1B region has been deleted (ONYX-O15) (Barker and Berk, 1987) and herpesviruses in which the ICP34.5 gene has been removed. ONYX-O15 has been tested in a variety of clinical situations with maximal success following repeated administration at high dose in head and neck cancer patients (Lamont *et al.*, 2000; Khuri *et al.*, 2000; Ganly *et al.*, 2000) or intrahepatic artery infusion into liver cancer patients (Reid *et al.*, 2001), both combined with chemotherapy.

This chapter aims to review work using HSV as an oncolytic agent, from the first preclinical work using HSV with mutations in the thymidine kinase (TK) gene, through clinical experiences with "first-generation" oncolytic HSV in which ICP34.5 or ICP34.5 and ICP6 are deleted, and finally to discuss newer (second- and third-generation) viruses with improved tumor-selective replication and/or that include the insertion of potentially therapeutic genes.

## II. Advantages of Oncolytic Cancer Therapy

As discussed above, oncolytic cancer therapy has a number of potential advantages over traditional approaches to cancer treatment. Side effects are likely to be less pronounced than is often the case with chemotherapy or radiotherapy and the use of an oncolytic virus might have advantages over surgery at sites in which surgery may be impractical. This includes areas around the head and neck and spinal cord where the risks associated with the surgery can be sufficient to outweigh any potential benefit.

Oncolytic agents also provide benefits over other gene-based approaches to cancer treatment in that any delivered therapeutic gene will be disseminated further through the tumor than using a nonreplicative gene delivery approach.

## III. Advantages of HSV as an Oncolytic Agent

HSV has a number of intrinsic properties making it highly attractive for development as an oncolytic therapy for cancer. These include the following:

1. HSV can infect and replicate in most actively dividing human cells such that oncolytic treatment with an appropriately engineered oncolytic version of HSV could allow treatment of most solid tumor types.

2. HSV also infects rodent cells effectively, meaning that immune-competent mouse models can be used, which is important for preclinical safety and efficacy testing (adenovirus, for example, infects mouse tumor cells poorly, requiring the use of immune-incompetent xenograph models).
3. HSV provides very robust lytic replication in permissive cells with a short replication time and as such provides a highly potent starting point for development as an oncolytic agent.
4. HSV is a large virus allowing large genetic inserts. Individual or multiple therapeutic genes can easily be incorporated into oncolytic versions of the virus. This is often not the case with other oncolytic agents where packaging constraints are often problematic.
5. HSV is a very well-characterized virus and effects in humans have been extensively studied. Use as an oncolytic agent should not therefore provide surprises when used in the clinic. This is not the case for oncolytic agents based on, e.g., nonhuman pathogens where effects in humans cannot easily be predicted.
6. Known mutations can supply tumor-selective replicative properties to the virus, rendering the virus nonpathogenic (see below).
7. HSV replication can be terminated with the drug acyclovir, probably the most effective antiviral drug currently available.

## IV. Attenuation of HSV to Provide Tumor-Selective Replication

A number of strategies have been used to attenuate HSV for oncolytic use. These are discussed below.

### A. Mutation of Thymidine Kinase (TK)

The first replicative versions of HSV tested for cancer treatment were mutated for the thymidine kinase (TK) gene (Markert *et al.*, 1992; Martuza *et al.*, 1991; Boviatsis *et al.*, 1994). TK is involved in optimizing nucleic acid metabolism for virus growth and is necessary for efficient replication in neurons. TK mutants are therefore highly neuroattenuated. TK(−) viruses were tested in intracranial mouse models of glioma and other nervous system-derived tumor types showing slowed tumor growth and prolonged survival. However, although these results were promising, TK mutants inoculated at a higher dose into test animals cause fatal encephalitis and therefore were concluded not to provide a sufficient level of safety for use in humans. The TK gene is also necessary for the activation of acyclovir, and so TK mutants are no longer sensitive to its action. Acyclovir could not therefore be used to terminate replication of such viruses if it were found to be necessary during clinical use.

### B. Mutation of ICP6

A second gene, like TK, required for the neurovirulence of HSV is the gene encoding ICP6, the large subunit of ribonucleotide reductase. ICP6 is necessary for the conversion of rNTPs to dNTPs in neurons, which are otherwise lacking but necessary for the synthesis of new viral DNA during virus replication. ICP6 mutants have also therefore been tested as replicative anticancer agents (Mineta *et al.*, 1994; Boviatsis *et al.*, 1994; Carroll *et al.*, 1996). ICP6 mutants are sensitive to acyclovir/gancyclovir. As a result ICP6 mutants have been extensively tested in combination with gancyclovir (Mineta *et al.*, 1994; Carroll *et al.*, 1997; Kramm *et al.*, 1996; Boviatsis *et al.*, 1994a), which when converted by TK into an active form blocks DNA replication in infected cells, which may be beneficial in cancer treatment. This showed enhanced killing of tumor cells *in vitro* (including a bystander effect) and showed improved survival of animals and/or enhanced effects in reducing tumor growth in a number of tumor models as compared to the use of the ICP6 mutant without gancycolvir. However, like TK mutants, ICP6 mutants can cause fatal encephalitis if used at sufficient dose, and were thus not thought to provide a sufficient margin of safety for testing in humans.

### C. Mutation of ICP34.5

ICP34.5, the so-called neurovirulence factor, has been found to be essential for pathogenicity of HSV, a phenotype initially identified through the work of Chou and Roizman (1992, 1994; Chou *et al.*, 1990, 1995; He *et al.*, 1997). Deletion of ICP34.5 provides the greatest degree of attenuation of any individual mutation where the virus can still replicate in actively dividing cells. As such, ICP34.5-deleted HSV provides a considerably improved safety profile as compared to the use of viruses mutated for ICP6 or TK. Viruses mutated in ICP34.5 have therefore been tested extensively in animal tumor models, initially by Markert *et al.* (1993), but then subsequently by others (e.g., Andreansky *et al.*, 1996, 1997; Chambers *et al.*, 1995; Kesari *et al.*, 1995; Randazzo *et al.*, 1995; see below). This work has demonstrated considerable antitumor efficacy, combined with a good safety profile, which will be discussed below. A number of ICP34.5-deleted versions of HSV are also currently in clinical trials, which will also be discussed below.

ICP34.5 has been studied extensively and appears to provide multiple functions in the virus life cycle, the removal of which provides the property of tumor-selective replication. One function of ICP34.5 is to block the arrest in translation, which usually occurs in virus-infected cells as an antiviral response preventing virus replication (Chou and Roizman, 1992, 1994). This effect is mediated through the cellular PKR kinase, which phosphorylates the translation initiation factor eIF2$\alpha$, thereby stopping translation. ICP34.5 recruits protein phosphatase 1$\alpha$ to rephosphorylate eIF2$\alpha$ allowing protein translation and continued virus replication (Chou *et al.*, 1995; He *et al.*, 1997a). Tumor cells often have an impaired PKR pathway and/or elevated levels of eIF2$\alpha$, which means that although ICP34.5-deleted HSV cannot grow in normal cells (which have a functional PKR pathway), they can replicate in tumor cells as inactivation of the PKR response is less critical for virus growth.

A second function of ICP34.5 appears to be in allowing new virions to become packaged and leave infected cells in a cell type-specific fashion (Brown *et al.*, 1994). Thus in nonpermissive cells in the absence of ICP34.5 nucleocapsids are retained in the nucleus and a productive infection cannot ensue. In permissive cells, such as tumor cells, ICP34.5-deleted HSV leaves the nucleus as normal and productive virus replication occurs. The mechanism behind this phenotype is not understood.

### D. ICP34.5/ICP6 Double Mutants

Following preclinical testing with oncolytic versions of HSV mutated for TK, ICP6, or ICP34.5, considerable data had been generated demonstrating promise for tumor treatment. Much of this work had been focused on the treatment of intracranial glioma and it was for this indication that initial clinical work was to be considered. However, to allow this to occur, it was decided to generate viruses with the greatest possible level of safety for first clinical use. Here double mutants mutated for both ICP34.5 and ICP6 were constructed (leaving TK intact to retain acyclovir sensitivity) (Mineta *et al.*, 1995; Kramm *et al.*, 1997). Although showing less effect in slowing tumor growth than the singly mutated viruses, effects on tumor growth in rodents were demonstrated and one of these viruses (G207) has been entered into clinical trials for glioma where a good safety profile has been demonstrated (see below).

### E. Use of HSV1/HSV2 Hybrids

A further approach to providing a replication-competent version of HSV, which is attenuated such that it cannot cause disease, initially for anti-HSV vaccine purposes, has been to replace portions of the HSV1 genome with the homologous regions from HSV2. Such viruses can be grown in culture but are often nonpathogenic *in vivo* for ill-defined reasons probably relating to the fact that HSV1 and HSV2 proteins do not function in the life cycle of the other at 100% efficiency. One such virus (R7020, also known as NV1020; Meignier *et al.*, 1988, 1990) has been tested in preclinical tumor models (Ebright *et al.*, 2002; Wong *et al.*, 2001; Cozzi *et al.*, 2001; McAuliffe *et al.*, 2000; Delman *et al.*, 2000) leading to a clinical trial by intrahepatic artery infusion in liver cancer patients. The results of this trial have not yet been reported. NV1020 has no specific disablement providing tumor selectivity or reduced pathogenicity, but has a portion of the unique long region of the viral genome encoding glycoprotein G, D, I, and E, extending into the long repeat region encoding ICP34.5, replaced by the homologous region from HSV2. This has resulted in one copy of ICP34.5 being inactivated, but still leaves a functional copy of ICP34.5 in the other long repeat region of the genome. Any level of tumor specificity is therefore not provided by a lack of ICP34.5.

## V. Effects of Preexisting Immunity to HSV

HSV is a human pathogen that is highly prevalent in the population. As such the effects of prior immunity to HSV on tumor treatment with oncolytic HSV are worthy of consideration. Additionally, immune effects may also affect repeat dosing, which will probably be required for effective tumor treatment.

A number of studies have assessed the effects of prior exposure to HSV on tumor treatment with oncolytic HSV (Lambright *et al.*, 2000; Chahlavi *et al.*, 1999; Delman *et al.*, 2000; Lui *et al.*, 2002). In each case it has been concluded that prior immunization with HSV does not affect subsequent treatment.

## VI. Effects of the Immune Response in Enhancing Tumor Therapy

Oncolytic treatment of tumors might be expected to be multimodal in its mode of action. Not only would tumor cells be expected to be killed directly by lytic virus replication, but also release of antigens from the tumor might be expected to induce a systemic antitumor immune effect. A number of studies have been performed to explore this possibility (Miller and Fraser, 2000; Toda *et al.*, 1999, 2002; Todo *et al.*, 1999; Lui *et al.*, 2002). Here cellular immune responses to tumor antigens in mouse models have been demonstrated and considerably enhanced antitumor effects shown in normal rather than SCID or lymphocyte-depleted animals. Moreover, antitumor effects can be seen in both injected and noninjected tumors in which tumors had been implanted in both flanks of mice, indicating an immune effect. Use of oncolytic versions of HSV therefore not only treats tumors by direct virus replication, but also induces a systemic antitumor immune response that may be useful in the treatment of disseminated disease. Building on this information, versions of oncolytic HSV with enhanced antitumor immune-stimulating properties have now been constructed, and these are further discussed later.

## VII. Clinical Experience with Oncolytic HSV

To date results of clinical trials with oncolytic versions of HSV have been reported (HSV strains G207 and 1716).

### A. G207

G207 is a version of HSV1 strain F, which is deleted for ICP34.5 and has an insertion of the *Escherichia coli LacZ* gene into the gene encoding ICP6 (Mineta *et al.*, 1995). This has been extensively tested for safety in preclinical models, including intracerebral injection in the *Aotus* monkey, which is highly sensitive to infection with wild-type HSV (Todo *et al.*, 2000). A phase I clinical trial has been conducted in which up to $3 \times 10^9$ pfu (at five sites in a volume of 0.2 ml/site) was administered by intratumoral injection in malignant glioma patients (Markert *et al.*, 2000). Twenty-one patients were treated, with three at the highest dose. In these 21 patients, no toxicity or virus-associated serious adverse events were documented and results were obtained possibly suggestive of slowed tumor progression in some cases. However, as biopsies could not be taken and all histological examination performed at autopsy, evidence of virus replication could not be obtained. These promising safety data were concluded to warrant further clinical evaluation of G207 in malignant glioma patients.

### B. 1716

1716 is a spontaneous mutant of HSV1 strain 17+, which was found to be nonneurovirulent in mice (MacLean *et al.*, 1991), although it can replicate in dividing cells in the brain such as ependymal cells (Kesari *et al.*, 1998). 1716 was found to be mutated in the long terminal repeat regions of the HSV genome, as for a number of other nonneurovirulent strains of HSV1 and HSV2 (Thompson *et al.*, 1983, 1989; Taha *et al.*, 1989a,b). The nonneurovirulent phenotype was

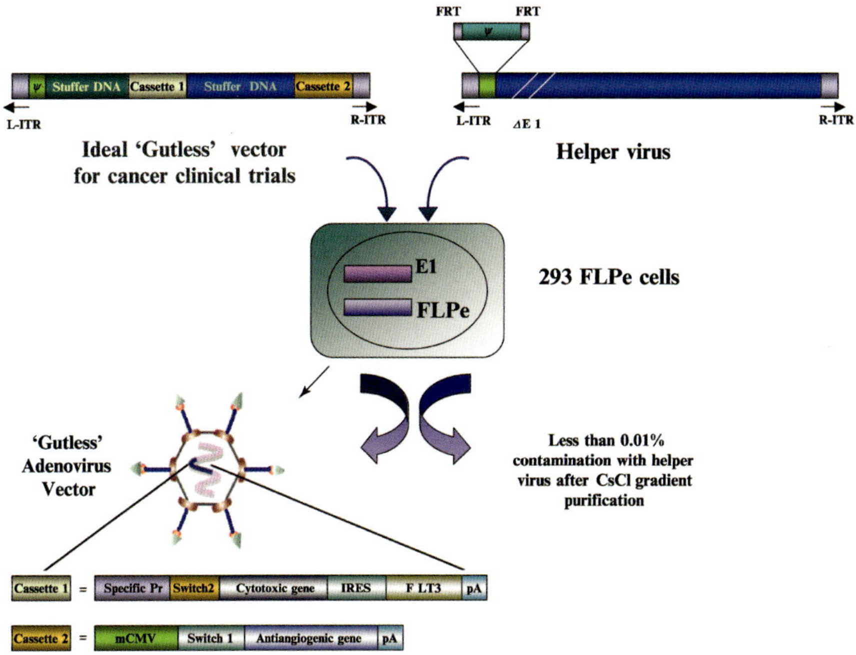


Lowenstein et al., FIG. 2. An E1-, E3-deleted helper adenovirus (FL helper) with a packaging signal sensitive to FLP3-mediated excision. (A) FL helper was generated by homologous recombination in 293 cells after cotransfection with plasmid pΔE1sp1A-2xFRT and pBHG10-CMVluc-I. The bottom half of the figure indicates the left-end region of the helper virus genome, before and after FLP3-mediated recombination; the figure also indicates the location of the PCR primers used to determine excision of the packaging signal, $\phi$ (flanked by black arrows). (B) Production of gutless adenoviral vectors. Diagram of an ideal gene delivery vector for neurological gene therapy using the "gutless" adenoviral vector. The packaging signal of the helper virus, flanked by FRT sites, is excised after growth in 293 cells expressing FLPe recombinase. The cloning capacity of this system (N30 kbp) will enable the simultaneous use of two or more therapeutic genes and also includes inducible promoter systems.

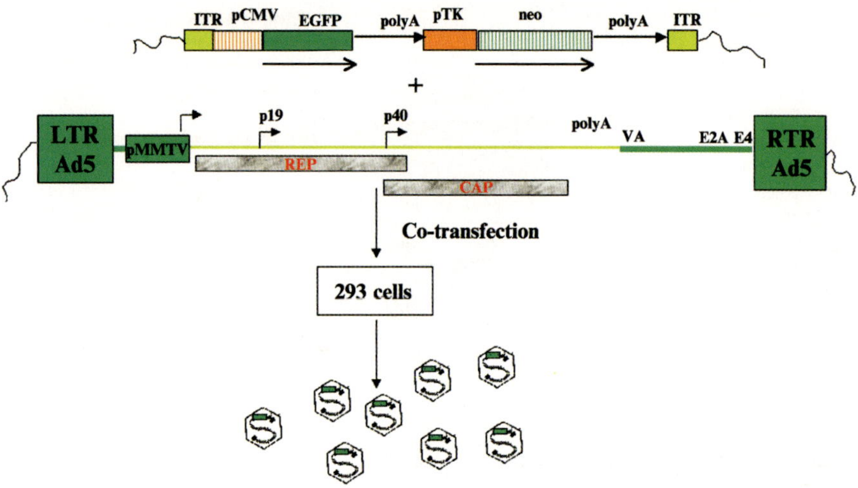


Lehtonen and Tenenbaum, FIG. 6. Production of rAAV by the cotransfection method. Recombinant AAV virus is obtained by cotransfection of adenovirus-immortalized human embryonic kidney (HEK-293) cells with the vector plasmid together with a helper/packaging plasmid (e.g., pDG) expressing AAV viral genes and the adenoviral genes required for AAV replication and encapsidation (Grimm *et al.*, 1998). The rAAV vector stock, devoid of adenovirus, is purified by cesium chloride gradient (Snyder *et al.*, 1996) or by iodixanol gradient and heparin agarose column chromatography (Zolotukhin *et al.*, 2000).

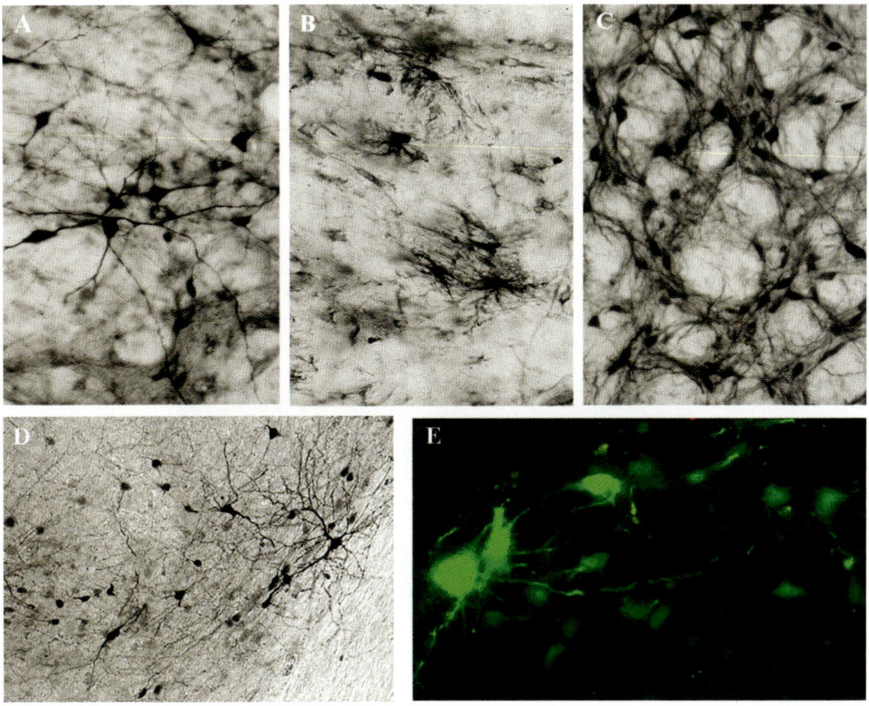


Lehtonen and Tenenbaum, FIG. 7. Tropism of rAAV vectors for specific cells in the brain. A purified viral suspension of rAAV2 expressing gfp-marker gene under the control of the CMV promoter was stereotaxically injected in various regions of the brain. One month after injection, transgene expression was detected either by observing native gfp fluorescence or by immunohistochemistry using peroxidase staining. Almost exclusively neurons are transduced in the striatum (A), globus pallidus (C), cortex (D), and substantia nigra pars compacta (E). In white matter, e.g., in the internal capsule (B), glial cells are efficiently transduced.

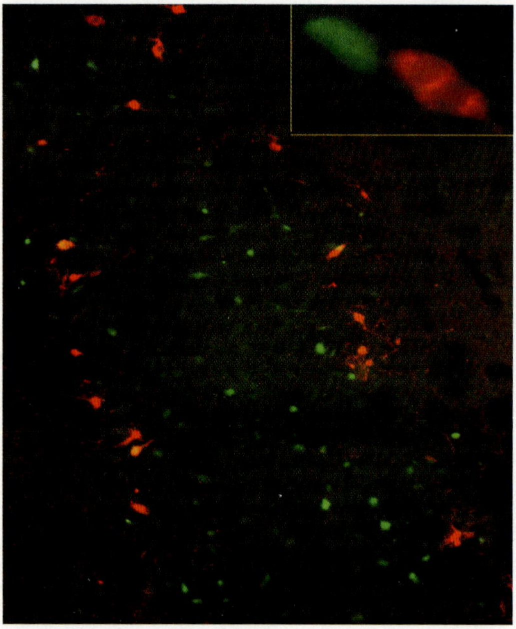

Lehtonen and Tenenbaum, FIG. 8. *Ex vivo* gene delivery to embryonic dopaminergic grafts using rAAV2 vectors. rAAV2–CMV–gfp and transplanted into the striatum of 6-hydroxydopamine-lesioned rats. Three months posttransplantation, brains were analyzed by double immunofluorescence for gfp (in green) and tyrosine hydroxylase (in red). Dopaminergic neurons are located at the border between the graft and the host striatum whereas gfp-positive cells are present all over the graft.

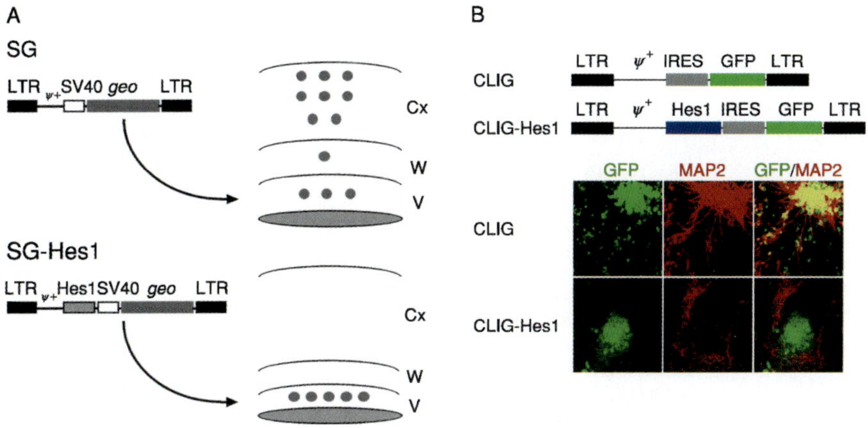


Kageyama et al., FIG. 6. Maintenance of neural precursor cells by *Hes1*-transducing retrovirus. (A) SG virus (produced from pLXSG) directs *geo* (fusion of *β-galactosidase* and *neo*) expression from the internal SV40 promoter whereas SG–Hes1 virus directs *Hes1* expression from the 5′-LTR promoter and *geo* expression from the internal SV40 promoter. One microliter of SG or SG–Hes1 retrovirus ($10^6$ cfu/ml) in 0.2 mg/ml of Polybrene was injected into lateral ventricles of mouse embryos at Days 12–13, and the fates of the virus-infected cells were examined by X-gal staining at postnatal Days 0–4. The majority of the cells infected with SG virus migrated out of the ventricular zone and differentiated into neurons and glial cells whereas those infected with SG–Hes1 virus remained in the ventricular zone. Cx, cortex; V, ventricular zone; W, white matter. (B) CLIG directs only GFP expression whereas CLIG–Hes1 directs Hes1 and GFP expression. Neural precursor cells prepared from mouse embryos were infected with CLIG or CLIG–Hes1 virus. Cells infected with CLIG virus differentiated into neurons with multiple neurites ($MAP2^+$) whereas those infected with CLIG–Hes1 did not.

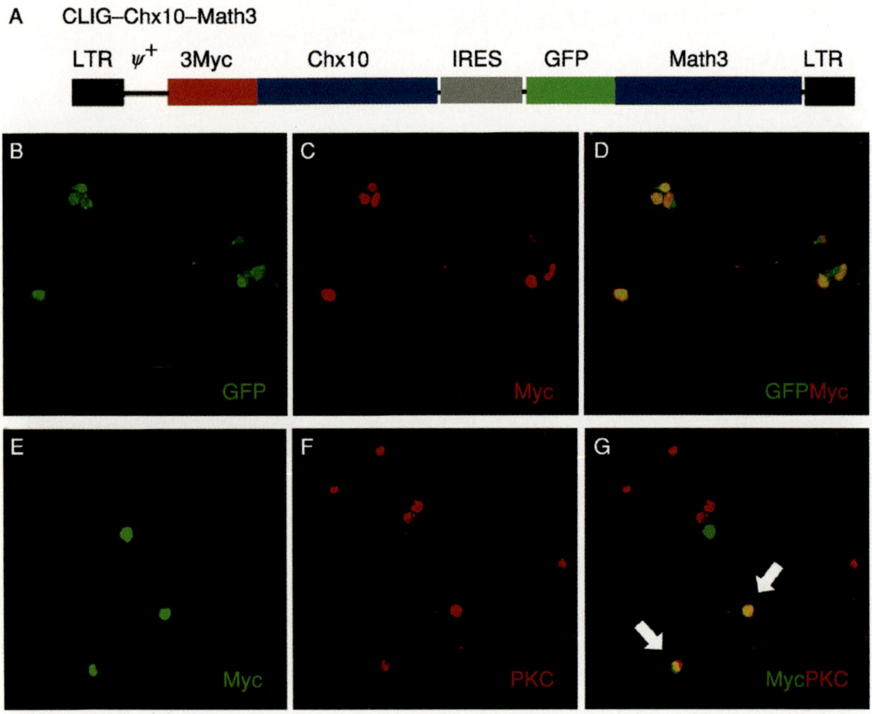


Kageyama et al., FIG. 7. Combinations of bHLH and homeobox genes promote specification of interneurons. (A) Structure of CLIG–Chx10–Math3. CLIG–Chx10–Math3 carries the homeobox gene *Chx10* and the bHLH gene *Math3*, which are under the control of the 5′-LTR. In this vector, Chx10 has three repeats of Myc tags whereas Math3 is fused with GFP. Dual gene expression can be monitored by Myc and GFP expression. (B–D) Neural precursor cells prepared from retina of mouse embryos were infected with CLIG–Chx10–Math3. Almost all cells infected with CLIG–Chx10–Math3 virus coexpressed GFP and Myc, indicating that this virus efficiently directs dual gene expression. (E–G) Many of the cells infected with CLIG–Chx10–Math3 ($Myc^+$) differentiated into bipolar cells ($PKC^+$, arrows in G). This retroviral system is useful for generation of specific neuronal subtypes.

later found to be due to mutation of ICP34.5, which is encoded in this region (McKay *et al.*, 1993).

1716 has been tested in a number of mouse models of cancer where, following direct intratumoral injection into either subcutaneous or intracranial tumors, therapeutic effects were seen (Kucharczuk *et al.*, 1997; Lasner *et al.*, 1996; Kesari *et al.*, 1995; Randazzo *et al.*, 1995, 1997).

Following these promising preclinical data, a Phase I clinical trial in glioma patients was performed (Rampling *et al.*, 2000). This was the first use of an oncolytic virus in humans. In this trial, in groups of three patients, $10^3$, $10^4$, or $10^5$ pfu of virus was injected into the tumor in a total volume of 1 ml. No adverse effects were seen and some suggestions of improved survival were demonstrated. In a further Phase I clinical trial in which 12 patients were injected with $10^5$ pfu of 1716 4–9 days prior to removal of the injected tumor, virus replication was demonstrated in two of the tumors (Papanastassiou *et al.*, 2002). 1716 therefore replicates in glioma tissue, without evidence of toxicity at the doses so far used. 1716 has also been tested in a Phase I clinical trial by intratumoral inoculation in melanoma patients (MacKie *et al.*, 2001). Here five patients with stage 4 melanoma were injected with either one, two, or three injections of $10^3$ pfu of 1716 in 1 ml. Even at this low virus dose, tumor flattening was seen in one patient and histopathological evidence of tumor necrosis was seen in three patients.

## VIII. Improving Oncolytic HSV

Initial Phase I clinical trials with oncolytic HSV have used either HSV1 strain F with ICP34.5 and ICP6 inactivated or HSV1 strain 17+ with ICP34.5 inactivated. These have provided clinical proof-of-principle, most importantly safety data. However, for optimal clinical efficacy there are a number of means by which potential improvements over these initial viruses might be made and that have been tested in preclinical models. Such second- and third-generation viruses are also now beginning to enter clinical trials in a number of cancer types.

### A. Incorporation of Therapeutic Genes

Oncolytic virus replication can produce a therapeutic effect, at least in preclinical models. However, combination of oncolytic therapy with the delivery of therapeutic genes may be particularly attractive, as discussed above, as virus replication will disseminate the therapeutic gene further through the tumor than is the case with nonreplicative gene delivery approaches. Moreover, if the therapeutic gene is chosen carefully, this may be synergistic with the antitumor effect of virus replication. Genes of a number of types have been delivered with oncolytic HSV, and effects in preclinical models assessed.

1. *Cytokines*

Cytokines and related molecules are involved in the modulation of immune responses, including the body's attempt to combat cancer. As such, molecules including a number of interleukins and interferons have been tested in gene delivery approaches to the treatment of cancer, generally using nonreplicative vectors (see Section I). However, incorporation of such genes into oncolytic viruses might be particularly attractive as oncolytic virus treatment will release tumor-associated antigens to which an immune response might be induced, which could then be enhanced by the immune stimulatory molecule.

Oncolytic versions of HSV that encode a number of immune stimulators have been tested in preclinical models. These molecules include IL-4 (Andreansky *et al.*, 1998), IL-12 (Parker *et al.*, 2000; Bennett *et al.*, 2001; Wong *et al.*, 2001), granulocyte–macrophage colony-stimulating factor (GM-CSF) (Wong *et al.*, 2001; Lui *et al.*, 2002), and B7.1 (Todo *et al.*, 2001). These potent immune stimulators were all found to improve survival and/or tumor shrinkage in xenograft models, and no safety issues were encountered. The use of cytokines and similar molecules in combination with oncolytic therapy might, therefore, be expected to improve therapeutic effects when used for tumor treatment in humans. Such an approach also has the advantage that the cytokine is delivered locally to the relevant site. This reduces the possibility of toxicity, which might be expected if cytokines were administered to patients systemically, which has been considered for cancer treatment.

2. *Prodrug activators*

Prodrug-activating genes encode enzymes that convert an inactive prodrug to an active form. The prodrug can thus be administered to a patient without the possibility of systemic toxicity from the active drug. This is locally converted in the tumor to produce a therapeutic effect. Prodrug activation has been attempted with a number of nonreplicative vectors, including clinical trials. However, as with all gene transfer approaches, the use of replicative vectors might be expected to be advantageous, and as a result a number of prodrug-activating systems have been tested in oncolytic HSV.

HSV encodes a natural prodrug-converting enzyme, TK, which can convert prodrugs such as gancyclovir to an active form, following which DNA synthesis is blocked. Oncolytic versions of HSV, in which TK is not deleted, have therefore been tested in combination with gancylovir with beneficial results on tumor treatment in rodent models (Todo *et al.*, 2000; Carroll *et al.*, 1997; Boviatsis *et al.*,

1994b). Other prodrug-activating systems have also been tested in oncolytic HSV, including cytosine deaminase conversion of 5-flurouracil, and cytochrome P-450 conversion of cyclophosphamide (Pawlik *et al.*, 2002; Aghi *et al.*, 1999; Rainov *et al.*, 1998; Chase *et al.*, 1998), including combination (Aghi *et al.*, 1998), again with beneficial effects.

### B. Improved Tumor-Selective Replication via US11 Upregulation

One function of ICP34.5 is to mediate dephosphorylation of the translation initiation factor eIF2$\alpha$ after phosphorylation by the cellular PKR kinase (He *et al.*, 1997b). In the absence of ICP34.5, cells respond to HSV infection by terminating translation by phosphorylation of eIF2$\alpha$, resulting in apoptosis and prevention of virus replication. ICP34.5 deletion mutants will not therefore grow effectively in cells with a functional PKR pathway, as is the case for most normal cells. Tumor cells often have a disabled PKR system and/or elevated levels of eIF2$\alpha$ as compared to normal cells. This allows productive infection of these cells by HSV ICP34.5 mutants. However, ICP34.5 mutants replicate with considerably reduced efficiency in most tumor cells as compared to wild-type HSV. Therefore, it would be advantageous to identify mutants of HSV that retained the tumor-selective replicative ability of ICP34.5 mutants, but that replicate with greater efficiency in tumor cells to improve the therapeutic effect.

To this end, an ICP34.5-deleted version of HSV was serially passaged on a tumor cell line in which replication was poor, and mutants with enhanced growth characteristics were isolated (Mohr and Gluzman, 1996). These "second site suppressor" mutants were also found to give improved antitumor activity *in vivo*, importantly without compromising safety (Mohr *et al.*, 2001). Following considerable further work, the mechanism behind this phenotype was identified.

The second site suppressor viruses were all found to have mutations in the region of the genome encoding ICP47 (responsible for blocking antigen presentation in HSV-infected cells). These mutations resulted in the positioning of the US11 gene, which is next to the US12 gene encoding ICP47, such that it was then expressed from the ICP47 promoter. The ICP47 promoter is active at immediate early times after infection as compared to late times when US11 is usually expressed. Thus, expression of US11 as an IE rather than an L gene results in the phenotype observed. US11 is an RNA-binding protein that blocks the phosphorylation of eIF2$\alpha$ in HSV-infected cells when expressed early in the viral replication cycle and as such allows continued translation of virus proteins and virus replication to occur more effectively (Cassady *et al.*, 1998; He *et al.*, 1997; Poppers *et al.*, 2000; Mulvey *et al.*, 1999).

A number of preclinical studies have been performed in rodent models demonstrating that combined ICP34.5 deletion/US11 upregulation provides a greatly enhanced antitumor effect (Lui *et al.*, 2002; Taneja *et al.*, 2001; Todo *et al.*, 2001). Additionally, as ICP47 is deleted, presentation of HSV antigens and tumor antigens on the tumor cells and on infiltrating antigen-presenting cells will also be enhanced. This may contribute to an improved antitumor immune response, particularly when combined with delivery of immune stimulatory genes to the tumor (Lui *et al.*, 2002).

### C. Use of Clinical Isolates of HSV

So far, most oncolytic versions of HSV have been based on serially passaged laboratory strains of HSV. These might be anticipated to have lost/reduced certain properties beneficial for an oncolytic version of HSV during serial passage over time. Such properties include the ability to replicate abundantly in human cells and to replicate and spread in solid tissue such as a tumor mass due to the laboratory culture conditions used. In an attempt to produce more potent versions of oncolytic HSV we tested a number of clinical isolates of HSV and have shown them to be considerably more effective at tumor cell killing than strain 17+ (Lui *et al.*, 2002). Such a virus has therefore been deleted for ICP34.5 and additionally deleted for ICP47, resulting in US11 upregulation (see above). Finally we have inserted the gene encoding human GM-CSF. This virus is designed to generate a potent antitumor immune response as well as direct tumor cell killing by lytic virus replication and is currently in a Phase I clinical trial in a number of tumor types.

### D. Combination with Other Therapies

Oncolytic anticancer therapy is likely to be used in combination with preexisting cancer treatments such as chemotherapy or radiotherapy. As a result, oncolytic HSV has on a number of occasions been tested in combination with chemotherapy or radiotherapy in a preclinical setting. Combination of radiotherapy with various versions of oncolytic HSV has been demonstrated to give improved effects as compared to either therapy alone in most cases in test mice (Blank *et al.*, 2002; Chung *et al.*, 2002; Bradley *et al.*, 1999; Advani *et al.*, 1998). However, in one case in a xenograft prostate cancer model, although both therapies were effective, combination of the two did not add to the effects (Jorgensen *et al.*, 2001).

In combination experiments with chemotherapy, cisplatin did not inhibit the cytopathic effect of G207 *in vitro*, and *in vivo* combination therapy resulted in improved antitumor effects in cisplatin-sensitive tumor models (Chahlavi *et al.*, 1999). In another study mitomycin C, cisplatin, methotrexate, and doxorubicin were tested (Toyoizumi *et al.*, 1999). Here the oncolytic effect of 1716 in combination with mitomycin C was synergistic, with evidence of an additive effect with the other drugs. In no case was antagonism observed. The *in vivo* effect of mitomycin C in combination with 1716 was then examined in a mouse model giving an additive effect.

Oncolytic therapy, in the models so far tested, is, therefore, generally compatible with the combined use of other therapies such as chemotherapy or radiotherapy, and in most cases combined therapy has provided improvements over either therapy alone. In no case has a reduction in the effectiveness of chemotherapy or radiotherapy been shown. These types of studies are important, as oncolytic agents will often be required to be tested or used in combination with current standard therapy; that they do not reduce the effectiveness of such therapy will be important in facilitating the clinical use of oncolytic viruses.

## IX. Conclusions

Oncolytic virus therapy is undoubtedly a highly promising new approach to the treatment of solid tumors, including those in the brain. A number of oncolytic viruses, including those based on HSV, have been tested in the clinic where they have generally proved to be well tolerated in the Phase I and II clinical trials conducted so far. Some indications of clinical efficacy have also been shown. However, it is evident that improvements in efficacy would be beneficial to increase effectiveness and also to reduce the doses of virus required. Reduction in dose would potentially benefit the patient, and would also minimize production issues during scale-up for later stage clinical testing and ultimate use in routine clinical practice.

In this chapter we have discussed preclinical and clinical work with first-generation oncolytic HSV. We have then discussed further developments in the field aimed at increasing the antitumor potency of the approach. Here both the tumor-selective replicative capabilities of the virus have been increased and/or potentially therapeutic genes have been inserted. In particularly promising approaches to the development of second- and third-generation oncolytic HSV the immune-stimulating properties of the virus have been enhanced by the inclusion of cytokine-encoding genes (Fig. 1). Such viruses, as well as having a direct oncolytic effect, aim to provide an *in situ*, patient-specific, antitumor vaccine, which, when administered into a tumor, would be expected to have effects on the treated tumor and also to induce an enhanced immune response, which may be beneficial in the treatment of metastatic disease.

To develop oncolytic HSV with greater tumor-selective replicative ability the use of clinical isolates of HSV has been considered. Here it was anticipated that

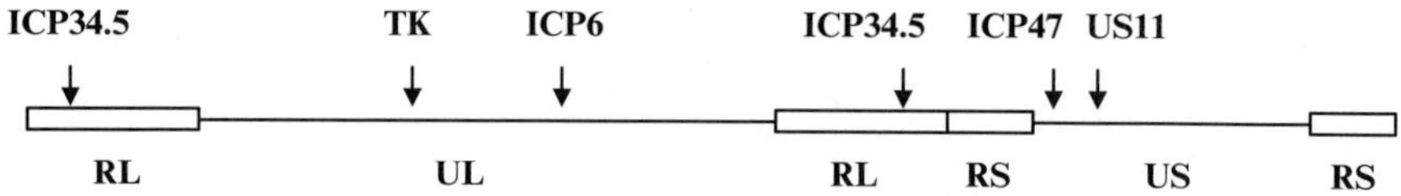


FIG. 1. The HSV genome and positions of the various genes mentioned in the text. The HSV genome consists of 152 kb of double-stranded DNA with a long unique region (UL) flanked by identical long repeated regions (RL), and a short unique region flanked by identical short repeated regions (RS). Genes encoded in the repeat regions are therefore present in two copies per viral genome.

recent clinical isolates might be more effective than the previously tested serially passaged laboratory strains. This was found to be the case. Tumor-selective replication has also been increased by a number of groups through the inclusion of a mutation resulting in the expression of the US11 gene as an IE rather than L gene (Taneja *et al.*, 2001). Increased expression of US11 is accomplished by deletion of the coding sequence for ICP47. Because ICP47 usually inhibits antigen presentation in HSV-infected cells (Hill *et al.*, 1995), this deletion would also be expected to increase levels of antigen presentation in infected tumor cells.

Overall, oncolytic viruses including HSV are highly promising for cancer treatment. Clinical work so far has demonstrated that such viruses are well tolerated with none of the side effects associated with the use of traditional anticancer drugs. Later stage clinical trials with these first-generation viruses will aim to demonstrate efficacy. More potent second- and third-generation viruses have also been developed. Bearing in mind that there is currently little effective therapy for a number of solid tumor types, particularly in the later stages of disease, it can be hoped that the use of oncolytic agents may in the future offer improved treatment for cancer patients.

## References

Advani, S. J., Sibley, G. S., Song, P. Y., Hallahan, D. E., Kataoka, Y., Roizman, B., and Weichselbaum, R. R. (1998). Enhancement of replication of genetically engineered herpes simplex viruses by ionizing radiation: A new paradigm for destruction of therapeutically intractable tumors. *Gene Ther.* **5,** 160–165.

Aghi, M., Kramm, C. M., Chou, T. C., Breakefield, X. O., and Chiocca, E. A. (1998). Synergistic anticancer effects of ganciclovir/thymidine kinase and 5-fluorocytosine/cytosine deaminase gene therapies. *J. Natl. Cancer Inst.* **90,** 370–380.

Aghi, M., Chou, T. C., Suling, K., Breakefield, X. O., and Chiocca, E. A. (1999). Multimodal cancer treatment mediated by a replicating oncolytic virus that delivers the oxazaphosphorine/rat cytochrome P450 2B1 and ganciclovir/herpes simplex virus thymidine kinase gene therapies. *Cancer Res.* **59,** 3861–3865.

Andreansky, S. S., He, B., Gillespie, G. Y., Soroceanu, L., Markert, J., Chou, J., Roizman, B., and Whitley, R. J. (1996). The application of genetically engineered herpes simplex viruses to the treatment of experimental brain tumors. *Proc. Natl. Acad. Sci. USA* **93,** 11313–11318.

Andreansky, S., Soroceanu, L., Flotte, E. R., Chou, J., Markert, J. M., Gillespie, G. Y., Roizman, B., and Whitley, R. J. (1997). Evaluation of genetically engineered herpes simplex viruses as oncolytic agents for human malignant brain tumors. *Cancer Res.* **57,** 1502–1509.

Andreansky, S., He, B., van Cott, J., McGhee, J., Markert, J. M., Gillespie, G. Y., Roizman, B., and Whitley, R. J. (1998). Treatment of intracranial gliomas in immunocompetent mice using herpes simplex viruses that express murine interleukins. *Gene Ther.* **5,** 121–130.

Barker, D. D., and Berk, A. J. (1987). Adenovirus proteins from both E1B reading frames are required for transformation of rodent cells by viral infection and DNA transfection. *Virology* **156,** 107–121.

Belldegrun, A., Tso, C. L., Zisman, A., Naitoh, J., Said, J., Pantuck, A. J., Hinkel, A., deKernion, J., and Figlin, R. (2001). Interleukin 2 gene therapy for prostate cancer: Phase I clinical trial and basic biology. *Hum. Gene Ther.* **12,** 883–892.

Bennett, J. J., Malhotra, S., Wong, R. J., Delman, K., Zager, J., St Louis, M., Johnson, P., and Fong, Y. (2001). Interleukin 12 secretion enhances antitumor efficacy of oncolytic herpes simplex viral therapy for colorectal cancer. *Ann. Surg.* **233,** 819–826.

Blank, S. V., Rubin, S. C., Coukos, G., Amin, K. M., Albelda, S. M., and Molnar-Kimber, K. L. (2002). Replication-selective herpes simplex virus type 1 mutant therapy of cervical cancer is enhanced by low-dose radiation. *Hum. Gene Ther.* **13,** 627–639.

Boviatsis, E. J., Park, J. S., Sena-Esteves, M., Kramm, C. M., Chase, M., Efird, J. T., Wei, M. X., Breakefield, X. O., and Chiocca, E. A. (1994a). Long-term survival of rats harboring brain neoplasms treated with ganciclovir and a herpes simplex virus vector that retains an intact thymidine kinase gene. *Cancer Res.* **54,** 5745–5751.

Boviatsis, E. J., Scharf, J. M., Chase, M., Harrington, K., Kowall, N. W., Breakefield, X. O., and Chiocca, E. A. (1994b). Antitumor activity and reporter gene transfer into rat brain neoplasms inoculated with herpes simplex virus vectors defective in thymidine kinase or ribonucleotide reductase. *Gene Ther.* **1,** 323–331.

Bradley, J. D., Kataoka, Y., Advani, S., Chung, S. M., Arani, R. B., Gillespie, G. Y., Whitley, R. J., Markert, J. M., Roizman, B., and Weichselbaum, R. R. (1999). Ionizing radiation improves survival in mice bearing intracranial high-grade gliomas injected with genetically modified herpes simplex virus. *Clin. Cancer Res.* **5,** 1517–1522.

Brown, S. M., MacLean, A. R., Aitken, J. D., and Harland, J. (1994). ICP34.5 influences herpes simplex virus type 1 maturation and egress from infected cells *in vitro*. *J. Gen. Virol.* **75**(pt. 12), 3679–3686.

Carroll, N. M., Chiocca, E. A., Takahashi, K., and Tanabe, K. K. (1996). Enhancement of gene therapy specificity for diffuse colon carcinoma liver metastases with recombinant herpes simplex virus. *Ann. Surg.* **224,** 323–329.

Carroll, N. M., Chase, M., Chiocca, E. A., and Tanabe, K. K. (1997). The effect of ganciclovir on herpes simplex virus-mediated oncolysis. *J. Surg. Res.* **69,** 413–417.

Cassady, K. A., Gross, M., and Roizman, B. (1998). The second-site mutation in the herpes simplex virus recombinants lacking the gamma 134.5 genes precludes shutoff of protein synthesis by blocking the phosphorylation of eIF-2alpha. *J. Virol.* **72,** 7005–7011.

Chahlavi, A., Rabkin, S., Todo, T., Sundaresan, P., and Martuza, R. (1999a). Effect of prior exposure to herpes simplex virus 1 on viral vector-mediated tumor therapy in immunocompetent mice. *Gene Ther.* **6,** 1751–1758.

Chahlavi, A., Todo, T., Martuza, R. L., and Rabkin, S. D. (1999b). Replication-competent herpes simplex virus vector G207 and cisplatin combination therapy for head and neck squamous cell carcinoma. *Neoplasia* **1,** 162–169.

Chambers, R., Gillespie, G. Y., Soroceanu, L., Andreansky, S., Chatterjee, S., Chou, J., Roizman, B., and Whitley, R. J. (1995). Comparison of genetically engineered herpes simplex viruses for the treatment of brain tumors in a scid mouse model of human malignant glioma. *Proc. Natl. Acad. Sci. USA* **92,** 1411–1415.

Chase, M., Chung, R. Y., and Chiocca, E. A. (1998). An oncolytic viral mutant that delivers the CYP2B1 transgene and augments cyclophosphamide chemotherapy. *Nat. Biotechnol.* **16,** 444–448.

Chen, Y., DeWeese, T., Dilley, J., Zhang, Y., Li, Y., Ramesh, N., Lee, J., Pennathur-Das, R., Radzyminski, J., Wypych, J., Brignetti, D., Scott, S., Stephens, J., Karpf, D. B., Henderson, D. R., and Yu, D. C. (2001). CV706, a prostate cancer-specific adenovirus variant, in combination with radiotherapy produces synergistic antitumor efficacy without increasing toxicity. *Cancer Res.* **61,** 5453–5460.

Chou, J., and Roizman, B. (1992). The gamma 1(34.5) gene of herpes simplex virus 1 precludes neuroblastoma cells from triggering total shutoff of protein synthesis characteristic of programed cell death in neuronal cells. *Proc. Natl. Acad. Sci. USA* **89,** 3266–3270.

Chou, J., and Roizman, B. (1994). Herpes simplex virus 1 gamma(1)34.5 gene function, which blocks the host response to infection, maps in the homologous domain of the genes expressed during growth arrest and DNA damage. *Proc. Natl. Acad. Sci. USA* **91,** 5247–5251.

Chou, J., Kern, E. R., Whitley, R. J., and Roizman, B. (1990). Mapping of herpes simplex virus-1 neurovirulence to gamma 134.5, a gene nonessential for growth in culture. *Science* **250,** 1262–1266.

Chou, J., Chen, J. J., Gross, M., and Roizman, B. (1995). Association of a M(r) 90,000 phosphoprotein with protein kinase PKR in cells exhibiting enhanced phosphorylation of translation initiation factor eIF-2 alpha and premature shutoff of protein synthesis after infection with gamma 134.5-mutants of herpes simplex virus 1. *Proc. Natl. Acad. Sci. USA* **92,** 10516–10520.

Chung, S. M., Advani, S. J., Bradley, J. D., Kataoka, Y., Vashistha, K., Yan, S. Y., Markert, J. M., Gillespie, G. Y., Whitley, R. J., Roizman, B., and Weichselbaum, R. R.(2002). The use of a genetically engineered herpes simplex virus (R7020) with ionizing radiation for experimental hepatoma. *Gene Ther.* **9,** 75–80.

Cozzi, P. J., Malhotra, S., McAuliffe, P., Kooby, D. A., Federoff, H. J., Huryk, B., Johnson, P., Scardino, P. T., Heston, W. D., and Fong, Y. (2001). Intravesical oncolytic viral therapy using attenuated, replication-competent herpes simplex viruses G207 and Nv1020 is effective in the treatment of bladder cancer in an orthotopic syngeneic model. *FASEB J.* **15,** 1306–1308.

Delman, K. A., Bennett, J. J., Zager, J. S., Burt, B. M., McAuliffe, P. F., Petrowsky, H., Kooby, D. A., Hawkins, W. G., Horsburgh, B. C., Johnson, P., and Fong, Y. (2000). Effects of preexisting immunity on the response to herpes simplex-based oncolytic viral therapy. *Hum. Gene Ther.* **11,** 2465–2472.

Ebright, M. I., Zager, J. S., Malhotra, S., Delman, K. A., Weigel, T. L., Rusch, V. W., and Fong, Y. (2002). Replication-competent herpes virus NV1020 as direct treatment of pleural cancer in a rat model. *J. Thorac. Cardiovasc. Surg.* **124,** 123–129.

Ferrantini, M., and Belardelli, F. (2000). Gene therapy of cancer with interferon: Lessons from tumor models and perspectives for clinical applications. *Semin. Cancer Biol.* **10,** 145–157.

Fujiwara, T., Kataoka, M., and Tanaka, N. (2000). Adenovirus-mediated p53 gene therapy for human cancer. *Mol. Urol.* **4,** 51–54.

Ganly, I., Kirn, D., Eckhardt, G., Rodriguez, G. I., Soutar, D. S., Otto, R., Robertson, A. G., Park, O., Gulley, M. L., Heise, C., Von Hoff, D. D., Kaye, S. B., and Eckhardt, S. G. (2000). A phase I study of Onyx-015, an E1B attenuated adenovirus, administered intratumorally to patients with recurrent head and neck cancer. *Clin. Cancer Res.* **6,** 798–806.

Hawkins, L. K., Lemoine, N. R., and Kirn, D. (2002). Oncolytic biotherapy: A novel therapeutic plafform. *Lancet Oncol.* **3,** 17–26.

He, B., Chou, J., Brandimarti, R., Mohr, I., Gluzman, Y., and Roizman, B. (1997a). Suppression of the phenotype of gamma(1)34.5-herpes simplex virus 1: Failure of activated RNA-dependent protein kinase to shut off protein synthesis is associated with a deletion in the domain of the alpha47 gene. *J. Virol.* **71,** 6049–6054.

He, B., Gross, M., and Roizman, B. (1997b). The gamma(1)34.5 protein of herpes simplex virus 1 complexes with protein phosphatase 1alpha to dephosphorylate the alpha subunit of the eukaryotic translation initiation factor 2 and preclude the shutoff of protein synthesis by double-stranded RNA-activated protein kinase. *Proc. Natl. Acad. Sci. USA* **94,** 843–848.

Hill, A., Jugovic, P., York, I., Russ, G., Bennink, J., Yewdell, J., Ploegh, H., and Johnson, D. (1995). Herpes simplex virus turns off the TAP to evade host immunity. *Nature* **375,** 411–415.

Jorgensen, T. J., Katz, S., Wittmack, E. K., Varghese, S., Todo, T., Rabkin, S. D., and Martuza, R. L. (2001). Ionizing radiation does not alter the antitumor activity of herpes simplex virus vector G207 in subcutaneous tumor models of human and murine prostate cancer. *Neoplasia* **3,** 451–456.

Kaushik, A. (2001). Leuvectin Vical Inc. *Curr. Opin. Investig. Drugs.* **2,** 976–981.

Kesari, S., Randazzo, B. P., Valyi-Nagy, T., Huang, Q. S., Brown, S. M., MacLean, A. R., Lee, V. M., Trojanowski, J. Q., and Fraser, N. W. (1995). Therapy of experimental human brain tumors using a neuroattenuated herpes simplex virus mutant. *Lab. Invest.* **73,** 636–648.

Kesari, S., Lasner, T. M., Balsara, K. R., Randazzo, B. P., Lee, V. M., Trojanowski, J. Q., and Fraser, N. W. (1998). A neuroattenuated ICP34.5-deficient herpes simplex virus type 1 replicates in ependymal cells of the murine central nervous system. *J. Gen. Virol.* **79**(pt. 3), 525–536.

Khuri, F. R., Nemunaitis, J., Ganly, I., Arseneau, J., Tannock, I. F., Romel, L., Gore, M., Ironside, J., MacDougall, R. H., Heise, C., Randlev, B., Gillenwater, A. M., Bruso, P., Kaye, S. B., Hong, W. K., and Kirn, D. H. (2000). A controlled trial of intratumoral ONYX-015, a selectively-replicating adenovirus, in combination with cisplatin and 5-fluorouracil in patients with recurrent head and neck cancer. *Nat. Med.* **6,** 879–885.

Kramm, C. M., Rainov, N. G., Sena-Esteves, M., Barnett, F. H., Chase, M., Herrlinger, U., Pechan, P. A., Chiocca, E. A., and Breakefield, X. O. (1996). Long-term survival in a rodent model of disseminated brain tumors by combined intrathecal delivery of herpes vectors and ganciclovir treatment. *Hum. Gene Ther.* **7,** 1989–1994.

Kramm, C. M., Chase, M., Herrlinger, U., Jacobs, A., Pechan, P. A., Rainov, N. G., Sena-Esteves, M., Aghi, M., Barnett, F. H., Chiocca, E. A., and Breakefield, X. O. (1997). Therapeutic efficiency and safety of a second-generation replication-conditional HSV1 vector for brain tumor gene therapy. *Hum. Gene Ther.* **8,** 2057–2068.

Kucharczuk, J. C., Randazzo, B., Chang, M. Y., Amin, K. M., Elshami, A. A., Sterman, D. H., Rizk, N. P., Molnar-Kimber, K. L., Brown, S. M., MacLean, A. R., Litzky, L. A., Fraser, N. W., Albelda, S. M., and Kaiser, L. R. (1997). Use of a "replication-restricted" herpes virus to treat experimental human malignant mesothelioma. *Cancer Res.* **57,** 466–471.

Lambright, E. S., Kang, E. H., Force, S., Lanuti, M., Caparrelli, D., Kaiser, L. R., Albelda, S. M., and Molnar-Kimber, K. L. (2000). Effect of preexisting anti-herpes immunity on the efficacy of herpes simplex viral therapy in a murine intraperitoneal tumor model. *Mol. Ther.* **2,** 387–393.

Lamont, J. P., Nemunaitis, J., Kuhn, J. A., Landers, S. A., and McCarty, T. M. (2000). A prospective phase II trial of ONYX-015 adenovirus and chemotherapy in recurrent squamous cell carcinoma of the head and neck (the Baylor experience). *Ann. Surg. Oncol.* **7,** 588–592.

Lasner, T. M., Kesari, S., Brown, S. M., Lee, V. M., Fraser, N. W., and Trojanowski, J. Q. (1996). Therapy of a murine model of pediatric brain tumors using a herpes simplex virus type-1

ICP34.5 mutant and demonstration of viral replication within the CNS. *J. Neuropathol. Exp. Neurol.* **55,** 1259–1269.

Lui, B. L., Robinson, M., Han, Z. Q., Branston, R. H., English, C., Reay, P., McGrath, Y., Thomas, S. K., Thornton, M., Bullock, P., Love, C. A., and Coffin, R. S. (2002). ICP34.5 deleted herpes simplex virus with enhanced oncolytic, immune stimulating, and anti-tumour properties. *Gene Ther.*

MacKie, R. M., Stewart, B., and Brown, S. M. (2001). Intralesional injection of herpes simplex virus 1716 in metastatic melanoma. *Lancet* **357,** 525–526.

MacLean, A. R., ul-Fareed, M., Robertson, L., Harland, J., and Brown, S. M. (1991). Herpes simplex virus type 1 deletion variants 1714 and 1716 pinpoint neurovirulence-related sequences in Glasgow strain 17+ between immediate early gene 1 and the 'a' sequence. *J. Gen. Virol.* **72**(pt. 3), 631–639.

Markert, J. M., Coen, D. M., Malick, A., Mineta, T., and Martuza, R. L. (1992). Expanded spectrum of viral therapy in the treatment of nervous system tumors. *J. Neurosurg.* **77,** 590–594.

Markert, J. M., Malick, A., Coen, D. M., and Martuza, R. L. (1993). Reduction and elimination of encephalitis in an experimental glioma therapy model with attenuated herpes simplex mutants that retain susceptibility to acyclovir. *Neurosurgery* **32,** 597–603.

Markert, J. M., Medlock, M. D., Rabkin, S. D., Gillespie, G. Y., Todo, T., Hunter, W. D., Palmer, C. A., Feigenbaum, F., Tornatore, C., Tufaro, F., and Martuza, R. L. (2000). Conditionally replicating herpes simplex virus mutant, G207 for the treatment of malignant glioma: Results of a phase I trial. *Gene Ther.* **7,** 867–874.

Martuza, R. L., Malick, A., Markert, J. M., Ruffner, K. L., and Coen, D. M. (1991). Experimental therapy of human glioma by means of a genetically engineered virus mutant. *Science* **252,** 854–856.

McAuliffe, P. F., Jarnagin, W. R., Johnson, P., Delman, K. A., Federoff, H., and Fong, Y. (2000). Effective treatment of pancreatic tumors with two multimutated herpes simplex oncolytic viruses. *J. Gastrointest. Surg.* **4,** 580–588.

McKay, E. M., McVey, B., Marsden, H. S., Brown, S. M., and MacLean, A. R. (1993). The herpes simplex virus type 1 strain 17 open reading frame RL1 encodes a polypeptide of apparent M(r) 37K equivalent to ICP34.5 of herpes simplex virus type 1 strain F. *J. Gen. Virol.* **74**(pt. 11), 2493–2497.

Meignier, B., Longnecker, R., and Roizman, B. (1988). In vivo behavior of genetically engineered herpes simplex viruses R7017 and R7020: Construction and evaluation in rodents. *J. Infect. Dis.* **158,** 602–614.

Meignier, B., Martin, B., Whitley, R. J., and Roizman, B. (1990). In vivo behavior of genetically engineered herpes simplex viruses R7017 and R7020. II. Studies in immunocompetent and immunosuppressed owl monkeys (Aotus trivirgatus). *J. Infect. Dis.* **162,** 313–321.

Miller, C. G., and Fraser, N. W. (2000). Role of the immune response during neuro-attenuated herpes simplex virus-mediated tumor destruction in a murine intracranial melanoma model. *Cancer Res.* **60,** 5714–5722.

Mineta, T., Rabkin, S. D., and Martuza, R. L. (1994). Treatment of malignant gliomas using ganciclovir-hypersensitive, ribonucleotide reductase-deficient herpes simplex viral mutant. *Cancer Res.* **54,** 3963–3966.

Mineta, T., Rabkin, S. D., Yazaki, T., Hunter, W. D., and Martuza, R. L. (1995). Attenuated multi-mutated herpes simplex virus-1 for the treatment of malignant gliomas. *Nat. Med.* **1,** 938–943.

Mohr, I., and Gluzman, Y. (1996). A herpesvirus genetic element which affects translation in the absence of the viral GADD34 function. *EMBO J.* **15,** 4759–4766.

Mohr, I., Sternberg, D., Ward, S., Leib, D., Mulvey, M., and Gluzman, Y. (2001). A herpes simplex virus type 1 gamma34.5 second-site suppressor mutant that exhibits enhanced growth in cultured glioblastoma cells is severely attenuated in animals. *J. Virol.* **75,** 5189–5196.

Mulvey, M., Poppers, J., Ladd, A., and Mohr, I. (1999). A herpesvirus ribosome-associated, RNA-binding protein confers a growth advantage upon mutants deficient in a GADD34-related function. *J. Virol.* **73,** 3375–3385.

Papanastassiou, V., Rampling, R., Fraser, M., Petty, R., Hadley, D., Nicoll, J., Harland, J., Mabbs, R., and Brown, M. (2002). The potential for efficacy of the modified (ICP 34.5(-)) herpes simplex virus HSV1716 following intratumoural injection into human malignant glioma: A proof of principle study. *Gene Ther.* **9,** 398–406.

Parker, J. N., Gillespie, G. Y., Love, C. E., Randall, S., Whitley, R. J., and Markert, J. M. (2000). Engineered herpes simplex virus expressing IL-12 in the treatment of experimental murine brain tumors. *Proc. Natl. Acad. Sci. USA* **97,** 2208–2213.

Pawlik, T. M., Nakamura, H., Mullen, J. T., Kasuya, H., Yoon, S. S., Chandrasekhar, S., Chiocca, E. A., and Tanabe, K. K. (2002). Prodrug bioactivation and oncolysis of diffuse liver metastases by a herpes simplex virus 1 mutant that expresses the CYP2B1 transgene. *Cancer* **95,** 1171–1181.

Poppers, J., Mulvey, M., Khoo, D., and Mohr, I. (2000). Inhibition of PKR activation by the proline-rich RNA binding domain of the herpes simplex virus type 1 Us11 protein. *J. Virol.* **74,** 11215–11221.

Rainov, N. G., Sena-Esteves, M., Fraefel, C., Dobberstein, K. U., Chiocca, E. A., and Breakefield, X. O. (1998). A chimeric fusion protein of cytochrome CYP4B1 and green fluorescent protein for detection of pro-drug activating gene delivery and for gene therapy in malignant glioma. *Adv. Exp. Med. Biol.* **451,** 393–403.

Rampling, R., Cruickshank, G., Papanastassiou, V., Nicoll, J., Hadley, D., Brennan, D., Petty, R., MacLean, A., Harland, J., McKie, E., Mabbs, R., and Brown, M. (2000). Toxicity evaluation of replication-competent herpes simplex virus (ICP 34.5 null mutant 1716) in patients with recurrent malignant glioma. *Gene Ther.* **7,** 859–866.

Randazzo, B. P., Kesari, S., Gesser, R. M., Alsop, D., Ford, J. C., Brown, S. M., MacLean, A., and Fraser, N. W. (1995). Treatment of experimental intracranial murine melanoma with a neuroattenuated herpes simplex virus 1 mutant. *Virology* **211,** 94–101.

Randazzo, B. P., Bhat, M. G., Kesari, S., Fraser, N. W., and Brown, S. M. (1997). Treatment of experimental subcutaneous human melanoma with a replication-restricted herpes simplex virus mutant. *J. Invest. Dermatol.* **108,** 933–937.

Reid, T., Galanis, E., Abbruzzese, J., Sze, D., Andrews, J., Romel, L., Hatfield, M., Rubin, J., and Kirn, D. (2001). Intra-arterial administration of a replication-selective adenovirus (dl1520) in patients with colorectal carcinoma metastatic to the liver: A phase I trial. *Gene Ther.* **8,** 1618–1626.

Ring, C. J. (2002). Cytolytic viruses as potential anti-cancer agents. *J. Gen. Virol.* **83,** 491–502.

Stewart, A. K., Lassam, N. J., Quirt, I. C., Bailey, D. J., Rotstein, L. E., Krajden, M., Dessureault, S., Gallinger, S., Cappe, D., Wan, Y., Addison, C. L., Moen, R. C., Gauldie, J., and Graham, F. L. (1999). Adenovector-mediated gene delivery of interleukin-2 in metastatic breast cancer and melanoma: Results of a phase 1 clinical trial. *Gene Ther.* **6,** 350–363.

Taha, M. Y., Clements, G. B., and Brown, S. M. (1989a). The herpes simplex virus type 2 (HG52) variant JH2604 has a 1488 bp deletion which eliminates neurovirulence in mice. *J. Gen. Virol.* **70**(pt. 11), 3073–3078.

Taha, M. Y., Clements, G. B., and Brown, S. M. (1989b). A variant of herpes simplex virus type 2 strain HG52 with a 1.5 kb deletion in RL between 0 to 0.02 and 0.81 to 0.83 map units is non-neurovirulent for mice. *J. Gen. Virol.* **70**(pt. 3), 705–716.

Taneja, S., MacGregor, J., Markus, S., Ha, S., and Mohr, I. (2001). Enhanced antitumor efficacy of a herpes simplex virus mutant isolated by genetic selection in cancer cells. *Proc. Natl. Acad. Sci. USA* **98,** 8804–8808.

Thompson, R. L., Wagner, E. K., and Stevens, J. G. (1983). Physical location of a herpes simplex virus type-1 gene function(s) specifically associated with a 10 million-fold increase in HSV neurovirulence. *Virology* **131,** 180–192.

Thompson, R. L., Rogers, S. K., and Zerhusen, M. A. (1989). Herpes simplex virus neurovirulence and productive infection of neural cells is associated with a function which maps between 0.82 and 0.832 map units on the HSV genome. *Virology* **172,** 435–450.

Toda, M., Rabkin, S. D., Kojima, H., and Martuza, R. L. (1999). Herpes simplex virus as an in situ cancer vaccine for the induction of specific anti-tumor immunity. *Hum. Gene Ther.* **10,** 385–393.

Toda, M., Iizuka, Y., Kawase, T., Uyemura, K., and Kawakami, Y. (2002). Immuno-viral therapy of brain tumors by combination of viral therapy with cancer vaccination using a replication-conditional HSV. *Cancer Gene Ther.* **9,** 356–364.

Todo, T., Rabkin, S. D., Sundaresan, P., Wu, A., Meehan, K. R., Herscowitz, H. B., and Martuza, R. L. (1999). Systemic antitumor immunity in experimental brain tumor therapy using a multimutated, replication-competent herpes simplex virus. *Hum. Gene Ther.* **10,** 2741–2755.

Todo, T., Feigenbaum, F., Rabkin, S. D., Lakeman, F., Newsome, J. T., Johnson, P. A., Mitchell, E., Belliveau, D., Ostrove, J. M., and Martuza, R. L. (2000a). Viral shedding and biodistribution of G207, a multimutated, conditionally replicating herpes simplex virus type 1, after intracerebral inoculation in Aotus. *Mol. Ther.* **2,** 588–595.

Todo, T., Rabkin, S. D., and Martuza, R. L. (2000b). Evaluation of ganciclovir-mediated enhancement of the antitumoral effect in oncolytic, multimutated herpes simplex virus type 1 (G207) therapy of brain tumors. *Cancer Gene Ther.* **7,** 939–946.

Todo, T., Martuza, R. L., Dallman, M. J., and Rabkin, S. D. (2001a). In situ expression of soluble B7-1 in the context of oncolytic herpes simplex virus induces potent antitumor immunity. *Cancer Res.* **61,** 153–161.

Todo, T., Martuza, R. L., Rabkin, S. D., and Johnson, P. A. (2001b). Oncolytic herpes simplex virus vector with enhanced MHC class I presentation and tumor cell killing. *Proc. Natl. Acad. Sci. USA* **98,** 6396–6401.

Toyoizumi, T., Mick, R., Abbas, A. E., Kang, E. H., Kaiser, L. R., and Molnar-Kimber, K. L. (1999). Combined therapy with chemotherapeutic agents and herpes simplex virus type 1 ICP34.5 mutant (HSV-1716) in human non-small cell lung cancer. *Hum. Gene Ther.* **10,** 3013–3029.

Wong, R. J., Kim, S. H., Joe, J. K., Shah, J. P., Johnson, P. A., and Fong, Y. (2001a). Effective treatment of head and neck squamous cell carcinoma by an oncolytic herpes simplex virus. *J. Am. Coll. Surg.* **193,** 12–21.

Wong, R. J., Patel, S. G., Kim, S., DeMatteo, R. P., Malhotra, S., Bennett, J. J., St Louis, M., Shah, J. P., Johnson, P. A., and Fong, Y. (2001b). Cytokine gene transfer enhances herpes oncolytic therapy in murine squamous cell carcinoma. *Hum. Gene Ther.* **12,** 253–265.

Xu, G., and McLeod, H. L. (2001). Strategies for enzyme/prodrug cancer therapy. *Clin. Cancer Res.* **7,** 3314–3324.

# RECOMBINANT RETROVIRUS VECTORS FOR TREATMENT OF MALIGNANT BRAIN TUMORS

Nikolai G. Rainov

Department of Neurological Science
University of Liverpool
and The Walton Centre for Neurology and Neurosurgery NHS Trust
Liverpool L9 7LJ, United Kingdom

Christof M. Kramm

Department of Paediatrics
Heinrich Heine University
Duesseldorf, Germany



## I. Introduction

The susceptibility of malignant brain tumors of glial origin (gliomas) to virus vector-mediated gene therapy has been investigated only since the early 1990s. The initially favored and best explored gene therapy approach included insertion of a gene into tumor cells that renders these and their clonal progeny differentially sensitive to drug treatment (Moolten, 1986, Moolten and Wells, 1990). The gene/vector system most widely utilized until now is the herpes simplex virus thymidine kinase (*HSV-tk*) gene transferred by a replication-incompetent retrovirus (RV) vector that is released *in situ* by fibroblast-derived retrovirus vector-producing cells (RV-VPC). The early work on RV vectors initiated current developments of highly sophisticated third- and fourth-generation virus vectors for human gene therapy. Although at present there is an impressive variety of different gene therapy strategies for experimental treatment of brain tumors




0074-7742/03 $35.00

(Kramm *et al.*, 1995; Rainov and Kramm, 2001; Lam and Brakefield, 2001), their therapeutic success is still very limited and highly dependent on the qualities and limitations of the vector systems available.

## II. Properties of Retrovirus Vectors and Virus Producer Cells

### A. Retrovirus

Retrovirus particles carry two identical RNA viral genome molecules. Adhesion to the host cell is mediated by the RV envelope protein through interaction with cell surface glycoproteins (Coffin, 1990; Albritton *et al.*, 1993). The host range of RV is determined by the type of envelope protein and the presence of the respective receptor protein on the target cells (Coffin, 1990; Chu *et al.*, 1994; Uckert and Walther, 1994). The ability of the envelope protein to determine the host range of the virion has been used to design recombinant RV with tissue-specific infectivity (Federspiel *et al.*, 1994; Kasahara *et al.*, 1994; Valsesia-Wittmann *et al.*, 1994). After entering the cell, RV genomic RNA is transcribed by the retroviral reverse transcriptase (RT) to result in a double-stranded DNA copy of the retroviral genome. After DNA synthesis has occurred, the RV nucleoprotein complex has to enter the nucleus to gain access to the genome (Bowerman *et al.*, 1989). In the case of the Moloney murine leukemia virus (MMLV), the cell has to go through mitosis with the consequent disintegration of the nuclear membrane, so that the viral DNA–protein complex comes into contact with the host cell genome (Lewis and Emerman, 1994). Therefore, this type of virus is not able to infect nondividing cells and this holds true for most RV types, except for human immunodeficiency virus (HIV) (Lewis *et al.*, 1992). When the preintegration complex comes in contact with the cell genome, the viral DNA integrates in a process mediated by RV integrase and cellular DNA repair enzymes. The integration event does not require any specific cellular DNA sequences and therefore occurs at random (Varmus and Swanstrom, 1982). After integration has occurred, transcription of the RV genome commences, using the 5′ long terminal repeat (LTR) as a promoter, leading to the production of new infectious viral particles that are released from the host cell without lysing it. Most RV vectors presently used for gene therapy are derived from MMLV. Its proviral DNA is approximately 8.5 kb with two 635-bp LTR at both ends. Next to the 5′ LTR there is a sequence known as $\Psi$ (psi) that directs packaging of the viral genome RNA, as well as any other RNA bearing this sequence (Linial *et al.*, 1978). There are three coding regions in every RV genome. The *gag* region codes for the structural GAG polyprotein, the *pol* region is translated from the same transcript as the *gag* region, and the pol polyprotein is processed to give rise to RV enzymes protease, reverse transcriptase, and integrase. The *env* region encodes the envelope protein that confers host specificity to the virus (Coffin, 1990).

The provirus DNA of MMLV has been cloned into plasmids, which allows for relatively simple manipulation of the RV genome using recombinant DNA techniques (Chumakov *et al.*, 1982). Transfection of cells with plasmids carrying the intact provirus results in the production of wild-type RV. To generate recombinant RV vectors, most of the viral genome can be deleted, except for both LTRs and the packaging signal $\psi$, to accommodate up to 8 kb of foreign DNA. The deletion of all RV genes renders the vector replication deficient and therefore safe for use in humans (Danos and Mulligan, 1988).

Propagation of recombinant RV vectors requires cells capable of complementing viral functions that are absent. Although several strategies have been used to develop these so-called packaging cell lines or virus vector producer cells, all of them have in common the fact that they provide in *trans* the viral proteins essential for the production of functional RV particles, while keeping production of wild-type RV to a minimum (Danos and Mulligan, 1988; Markowitz *et al.*, 1988).

Different designs have been used to create recombinant RV vectors capable of strong and long-term sustained expression of transgenes. Along with the request for transgene-driving promoters with variable characteristics to suit different applications, the viral backbone has also been subjected to changes that resulted in better expression and function (Hantzopoulos *et al.*, 1989; Ghattas *et al.*, 1991; Chakraborty *et al.*, 1993; Zitvogel *et al.*, 1994).

Retrovirus-mediated gene transfer is particularly applicable to brain tumors because of the selectivity of RV for mitotic cells. In the brain, such are essentially only the cells in a rapidly growing tumor, the proliferating activated endothelial cells in the neoplastic capillaries, and some reactive astrocytes (Shapiro and Shapiro, 1998), and only these cell types can be affected by RV-mediated gene transfer and therapy (Oldfield *et al.*, 1993). When compared to other types of vectors, e.g., herpes simplex type 1 (HSV) or adenovirus (AV), RV vectors are more selective in terms of transgene delivery to brain tumor cells, and the transgene is inherited by all progeny cells of the infected tumor cells (Boviatsis *et al.*, 1994; Tamiya *et al.*, 1995). Retrovirus vectors also show low toxicity to normal brain tissue (Ram *et al.*, 1993a,b; Long *et al.*, 1998; Shand *et al.*, 1999). Despite these advantages, low titers (typically $1 \times 10^5$–$10^7$ infectious RV particles/ml) and instability of the virus particles have limited the use of RV vectors for gene delivery *in vivo* (Boris-Lawrie and Temin, 1994) until RV producer cell lines were developed and routinely used.

### B. Retrovirus-Producing Cell Lines

Most current packaging cell lines are derived from murine fibroblasts and are ecotropic or amphotropic, depending on which type of Env protein they express

(Mann *et al.*, 1983). In the most basic version of a packaging cell line, cells are stably transfected with RV sequences that are wild type except for a mutation in the packaging ($\psi$) sequence. This allows for expression of all RV proteins, but not for packaging of retrovirus RNA into particles. Following transfection and stable integration of a plasmid construct, vector RNA is packaged into particles by the $\psi$ sequence it bears (Miller, 1990). In this basic system there is a relatively high frequency of recombination between the necessary helper virus genome and recombinant vector sequences, and unwanted wild-type RV particles are generated.

Later generations of RV packaging cells feature a division of the helper virus genome into two or more segments bearing additional mutations and integrated at separate sites in the host cell genome. Multiple recombination events would be needed to reconstitute the wild-type RV genome and therefore the chance of generating wild-type virus is essentially zero (Danos and Mulligan, 1988; Markowitz *et al.*, 1988). Such helper-free RV packaging cell lines (RV-VPC) usually produce vectors with an amphotropic range, and are able to release infectious RV particle in relatively high titers (up to $1 \times 10^7$ cfu) in culture or after intratumoral grafting (Lyons *et al.*, 1995).

Intratumoral grafting of RV-VPC was proven in most studies to be more efficient than direct injection of RV suspension (Short *et al.*, 1990; Ram *et al.*, 1993a; Rainov *et al.*, 1996, Kaloss *et al.*, 1999). When VPC are implanted into a brain tumor, RV vectors are continuously generated and released *in situ*. Xenogeneic or allogeneic VPC, however, survive for only a relatively short time (up to 14 days) in the brain of immunocompetent hosts (Shand *et al.*, 1999; Long *et al.*, 1999; Rainov *et al.*, 2000a). Cellular immunity directed toward foreign major histocompatibility class (MHC) molecules and RV antigens is probably responsible for the elimination of VPC (Barba *et al.*, 1994; Rainov *et al.*, 2000a). Another disadvantage preventing sufficient transduction efficiency might be the negligible migration potential of VPC derived from murine fibroblasts. These commonly used VPC do not migrate away from the site of implantation and therefore are not able to spread RV vectors over a considerable area of tumor and tumor-infiltrated brain. Released RV vectors can diffuse only a few hundred micrometers in the interstitium of a highly cellular brain tumor (Rainov *et al.*, 1996; Ram *et al.*, 1997; Puumalainen *et al.*, 1998). Thus, despite continuous vector production, the transduction efficiency is spatially limited and the gene delivery rate is low (Boris-Lawrie and Temin, 1994). Another major limitation of RV transduction efficiency is the inability of most recombinant RV vectors to replicate, which on the other hand is a safety feature and therefore deemed essential in human clinical trials (Long *et al.*, 1998, 1999; Shand *et al.*, 1999; Rainov *et al.*, 2000b). Each recombinant RV particle can infect only one tumor cell and does not generate progeny capable of secondary spread and infection of neighboring cells, unless wild-type RV is also present. Recently, a novel approach was studied that may allow for combining the biosafety of nonreplicating recombinant RV with the high transduction rate of herpes simplex virus type 1 (HSV)-based amplicon vectors. Sena-Esteves *et al.* (2002) cloned RV sequences into minimal HSV/AAV hybrid amplicons. Helper virus-free amplicon vectors were used to coinfect glioma cells in culture and generated a stable RV packaging population of tumor cells that produced retrovirus titers of approx. $1 \times 10^5$ cfu/ml. The authors suggested that this RV/amplicon vector system may facilitate the generation of a new type of RV-VPC. Infection by amplicons of cells with migratory properties such as glioma cells could result in secondary expansion and better spatial distribution of RV in an *in vivo* brain tumor environment (Sena-Esteves *et al.*, 2000).

## III. RV-Mediated Brain Tumor Therapy in Animal Models

In some early animal studies, Short *et al.* (1990) and Ezzeddine *et al.* (1991) infected C6 rodent glioma cells with RV to make them express the bacterial marker gene *lacZ* or the HSV-TK enzyme, respectively (Table I). Growth of subcutaneous C6 transgene-expressing cells in nude mice was completely blocked when tumor-bearing animals were treated with ganciclovir (GCV) (Ezzeddine *et al.*, 1991). Takamiya *et al.* (1992) used a C6 glioma cell line constitutively expressing the *lacZ* reporter gene to demonstrate strong inhibition of cell growth when co-cultured with HSV-TK expressing C6 cells in GCV-containing medium. These authors suggested that direct cell contact with passage of toxic metabolites, possibly through gap junctions, was necessary for this bystander effect. The importance of the direct bystander effect was confirmed independently by other investigators (Bi *et al.*, 1993; Freeman *et al.*, 1993; Ishii-Morita *et al.*, 1997; Burrows *et al.*, 2002). In another study, an RV-VPC line bearing the *HSV-tk* gene was implanted subcutaneously in nude mice mixed 1:1 with native C6 glioma cells (Takamiya *et al.*, 1993). The resulting tumors regressed under GCV treatment, while tumors cografted with control cells did not. Other groups demonstrated growth inhibition of 9L gliosarcoma in rat brain by intratumoral implantation of RV-VPC and GCV treatment (Ram *et al.*, 1993a; Rainov *et al.*, 1996). Because these studies used immunocompetent animals, the combination of HSV-TK/GCV therapy coupled to an immune response to mouse fibroblast antigens, retroviral glycoproteins displayed on the surface of the intratumoral RV-VPC, and possibly to HSV-TK itself may have contributed to tumor regression. Other studies suggested indeed that a systemic immune response against the tumor itself may be elicited by RV-VPC-mediated destruction of tumor cells (Barba *et al.*, 1994; Tapscott *et al.*, 1994; Vile *et al.*, 1997; Mullen *et al.*, 1998; Rainov *et al.*, 2000a).

Table I
ANIMAL MODELS OF RV-VPC GENE THERAPY[a]

| Authors | Year | Animal model | Tumor cell line | Main transgene system | Presence of antitumor effect | Additional treatment |
|---|---|---|---|---|---|---|
| Short *et al.* | 1990 | Rat | C6 | lacZ | N/A | — |
| Ezzeddine *et al.* | 1991 | Nude mouse | C6 | TK/GCV | Yes | — |
| Culver *et al.* | 1992 | Rat | 9L | TK/GCV | Yes | — |
| Takamiya *et al.* | 1992 | Nude mouse | C6 | TK/GCV | Yes | — |
| Takamiya *et al.* | 1993 | Nude mouse | C6 | TK/GCV | Yes | Wild-type RV |
| Ram *et al.* | 1993a | Rat | 9L | TK/GCV | Yes | — |
| Ram *et al.* | 1994 | Rat | 9L | TK/GCV | Yes | IL-2 |
| Barba *et al.* | 1994 | Rat | 9L | TK/GCV | Yes | — |
| Tamiya *et al.* | 1995 | Mouse | C6 | TK/GCV | N/A | — |
| Izquierdo *et al.* | 1995 | Rat | C6 | TK/GCV | Yes | — |
| Rainov *et al.* | 1996 | Rat | 9L | TK/GCV | Yes | — |
| Manome *et al.* | 1996 | Rat | 9L | CYP2B1/CPA | Yes | — |
| Benedetti *et al.* | 1997 | Nude mouse | U87, C6 | TK/GCV | Yes | IL-4 |
| Kruse *et al.* | 1997 | Rat | 9L | TK/GCV | Yes | — |
| Ono *et al.* | 1997 | Nude mouse | C6 | GPT/6-TX | Yes | — |
| Izquierdo *et al.* | 1997 | Rat | C6 | TK/GCV | Yes | — |
| Sandmair *et al.* | 1999 | Rat | BT4C | TK/GCV | | — |
| Sasaki *et al.* | 1999 | Rat | 9L | Antisense VEGF | Yes | — |
| Saleh *et al.* | 1999 | Rat, nude mouse | C6 | IL4 | Yes | None |

[a]Note that most of the studies use rodent glioma cells (C6, 9L, BT4C) and are able to demonstrate a significant antitumor effect of the treatment. TK/GCV, HSV1 thymidine kinase and ganciclovir; CYP2B1/CPA, cytochrome 2B1 and cyclophosphamide; GPT/6-TX, guaninephosphotransferase and 6-thioxanthine; VEGF, vascular endothelial growth factor; N/A, not available.

In further animal studies of similar design (Table I), the differences in survival rates seemed to reflect the different numbers of RV-VPC injected into the tumors. Implantation of RV-VPC in an amount representing a 50- to 125-fold of the initial tumor cell number resulted in significantly increased long-term survival (Culver *et al.*, 1992; Ram *et al.*, 1993a), but would not be realistic in human glioma patients. In other studies (Barba *et al.*, 1993; Rainov *et al.*, 1996), RV-VPC carrying the *HSV-tk* gene were grafted at much lower ratios to glioma cells (1:1 to 5:1). Survival in the latter studies was significantly prolonged in treated animals compared to controls, although the effect was not as dramatic as noted with the former high ratios of RV-VPC to tumor cells. In respect to clinical trials, a ratio of 1:1 may be achievable in patients with brain tumors if RV-VPC are implanted in the walls of the tumor resection cavity after surgical removal of the main tumor mass (Shand *et al.*, 1999; Rainov *et al.*, 2000b). More recent animal studies shifted their focus toward the testing of new generations of RV vectors. Lee *et al.* (2001) used an MMLV-based RV vector pseudotyped with a vesicular stomatitis virus G protein. The transduction efficiency of the pseudotyped RV was significantly higher in brain tumor cells compared with that of other retroviruses, such as PA317 or PG13 derived. An average of 10% gene expression was routinely obtained in brain tumors when pseudotyped RV was concentrated by ultracentrifugation and directly administrated to preestablished tumors in mice or rats. Tamura *et al.* (2001) also concentrated RV in VPC-free suspension and used the free RV particles to transduce the *HSV-tk* gene into a brain tumor in a mouse model. The transduction efficiency achieved with RV suspension was nearly as high as that observed after implantation of high-titer RV-VPC. Eighty percent of tumor-bearing mice were completely cured after administration of GCV (Tamura *et al.*, 2001). On the other hand, Sandmair *et al.* (1999) injected PA317 RV-VPC in a syngeneic rat BT4C rat glioma model and were unable to achieve prolonged survival in treated animals. These authors attributed the lack of efficacy to low gene transfer efficiency and limited life span of the RV-VPC. The same group continued to study factors limiting the efficacy of RV-VPC gene therapy in experimental brain tumors. Intracerebral tumors in syngeneic rats were produced by mixing stably transfected HSV-TK expressing BT4C rat glioma cells with parental tumor cells in various ratios. After 14 days of GCV treatment, tumors containing 10% or more HSV-TK-expressing cells showed a significant reduction in tumor size and a prolonged survival time. The authors concluded that at least a 10% RV transduction rate and even dispersion of expressing tumor cells throughout a large glioma are necessary for achieving a significant antitumor effect (Sandmair *et al.*, 2000b).

## IV. Clinical Studies with RV Vectors

### A. Early Clinical Studies

The first clinical study with stereotactic intratumoral inoculation of RV-VPC in patients with recurrent malignant brain tumors achieved some promising results (Ram *et al.*, 1997), although it was not a controlled and randomized protocol, and the end points were toxicity and biosafety of RV rather then tumor response (Table II). In two subsequent studies in patients with recurrent

TABLE II
CLINICAL TRIALS WITH RV-VPC GENE THERAPY[a]

| Authors | Year | Tumor type | Number of patients | Study phase | Significant tumor response |
|---|---|---|---|---|---|
| Izquierdo *et al.* | 1996 | Recurrent GBM | 5 | I | N/A |
| Izquierdo *et al.* | 1997 | Recurrent GBM | 2 | I | Yes |
| Colombo *et al.*, | 1997 | Recurrent GBM | 4 | I | N/A |
| Ram *et al.* | 1997 | Recurrent GBM | 15 | I/II | Yes |
| Puumalainen *et al.* | 1998 | Recurrent GBM | 10 | I | N/A |
| Klatzman *et al.* | 1998 | Recurrent GBM | 12 | I | Yes |
| Palu *et al.* | 1999 | Recurrent GBM | 4 | I | N/A |
| Shand *et al.* | 1999 | Recurrent GBM | 48 | II | Yes |
| Packer *et al.* | 2000 | Recurrent GBM | 12 | I | N/A |
| Rainov *et al.* | 2000b | Primary GBM | 248 | III | None |
| Harsh *et al.* | 2000 | Recurrent GBM | 5 | I | N/A |
| Sandmair *et al.* | 2000a | Primary GBM | 7 | I | None |
| Kramm *et al.* | 2002 | Recurrent ependymoma | 1 | Case study | N/A |
| Valery *et al.* | 2002 | Recurrent GBM | 1 | Case study | Yes |

[a]Note that the majority of studies are in patients with recurrent glioblastoma multiforme and that most trials are prospective uncontrolled phase I or II protocols. Tumor response is therefore a mostly subjective assessment not based on statistical comparison with a control group. GBM, glioblastoma multiforme; N/A, not available.

glioblastoma multiforme (GBM), HSV-TK/GCV gene therapy was performed by locally administered RV-VPC inoculated manually during open surgical resection of tumor (Klatzmann *et al.*, 1998; Shand *et al.*, 1999). In both studies, there were individual cases of extended recurrence-free periods, including one patient each with apparently arrested tumor progression for 2 years or longer. The authors of both reports independently concluded that the RV-VPC gene therapy strategy appeared to be adequately safe and effective to allow further clinical studies. The same conclusions were drawn by a separate group of investigators studying RV-VPC application in a total of seven patients with recurrent GBM (Izquierdo *et al.*, 1996, 1997). A phase I study using intraoperative RV-VPC administration was carried out in pediatric cases with recurrent malignant glioma (Packer *et al.*, 2000). Because there was no control group included in the study, it was impossible to provide evidence for a significant gene therapy-mediated antitumor effect, although Packer *et al.* (2000) and later Kramm *et al.* (2002) described single cases of long-term survivors after gene therapy. A common and severe drawback of all of the above clinical protocols was the fact that the overall transduction rate of tumor cells was very low, which was also confirmed in pilot studies in human GBM employing marker genes such as *lácz* (Puumalainen *et al.*, 1998). To overcome this limitation and to achieve a significant effect against tumors *in vivo*, Palu *et al.* (1999) used RV vectors in a combined HSV-TK and cytokine [interleukin (IL)-2] gene therapy approach, speculating that IL-2 transgene expression even in a few tumor cells may elicit a systemic antitumor response and could result in major antitumor toxicity mediated by the host cellular immune response. The study was not controlled and therefore it proved impossible to provide reliable data for the improved antitumor effect of this approach.

## B. Advanced Clinical Studies

An evaluation of the efficacy of RV-VPC gene therapy would be possible only by means of a large prospective, randomized, and controlled trial, and therefore such a trial has been planned since the mid-1990s, after RV-VPC were administered to human patients without serious side effects and with some alleged efficacy (Ram *et al.*, 1997; Shand *et al.*, 1999). In a phase III clinical study, RV-VPC-mediated gene therapy was used as an adjuvant to the standard therapy of maximum surgical resection and irradiation for newly diagnosed, previously untreated GBM (Rainov *et al.*, 2000b). Primary tumors were expected to respond better to gene therapy than most recurrent tumors investigated in previous studies. The treatment groups were well balanced with respect to baseline demographic characteristics, confirmed diagnosis, and recognized prognostic factors, allowing a valid comparison. Radiotherapy and gene therapy (GCV administration) overlapped partly in the protocol, and there were good reasons for this overlapping at the time the study protocol was drafted. On the one hand, the desire was to modify as little as possible the established standard algorhithm for treatment of GBM; on the other hand, there were experimental studies to suggest that HSV-TK/GCV gene therapy acts synergistically with radiation to enhance tumor cell killing (Boothman *et al.*, 1994; Kim *et al.*, 1994). After 4 years of clinical and laboratory follow-up of the 248 patients with primary GBM divided in a gene therapy arm and a control arm, survival analysis was not able to demonstrate that the addition of gene therapy provided any advantage in terms of time to tumor progression or overall survival time (Rainov *et al.*, 2000b).

### 1. *Biosafety Issues*

Even without proven benefit to GBM patients, gene therapy itself, e.g., grafting of RV-VPC in the walls of the tumor resection cavity, was not associated with major side effects or survival disadvantages. No increased incidence of postoperative neurological complications was noted, and no acute inflammation or meningitis was observed in any of the gene therapy cases. There also was no evidence for severe neurological side effects at the time of GCV administration, when the local gene therapy-induced cell destruction should occur. These safety findings compared favorably with those of previous studies using the same gene therapy paradigm but in which the safety assessment suffered from lack of controls (Ram *et al.*, 1997; Klatzman *et al.*, 1998; Shand *et al.*, 1999; Rainov *et al.*, 2000b).

Packer *et al.* (2000) treated 12 children with recurrent brain tumors and reported transient adverse events after RV-VPC injection in four patients (seizures and signs of elevated intracerebral pressure). In three of these patients, the adverse events occurred during GCV administration and were considered to be possibly due to the gene therapy. Previous phase I/II trials in adult patients with recurrent GBM showed, however, that the treatment is well tolerated. There were only two adverse events (seizures, hydrocephalus) possibly due to the gene therapy in a series of 48 adults (Shand *et al.*, 1999) and none in another series of 12 adults (Klatzman *et al.*, 1998). Harsh *et al.* (2000) administered VPC by stereotactic biopsy in five patients with recurrent GBM, resecting the tumors 5 days later while reinjecting the tumor bed with RV-VPC. These authors described no gene therapy-related adverse events, although one patient developed a lethal brain abscess after open surgery. The published results with murine RV-VPC seem to confirm the hypothesis that injection of foreign cells in the tumor margin and its surrounding, mostly normal brain tissue, does not cause significant clinical morbidity within the follow-up time frame (1–2 years). Compared with chemotherapy for malignant brain tumors, HSV-TK/GCV gene therapy appears to offer a better clinical safety profile as regards the incidence of adverse events with chemotherapy (Kramer *et al.*, 1997; Prados and Russo, 1998).

Biosafety monitoring is an important part of every clinical study with virus vectors. So far, follow-up assays have been able to exclude any gain of replication competence through recombination of the vector with wild-type RV. The presence of transduced blood cells in the systemic blood circulation in a small number of patients presents a mostly theoretical risk and should therefore not be considered an unacceptable safety risk in the treatment of an otherwise rapidly fatal cancer (Long *et al.*, 1998, 1999; Shand *et al.*, 1999). Absence of a humoral response to the vector core protein in a phase III clinical study (Rainov *et al.*, 2000b) suggests that the RV vector itself is not entering the systemic circulation, although antibodies to RV core protein have been shown previously in a small number of patients, particularly in those receiving multiple administrations of RV-VPC suspension (Ram *et al.*, 1997; Long *et al.*, 1998).

2. *Transduction Efficiency of RV Vectors*

The principal requirement for the success of virus-mediated gene therapy with replication-incompetent RV vectors seems to be the achievement of adequate transduction levels of a functional transgene in the target tumor cells. It has been shown that because of the direct pharmacological bystander effect and the indirect systemic antitumor immune response, transduction of all existing tumor cells is not a prerequisite for successful tumor ablation (Freeman *et al.*, 1993). Despite this fact, threshold transduction levels certainly do exist and are evident at least in animal experiments. An animal study using RV-VPC had been successful in a murine brain tumor model at minimum transduction levels of 10% (Lyons *et al.*, 1995), and the threshold figure of 10% has been observed by other investigators in similar experiments (Culver *et al.*, 1992; Ram *et al.*, 1993a; Freeman *et al.*, 1993; Sandmair *et al.*, 2000b).

Despite considerable numbers of brain tumor patients receiving RV-VPC gene therapy (Table II), surprisingly little is known about the tumor transduction rate *in vivo*, except that it is undoubtedly very low. Harsh *et al.* (2000) investigated brain tumor tissue by quantitation of transgene (*neo*) levels in seven patients, including a patient autopsied only 3 weeks after VPC injection. These authors reported consistent tumor transduction rates below 0.002% in all investigated cases. Stereotactic injection of the RV-VPC was carried out 5 days before open surgical resection of the tumor. There was no evidence for production of transgene HSV-TK protein in resected tumor tissue, as investigated by immunohistochemistry with an anti-HSV-TK antibody. Elevated HSV-TK enzyme activity was noted in RV-VPC-injected tumor samples from two patients, as measured by a more sensitive radioactive kinase assay employing $^3$H-labeled GCV. The investigators considered this, however, to be more likely to originate from VPC than from transduced tumor cells, because of the greater density of VPC in the assayed area as demonstrated by immunohistochemistry. In any case, HSV-*tk* activity in human GBM samples was rather low compared with transduction rates reported in successful control animal experiments (Harsh *et al.*, 2000).

In a previous series that included 51 post-VPC brain tumor biopsy or autopsy samples, 39 of which were proven positive for vector DNA by PCR, the highest transduction level (*neo*-quantitation) found in two patients was 2.6% (Long *et al.*, 1999). Most of the tumor specimens contained less than 0.03% transduced tumor cells. Other investigators have reported transgene expression for up to 7 days after intratumoral RV-VPC injection, but were unable to differentially describe percentages of transduced tumor cells (Ram *et al.*, 1997; Puumalainen *et al.*, 1998; Rainov *et al.*, 2000b).

In a gene marking study in human GBM patients, Puumalainen *et al.* (1998) injected RV or adenovirus (AV) vectors in suspension in 10 patients with malignant glioma via a catheter inserted into the tumor. RV was repeatedly injected on 3 consecutive days into the tumor, followed by its resection a few days later. Gene transfer efficiency varied between <0.01 and 4% with RV, and between <0.01 and 11% with AV. The transgene activity with both virus vectors was not evenly distributed throughout the tumor. Both glioma cells and endothelium in the neoplastic blood vessels were transduced with RV, but not with AV (Puumalainen *et al.*, 1998).

3. *Importance of Vector Delivery Modes to RV-VPC Gene Therapy*

Efficient gene delivery is the first and most important requirement for successful treatment of brain tumors using the virus vector paradigm. All previous clinical protocols called for direct manual or stereotactic injection of small volumes of RV-VPC or RV particle suspension into the walls of the tumor resection cavity or into the tumor mass (Izquierdo *et al.*, 1996; Klatzman *et al.*, 1998; Shand *et al.*, 1999). Besides the technically virtually impossible task of standardizing distances between injection tracks and applied volumes, reflux of vector suspension from needle tracks was invariably encountered with the manual technique (Rainov and Kramm, 2001). The actual volumes delivered to brain parenchyma were therefore presumably lower than those injected. Lack of dissemination of vectors away from the injection sites has also been noted (Ram *et al.*, 1997; Puumalainen *et al.*, 1998; Sandmair *et al.*, 2000a), confirming previous observations from animal studies that murine VPC are nonmigrating and remain mostly at the injection site (Ram *et al.*, 1993a; Rainov *et al.*, 1996). Also there is only a relatively small fraction of synchronously mitotic tumor cells in a large human glioma (Shapiro and Shapiro, 1998), and RV supply or production by VPC cannot be maintained for a period of time that is long enough to ensure infection of the major part of the tumor. Rapid inactivation and destruction of free RV vectors by complement (Russell *et al.*, 1995) are other negative factors limiting vector spread and tumor cell transduction.

4. *Lessons Learned from Clinical Trials with RV Vectors*

The greatest shortcoming of all RV-VPC-based clinical protocols in malignant brain tumors seems to be low efficiency of *in vivo* gene transfer to tumor cells. Our present inability to noninvasively measure transduction levels or functional transgene activity prevents a meaningful correlation of transduction rates with vector delivery techniques and with the neuroradiological and clinical changes after administration of a prodrug such as GCV. To quantify vector-mediated tumor transduction, minimally invasive gene transfer imaging must be included in future protocols, and there are already some experimental studies to suggest that positron-emission tomography (PET) imaging and quantitation of HSV-TK metabolism are feasible *in vivo* (Blasberg and Tjuvajev, 1999; Gambhir *et al.*, 2000). Quantitative or semiquantitative imaging of transgene function in areas surrounding resected GBM would ensure that the expected enzymatic activity is indeed present, and that there is a sufficient amount of vector distributed throughout areas neighboring the resection cavity and surrounding brain tissue. Further optimization of RV vectors and RV-VPC may also be essential for improvement of clinical efficacy. Recent developments such as hybrid vectors, e.g., based on HSV amplicons and retaining RV features, may dramatically increase vector availability in a large size human tumor and allow for controlled replication of RV vector in tumor cells and therefore for a secondary spread of infection throughout the tumor (Sena-Esteves *et al.*, 2002).

Combination of gene therapy with currently existing and established antineoplastic treatments, such as chemotherapy or radiation, may considerably increase the tumor-selective toxicity without potentiation of side effects or unwanted toxicity to normal cells (Rainov *et al.*, 2001). Furthermore, diffusion of prodrugs such as GCV into a brain tumor may vary according to the permeability of the tumor and the peritumoral tissue. Administration of GCV along a blood–tumor barrier modifier such as the selective B2 bradykinin agonist labradimil (Cereport, Alkermes, Inc., Cambridge, MA) could result in a significantly higher intra- and peritumoral GCV concentration and subsequent elimination of persisting transduced cells (Inamura *et al.*, 1994; Lemay *et al.*, 1998).

The current mode of administration of RV-VPC and their nonmigratory fibroblast phenotype limit the distribution of these cells to the immediate vicinity of the needle tract used for grafting. The employment of replication-deficient RV vectors further narrows the range of local gene therapy actions. Use of highly migratory VPC, which are able to "track" foci and even single tumor cells invading the surrounding normal brain (Herrlinger *et al.*, 2000), may solve the spatial distribution problem. Furthermore, the present manual application mode of RV-VPC may be improved by employing automated slow-speed infusion devices (Lieberman *et al.*, 1995), as it is virtually impossible to standardize manual intraparenchymal injection of small volumes of RV-VPC for equal spacing between injection tracks and for minimum reflux of the suspension.

Investigators have also claimed that the RV-based HSV-TK/GCV approach is accompanied by a systemic immune response against the RV vector, the grafted VPC, and even against autologous tumor cells, which involves T cells and natural killer (NK) cells (Barba *et al.*, 1994; Vile *et al.*, 1997; Mullen *et al.*, 1998; Rainov *et al.*, 2000a; Parsa *et al.*, 2001). This immunological effect could be enhanced through transfer of additional genes coding for immunostimulatory cytokines such as IL-2 (Palu *et al.*, 1999) or IL-12 (DiMeco *et al.*, 2000), and RV gene therapy approaches may indeed benefit from combining various concepts such as HSV-TK/GCV or other toxic gene/prodrug systems with simultaneous intratumoral cytokine gene transfer.

## V. Conclusions

Gene therapy is an exciting and highly sophisticated experimental paradigm for treatment of brain tumors. Despite more than a decade of experience from animal studies and clinical trials, it still seems to be at its very early stages of development. RV have been one of the earliest vectors used for gene transfer in the brain and employed as a tool for targeted and selective destruction of tumor cells. These vectors seem to be well tolerated and exceptionally biologically safe, although apparently not able to significantly improve survival of the patients in combined treatment protocols. RV clinical trials in glioma have nevertheless produced a substantial amount of data and have greatly contributed toward identification of the most serious shortcomings of the nonreplicative local-intratumoral virus vector gene therapy strategy. Moreover, these trials are instrumental in designing novel types of therapeutic virus vector systems and searching for more efficient modes of vector delivery.

## References

Albritton, L. M., Kim, J. W., Tseng, L., and Cunningham, J. M. (1993). Envelope-binding domain in the cationic amino acid transporter determines the host range of ecotropic murine retroviruses. *J. Virol.* **67,** 2091–2096.

Barba, D., Hardin, J., Ray, J., and Gage, F. H. (1993). Thymidine kinase-mediated killing of rat brain tumors. *J. Neurosurg.* **79,** 729–735.

Barba, D., Hardin, J., Sadelain, M., and Gage, F. H. (1994). Development of anti-tumor immunity following thymidine kinase-mediated killing of experimental brain tumors. *Proc. Natl. Acad. Sci. USA* **91,** 4348–4352.

Benedetti, S., Dimeco, F., Pollo, B., Cirenei, N., Colombo, B. M., Bruzzone, M. G., Cattaneo, E., Vescovi, A., Didonato, S., Colombo, M. P., and Finocchiaro, G. (1997). Limited efficacy of the HSV-TK/GCV system for gene therapy of malignant gliomas and perspectives for the combined transduction of the interleukin-4 gene. *Hum. Gene Ther.* **8,** 1345–1353.

Bi, W. L., Parysek, L. M., Warnick, R., and Stambrook, P. J. (1993). In vitro evidence that metabolic cooperation is responsible for the bystander effect observed with HSV-tk retroviral gene therapy. *Hum. Gene Ther.* **4,** 725–731.

Blasberg, R. G., and Tjuvajev, J. G. (1999). Herpes simplex virus thymidine kinase as a marker/reporter gene for PET imaging of gene therapy. *Q. J. Nucl. Med.* **43,** 163–169.

Boothman, D. A., Davis, T. W., and Sahijdak, W. M. (1994). Enhanced expression of thymidine kinase in human cells following ionizing radiation. *Int. J. Radiat. Oncol. Biol. Phys.* **30,** 391–398.

Boris-Lawrie, K., and Temin, H. M. (1994). The retroviral vector. Replication cycle and safety considerations for retrovirus-mediated gene therapy. *Ann. N.Y. Acad. Sci.* **716,** 59–70.

Boviatsis, E. J., Chase, M., Wei, M., Tamiya, T., Hurford, R. K., Jr., Kowall, N. W., Tepper, R. I., Breakefield, X. O., and Chiocca, E. A. (1994). Gene transfer into experimental brain tumors mediated by adenovirus, herpes-simplex-virus (HSV), and retrovirus vectors. *Hum. Gene Ther.* **5,** 183–191.

Bowerman, B., Brown, P. O., Bishop, J. M., and Varmus, H. E. (1989). A nucleoprotein complex mediates the integration of retroviral DNA. *Gen. Dev.* **3,** 469–478.

Burrows, F. J., Gore, M., Smiley, W. R., Kanemitsu, M. Y., Jolly, D. J., Read, S. B., Nicholas, T., and Kruse, C. A. (2002). Purified herpes simplex virus thymidine kinase retroviral particles: III. Characterization of bystander killing mechanisms in transfected tumor cells. *Cancer Gene Ther.* **9,** 87–95.

Chakraborty, A. K., Zink, M. A., Boman, B. M., and Hodgson, C. P. (1993). Synthetic retrotransposon vectors for gene therapy. *FASEB J.* **7,** 971–977.

Chu, T. H., Martinez, I., Sheay, W., and Dornburg, R. (1994). Cell targeting with retroviral vector particles containing antibody-envelope fusion proteins. *Gene Ther.* **1,** 292–299.

Chumakov, I., Stuhlmann, H., Harbers, K., and Jaenisch, R. (1982). Cloning of two genetically transmitted moloney leukemia proviral genomes: Correlation between biological activity of the cloned DNA and viral genome activation in the animal. *J. Virol.* **42,** 1088–1098.

Coffin, J. M. (1990). Retroviridae and their replication. *In* "Virology" (J. M. Coffin, Ed.), pp. 1437–1500. Raven Press, New York.

Colombo, F., Zanusso, M., Casentini, L., Cavaggioni, A., Franchin, E., Calvi, P., and Palu, G. (1997). Gene stereotactic neurosurgery for recurrent malignant gliomas. *Stereotact. Funct. Neurosurg.* **68,** 245–251.

Culver, K. W., Ram, Z., Wallbridge, S., Ishii, H., Oldfield, E. H., and Blaese, R. M. (1992). In vivo gene transfer with retroviral vector-producer cells for treatment of experimental brain tumors. *Science* **256,** 1550–1552.

Danos, O., and Mulligan, R. C. (1988). Safe and efficient generation of recombinant retroviruses with amphotropic and ecotropic host ranges. *Proc. Natl. Acad. Sci. USA* **85,** 6460–6464.

Dimeco, F., Rhines, L. D., Hanes, J., Tyler, B. M., Brat, D., Torchiana, E., Guarnieri, M., Colombo, M. P., Pardoll, D. M., Finocchiaro, G., Brem, H., and Olivi, A. (2000). Paracrine delivery of IL-12 against intracranial 9L gliosarcoma in rats. *Neurosurgery* **92,** 419–427.

Ezzeddine, Z. D., Martuza, R. L., Platika, D., Short, M. P., Malick, A., Choi, B., and Breakefield, X. O. (1991). Selective killing of glioma cells in culture and in vivo by retrovirus transfer of the herpes simplex virus thymidine kinase gene. *New Biol.* **3,** 608–614.

Federspiel, M. J., Bates, P., Young, J. A. T., and Varmus, H. E. (1994). A system for tissue-specific gene targeting: Transgenic mice susceptible to subgroup A avian leukosis virus-based retroviral vectors. *Proc. Natl. Acad. Sci. USA* **91,** 11241–11245.

Freeman, S. M., Abboud, C., Whartenby, K., Packman, C., Koeplin, D., Moolten, F., and Abraham, G. (1993). The "bystander effect": Tumor regression when a fraction of the tumor mass is genetically modified. *Cancer Res.* **53,** 5274–5283.

Gambhir, S. S., Herschman, H. R., Cherry, S. R., Barrio, J. R., Satyamurthy, N., Toyokuni, T., Phelps, M. E., Larson, S. M., Balatoni, J., Finn, R., Sadelain, M., Tjuvajev, J., and Blasberg, R. (2000). Imaging transgene expression with radionuclide imaging technologies. *Neoplasia* **2,** 118–138.

Ghattas, I. R., Sanes, J. R., and Majors, J. E. (1991). The encephalomyocarditis virus internal ribosome entry site allows efficient coexpression of two genes from a recombinant provirus in cultured cells and embryos. *Mol. Cell Biol.* **11,** 5848–5859.

Hantzopoulos, P. A., Sullenger, B. A., Ungers, G., and Gilboa, E. (1989). Improved gene expression upon transfer of the adenosine deaminase minigene outside the transcriptional unit of a retrovirus vector. *Proc. Natl. Acad. Sci. USA* **86,** 3519–3523.

Harsh, G. R., Deisboeck, T. S., Louis, D. N., Colvin, M., Silver, J. S., Qureshi, N. H., Kracher, J., Finkelstein, D., Chiocca, E. A., and Hochberg, F. H. (2000). Thymidine Kinase activation of ganciclovir in recurrent glimas: a gene-marking and neuropathological study. *J. Neurosurg.* **92,** 804–811.

Herrlinger, U., Woiciechowski, C., Sena-Esteves, M., Aboody, K. S., Jacobs, A. H., Rainov, N. G., Snyder, E., and Breakefield, X. O. (2000). Neural precursor cells for delivery of replication-conditional HSV-1 vectors to intracerebral gliomas. *Mol. Ther.* **1,** 347–357.

Inamura, T., Nomura, T., Bartus, R. T., and Black, K. L. (1994). Intracarotid infusion of RMP-7, a bradykinin analog: A method for selective drug delivery to brain tumors. *J. Neurosurg.* **81,** 752–758.

Ishii-Morita, H., Agbaria, R., Mullen, C. A., Hirano, H., Koeplin, D. A., Ram, Z., Oldfield, E. H., Johns, D. G., and Blaese, R. M. (1997). Mechanism of 'bystander effect' killing in the herpes simplex thymidine kinase gene therapy model of cancer. *Gene Ther.* **4,** 244–251.

Izquierdo, M., Cortes, M., De Felipe, P., Martin, V., Diez-Guerra, J., Talavera, A., and Perez-Higueras, A. (1995). Long-term rat survival after malignant brain tumour regression by retroviral gene therapy. *Gene Ther.* **2,** 66–69.

Izquierdo, M., Martin, V., de Felipe, P., Izquierdo, J. M., Perez-Higueras, A., Cortes, M. L., Paz, J. F., Isla, A., and Blazquez, M. G. (1996). Human malignant brain tumor response to herpes simplex thymidine kinase (HSVtk)/ganciclovir gene therapy. *Gene Ther.* **3,** 491–495.

Izquierdo, M., Cortes, M. L., Martin, V., de Felipe, P., Izquierdo, J. M., Perez-Higueras, A., Paz, J. F., Isla, A., and Blazquez, M. G. (1997). Gene therapy in brain tumours: Implications of the size of glioblastoma on its curability. *Acta Neurochir. Suppl.* **68,** 111–117.

Kaloss, M., Linscott, M., Wey, C., Lu, P., Long, Z., McGarrity, G. J., Otto, E., and Lyons, R. M. (1999). Distribution of retroviral vectors and vector producer cells using two routes of administration in rats. *Gene Ther.* **6,** 1389–1396.

Kasahara, N., Dozy, A. M., and Kan, Y. W. (1994). Tissue-specific targeting of retroviral vectors through ligand-receptor interactions. *Science* **266,** 1373–1376.

Kim, J. H., Kim, S. H., Brown, S. L., and Freytag, S. O. (1994). Selective enhancement by an antiviral agent of the radiation-induced cell killing of human glioma cells transduced with HSV-tk gene. *Cancer Res.* **54,** 6053–6056.

Klatzmann, D., Valery, C. A., Bensimon, G., Marro, B., Boyer, O., Mokhtari, K., Diquet, B., Salzmann, J. L., and Philippon, J. (1998). A phase I/II study of herpes simplex virus type 1 thymidine kinase "suicide" gene therapy for recurrent glioblastoma. *Hum. Gene Ther.* **9,** 2595–2604.

Kramer, E. D., Packer, R. J., Ginsberg, J., Goldman, S., Thompson, S., Bayer, L. A., Shen, V., Harris, R., Khan, S., and Finlay, J. L. (1997). Acute neurologic dysfunction associated with high-dose chemotherapy and autologous bone marrow rescue for primary malignant brain tumors. *Pediatr. Neurosurg.* **27,** 230–237.

Kramm, C. M., Sena-Esteves, M., Barnett, F. H., Rainov, N. G., Schuback, D. E., Yu, J. S., Pechan, P. A., Paulus, W., Chiocca, E. A., and Breakefield, X. O. (1995). Gene therapy for brain tumors. *Brain Pathol.* **5,** 345–381.

Kramm, C. M., Korholz, D., Rainov, N. G., Niehues, T., Fischer, U., Steffens, S., Frank, S., Banning, U., Horneff, G., Schroten, H., and Burdach, S. (2002). Systemic activation of the immune system during ganciclovir treatment following intratumoral herpes simplex virus type 1 thymidine kinase gene transfer in an adolescent ependymoma patient. *Neuropediatrics* **33,** 6–9.

Lam, P. Y., and Breakefield, X. O. (2001). Potential of gene therapy for brain tumors. *Hum. Mol. Genet.* **10,** 777–787.

Lee, H., Song, J. J., Kim, E., Yun, C. O., Choi, J., Lee, B., Kim, J., Chang, J. W., and Kim, J. H. (2001). Efficient gene transfer of VSV-G pseudotyped retroviral vector to human brain tumor. *Gene Ther.* **8,** 268–273.

Lemay, D. R., Kittaka, M., Gordon, E. M., Gray, B., Stins, M. F., Mccomb, J. G., Jovanovic, S., Tabrizi, P., Weiss, M. H., Bartus, R., Anderson, W. F., and Zlokovic, B. V. (1998). Intravenous RMP-7 increases delivery of ganciclovir into rat brain tumors and enhances the effects of herpes simplex virus thymidine kinase gene therapy. *Hum. Gene Ther.* **9,** 989–995.

Lewis, P. F., and Emerman, M. (1994). Passage through mitosis is required for oncoretroviruses but not for the human immunodeficiency virus. *J. Virol.* **68,** 510–516.

Lewis, P. F., Hensel, M., and Emerman, M. (1992). Human immunodeficiency virus infection of cells arrested in the cell cycle. *EMBO J.* **11,** 3053–3058.

Lieberman, D. M., Laske, D. W., Morrison, P. F., Bankiewicz, K. S., and Oldfield, E. H. (1995). Convection-enhanced distribution of large molecules in gray matter during interstitial drug infusion. *J. Neurosurg.* **82,** 1021–1029.

Linial, M., Medeiros, E., and Hayward, W. S. (1978). An avian oncovirus mutant (SE 21Q1b) deficient in the genomic RNA: Biological and biochemical characterization. *Cell* **15,** 1371–1381.

Long, Z., Li, L. P., Grooms, T., Lockey, C., Nader, K., Mychkovsky, I., Mueller, S., Burimski, I., Ryan, P., Kikuchi, G., Ennist, D., Marcus, S., Otto, E., and McGarrity, G. (1998). Biosafety monitoring of patients receiving intracerebral injections of murine retroviral vector producer cells. *Hum. Gene Ther.* **9,** 1165–1172.

Long, Z., Lu, P., Grooms, T., Mychkovsky, I., Westley, T., Fitzgerald, T., Sharma-Chibber, S., Shand, N., McGarrity, G., and Otto, E. (1999). Molecular evaluation of biopsy and autopsy specimens from patients receiving *in vivo* retroviral gene therapy. *Hum. Gene Ther.* **10,** 733–740.

Lyons, R. M., Forry-Schaudies, S., Otto, E., Wey, C., Patil-Koota, V., Kaloss, M., McGarrity, G. J., and Chiang, Y. L. (1995). An improved retroviral vector encoding the herpes simplex virus thymidine kinase gene increases antitumor efficacy in vivo. *Cancer Gene Ther.* **2,** 273–280.

Mann, R., Mulligan, R. C., and Baltimore, D. (1983). Construction of a retrovirus packaging mutant and its use to produce helper-free defective retrovirus. *Cell* **33,** 153–159.

Manome, Y., Wen, P. Y., Chen, L., Tanaka, T., Dong, Y., Yamazoe, M., Hirshowitz, A., Kufe, D. W., and Fine, H. A. (1996). Gene therapy for malignant gliomas using replication incompetent retroviral and adenoviral vectors encoding the cytochrome P450 2B1 gene together with cyclophosphamide. *Gene Ther.* **3,** 513–520.

Markowitz, D., Goff, S., and Bank, A. (1988). Construction and use of a safe and efficient amphotropic packaging cell line. *Virology* **167,** 400–406.

Miller, A. D. (1990). Retrovirus packaging cells. *Hum. Gene Ther.* **1,** 5–14.

Moolten, F. L. (1986). Tumor chemosensitivity conferred by inserted herpes thymidine kinase genes: A paradigm for a prospective cancer control strategy. *Cancer Res.* **46,** 5276–5281.

Moolten, F. L., and Wells, J. M. (1990). Curability of tumors bearing herpes thymidine kinase genes transferred by retroviral vectors. *J. Natl. Cancer Inst.* **82,** 297–300.

Mullen, C. A., Anderson, L., Woods, K., Nishino, M., and Petropoulos, D. (1998). Ganciclovir chemoablation of herpes thymidine kinase suicide gene-modified tumors produces tumor necrosis and induces systemic immune responses. *Hum. Gene Ther.* **9,** 2019–2030.

Oldfield, E. H., Ram, Z., Culver, K. W., Blaese, R. M., DeVroom, H. L., and Anderson, W. F. (1993). Gene therapy for the treatment of brain tumors using intra-tumoral transduction with the thymidine kinase gene and intravenous ganciclovir. *Hum. Gene Ther.* **4,** 39–69.

Ono, Y., Ikeda, K., Wei, M. X., Harsh, G. R., Tamiya, T., and Chiocca, E. A. (1997). Regression of experimental brain tumors with 6-thioxanthine and Escherichia coli gpt gene therapy. *Hum. Gene Ther.* **8,** 2043–2055.

Packer, R. J., Raffel, C., Villablanca, J. G., Tonn, J. C., Burdach, S. E., Burger, K., LaFond, D., McComb, J. G., Cogen, P. H., Vezina, G., and Kapcala, L. P. (2000). Treatment of progressive or recurrent pediatric malignant supratentorial brain tumors with herpes simplex virus thymidine kinase gene vector-producer cells followed by intravenous ganciclovir administration. *J. Neurosurg.* **92,** 249–254.

Palu, G., Cavaggioni, A., Calvi, P., Franchin, E., Pizzato, M., Boschetto, R., Parolin, C., Chilosi, M., Ferrini, S., Zanusso, A., and Colombo, F. (1999). Gene therapy of glioblastoma multiforme via combined expression of suicide and cytokine genes: A pilot study in humans. *Gene Ther.* **6,** 330–337.

Parsa, A. T., Chi, J. H., Hurley, P. T., Jeyapalan, S. A., and Bruce, J. N. (2001). Immunomodulation of glioma cells after gene therapy: Induction of major histocompatibility complex class I but not class II antigen in vitro. *Neurosurgery* **49,** 681–689.

Prados, M. D., and Russo, C. (1998). Chemotherapy of brain tumors. *Semin. Surg. Oncol.* **14,** 88–95.

Puumalainen, A. M., Vapalahti, M., Agrawal, R. S., Kossila, M., Laukkanen, J., Lehtolainen, P., Viita, H., Paljärvi, L., Vanninen, R., and Yla-Herttuala, S. (1998). Beta-galactosidase gene transfer to human malignant glioma in vivo using replication-deficient retroviruses and adenoviruses. *Hum. Gene Ther.* **9,** 1769–1774.

Rainov, N. G., and Kramm, C. M. (2001). Vector delivery methods and targeting strategies for gene therapy of brain tumors. *Current Gene Ther.* **1,** 367–383.

Rainov, N. G., Kramm, C. M., Aboody-Guterman, K., Chase, M., Ueki, K., Louis, D. N., Harsh, G. R., Chiocca, A., and Breakefield, X. O. (1996). Retrovirus-mediated gene therapy of experimental brain neoplasms using the herpes simplex virus-thymidine kinase/ganciclovir paradigm. *Cancer Gene Ther.* **3,** 99–106.

Rainov, N. G., Kramm, C. M., Banning, U., Riemann, D., Holzhausen, H. J., Heidecke, V., Burger, K. J., Burkert, W., and Korholz, D. (2000a). Immune response induced by retrovirus-mediated HSV-tk/GCV pharmacogene therapy in patients with glioblastoma multiforme. *Gene Ther* **7,** 1853–1858.

Rainov, N. G., and the GLI328 International Study Group (2000b). A phase III clinical evaluation of herpes simplex virus type 1 thymidine kinase and ganciclovir gene therapy as an adjuvant to surgical resection and radiation in adults with previously untreated glioblastoma multiforme. *Hum. Gene Ther.* **11,** 2389–2401.

Rainov, N. G., Fels, C., Droege, J. W., Schafer, C., Kramm, C. M., and Chou, T. C. (2001). Temozolomide enhances herpes simplex virus thymidine kinase/ganciclovir therapy of malignant glioma. *Cancer Gene Ther.* **8,** 662–668.

Ram, Z., Culver, K. W., Walbridge, S., Blaese, R. M., and Oldfield, E. H. (1993a). In situ retroviral-mediated gene transfer for the treatment of brain tumors in rats. *Cancer Res.* **53,** 83–88.

Ram, Z., Culver, K. W., Walbridge, S., Frank, J. A., Blaese, M. R., and Oldfield, E. H. (1993b). Toxicity studies of retroviral-mediated gene transfer for the treatment of brain tumors. *J. Neurosurg.* **79,** 400–407.

Ram, Z., Walbridge, S., Heiss, J. D., Culver, K. W., Blaese, R. M., and Oldfield, E. H. (1994). In vivo transfer of the human interleukin-2 gene: Negative tumoricidal results in experimental brain tumors. *J. Neurosurg.* **80,** 535–540.

Ram, Z., Culver, K. W., Oshiro, E. M., Viola, J. J., DeVroom, H. L., Otto, E., Long, Z., Chiang, Y., McGarrity, G. J., Muul, L. M., Katz, D., Blaese, R. M., and Oldfield, E. H. (1997). Therapy of malignant brain tumors by intratumoral implantation of retroviral vector-producing cells. *Nat. Med.* **3,** 1354–1361.

Russel, D. W., Berger, M. S., and Miller, A. D. (1995). The effects of human serum and cerebrospinal fluid on retroviral vectors and packaging cell lines. *Hum. Gene Ther.* **6,** 635–641.

Saleh, M., Wiegmans, A., Malone, Q., Stylli, S. S., and Kaye, A. H. (1999). Effect of in situ retroviral interleukin-4 transfer on established intracranial tumors. *J. Natl. Cancer Inst.* **91,** 438–445.

Sandmair, A. M., Loimas, S., Poptani, H., Vainio, P., Vanninen, R., Turunen, M., Tyynela, K., Vapalahti, M., and Yla-Herttuala, S. (1999). Low efficacy of gene therapy for rat BT4C malignant glioma using intra-tumoural transduction with thymidine kinase retrovirus packaging cell injections and ganciclovir treatment. *Acta Neurochir.* **141,** 867–873.

Sandmair, A. M., Loimas, S., Puranen, P., Immonen, A., Kossila, M., Puranen, M., Hurskainen, H., Tyynela, K., Turunen, M., Vanninen, R., Lehtolainen, P., Paljarvi, L., Johansson, R., Vapalahti, M., and Yla-Herttuala, S. (2000a). Thymidine kinase gene therapy for human malignant glioma, using replication-deficient retroviruses or adenoviruses. *Hum. Gene Ther.* **11,** 2197–2205.

Sandmair, A. M., Turunen, M., Tyynela, K., Loimas, S., Vainio, P., Vanninen, R., Vapalahti, M., Bjerkvig, R., Janne, J., and Yla-Herttuala, S. (2000b). Herpes simplex virus thymidine kinase gene therapy in experimental rat BT4C glioma model: Effect of the percentage of thymidine kinase-positive glioma cells on treatment effect, survival time, and tissue reactions. *Cancer Gene Ther.* **7,** 413–421.

Sasaki, M., Wizigmann-Voos, S., Risau, W., and Plate, K. H. (1999). Retrovirus producer cells encoding antisense VEGF prolong survival of rats with intracranial GS9L gliomas. *Int. J. Dev. Neurosci.* **17,** 579–591.

Sena-Esteves, M., Hampl, J. A., Camp, S. M., and Breakefield, X. O. (2002). Generation of stable retrovirus packaging cell lines after transduction with herpes simplex virus hybrid amplicon vectors—double dagger. *J. Gene Med.* **4,** 229–239.

Shand, N., Weber, F., Mariani, L., Bernstein, M., Gianella-Borradori, A., Long, Z., Sorensen, A. G., and Barbier, N. (1999). A phase 1-2 clinical trial of gene therapy for recurrent glioblastoma multiforme by tumor transduction with the herpes simplex thymidine kinase gene followed by ganciclovir. *Hum. Gene Ther.* **10,** 2325–2335.

Shapiro, W. R., and Shapiro, J. R. (1998). Biology and treatment of malignant glioma. *Oncology* **12,** 233–240.

Short, M. P., Choi, B. C., Lee, J. K., Malick, A., Breakefield, X. O., and Martuza, R. L. (1990). Gene delivery to glioma cells in rat brain by grafting of a retrovirus packaging cell line. *J. Neurosci. Res.* **27,** 427–439.

Takamiya, Y., Short, M. P., Ezzeddine, Z. D., Moolten, F. L., Breakefield, X. O., and Martuza, R. L. (1992). Gene therapy of malignant brain tumors: A rat glioma line bearing the herpes simplex virus type 1-thymidine kinase gene and wild type retrovirus kills other tumor cells. *J. Neurosci. Res.* **33,** 493–503.

Takamiya, Y., Short, P., Moolten, F. L., Fleet, C., Mineta, T., Breakefield, X. O., and Martuza, R. L. (1993). An experimental model of retrovirus gene therapy for malignant brain tumors. *J. Neurosurg.* **79,** 104–110.

Tamiya, T., Wei, M. X., Chase, M., Ono, Y., Lee, F., Breakefield, X. O., and Chiocca, E. A. (1995). Transgene inheritance and retroviral infection contribute to the efficiency of gene expression in solid tumors inoculated with retroviral vector producer cells. *Gene Ther.* **2,** 531–538.

Tamura, K., Tamura, M., Ikenaka, K., Yoshimatsu, T., Miyao, Y., Nanmoku, K., and Shimizu, K. (2001). Eradication of murine brain tumors by direct inoculation of concentrated high titer-recombinant retrovirus harboring the herpes simplex virus thymidine kinase gene. *Gene Ther.* **8,** 215–222.

Tapscott, S. J., Miller, A. D., Olson, J. M., Berger, M. S., Groudine, M., and Spence, A. M. (1994). Gene therapy of rat 9L gliosarcoma tumors by transduction with selectable genes does not require drug selection. *Proc. Natl. Acad. Sci. USA* **91,** 8185–8189.

Uckert, W., and Walther, W. (1994). Retrovirus-mediated gene transfer in cancer therapy. *Pharmacol. Ther.* **63,** 323–347.

Valsesia-Wittmann, S., Drynda, A., Deleage, G., Aumailley, M., Heard, J. M., Danos, O., Verdier, G., and Cosset, F. L. (1994). Modifications in the binding domain of avian retrovirus envelope protein to redirect the host range of retroviral vectors. *J. Virol.* **68,** 4609–4619.

Varmus, H. E., and Swanstrom, R. (1982). Replication of retroviruses. *In* "RNA Tumor Viruses" (H. E. Varmus and R. Swanstrom, Eds.), pp. 369–512. Cold Spring Harbour Press, Cold Spring Harbour, New York.

Vile, R. G., Castleden, S., Marshall, J., Camplejohn, R., Upton, C., and Chong, H. (1997). Generation of an anti-tumour immune response in a non-immunogenic tumor: HSVtk killing in vivo stimulates a mononuclear cell infiltrate and a Th1-like profile of intratumoural cytokine expression. *Int. J. Cancer.* **71,** 267–274.

Zitvogel, L., Tahara, H., Cai, Q., Storkus, W. J., Muller, G., Wolf, S. F., Gately, M., Robbins, P. D., and Lotze, M. T. (1994). Construction and characterization of retroviral vectors expressing biologically active human interleukin-12. *Hum. Gen Ther.* **5,** 1493–1506.

# ADENO-ASSOCIATED VIRAL VECTORS FOR PARKINSON'S DISEASE

Shin-ichi Muramatsu,[*] Lijun Wang,[*,†] Kunihiko Ikeguchi,[*] Ken-ichi Fujimoto,[*] Imaharu Nakano,[*] Takashi Okada,[†] Hiroaki Mizukami,[†] Yutaka Hanazono,[†] Akihiro Kume,[†] Imaharu Nakano,[*] and Keiya Ozawa[†]

[*]Division of Neurology and
[†]Division of Genetic Therapeutics
Jichi Medical School
Tochigi 329-0498, Japan



## I. Introduction

Parkinson's disease (PD) is a common neurodegenerative disorder among the elderly, with an estimated 1% of the population over 60 years old suffering from PD. Progressive loss of nigrostriatal dopamine (DA) neurons leads to a severe decrease in the DA content of the striatum. Cardinal motor symptoms, including resting tremor, muscular rigidity, and bradykinesia, are produced by the reduction of striatal DA to about 20% of normal levels. Replacement of DA is therefore important to alleviate the motor symptoms of the disease. The introduction of the DA precursor, L-3,4-dihydroxyphenylalanine (L-DOPA), in the late 1960s represented a major therapeutic advance in the management of PD, providing clinical benefit to virtually all patients and reducing mortality. Since then, oral administration of L-DOPA has been the mainstay in pharmacotherapy for PD. However, as the disease progresses, L-DOPA is less efficiently converted to DA in the striatum and loses efficacy. Frequent systemic administration of high-dose L-DOPA causes oscillations in motor performance (Jenner, 2000; Langston *et al.*, 2000) and some deleterious complications such as hallucinations due to dopaminergic




0074-7742/03 $35.00

stimulation in the mesolimbic system (Carey *et al.*, 1995). Novel therapeutic interventions substituting for oral L-DOPA administration are therefore required.

PD is regarded as a suitable candidate for gene therapy. Unlike other neurological disorders that affect broad regions of brain, such as Alzheimer's disease, lipid storage disease, and multiple sclerosis, PD is primarily confined to the well-defined nigrostriatal dopaminergic system. Therapeutic genes do not need to be delivered to the entire brain, but can be targeted to portions of the basal ganglia. Stereotactic surgical techniques for approaching the basal ganglia are already established in clinical practice. With the development of microelectrode recording for neural activity, a number of surgical treatments such as pallidotomy, thalmotomy, and deep brain stimulation in the subthalamic nucleus have been employed in PD over the past century. In addition, well-characterized rodent and primate PD models are available for testing novel therapeutic interventions.

Apart from genetic mutations described in some familial cases, the causes of PD remain largely unknown. Purely genetic causes probably account for only a small number of PD patients, and multiple factors including environmental factors may contribute to the development of sporadic PD. For this reason, the current strategy of gene therapy for PD does not pursue the correction of specific genetic abnormalities, but rather aims at either restoring local production of DA by delivering genes for DA-synthesizing enzymes to the striatum (Kaplitt *et al.*, 1994; During *et al.*, 1998; Fan *et al.*, 1998; Mandel *et al.*, 1998; Leff *et al.*, 1999; Bankiewicz *et al.*, 2000; Shen *et al.*, 2000; Sanchez-Pernaute *et al.*, 2001; Kirik *et al.*, 2002; Muramatsu *et al.*, 2002), or providing neuroprotection to block or slow ongoing degenerative processes by providing genes for growth factors (Klein *et al.*, 1999; Mandel *et al.*, 1997, 1999; Kirik *et al.*, 2000; Wang *et al.*, 2002), antioxidant molecules, or antiapoptotic substances (Mochizuki *et al.*, 2001).

## II. Recombinant Adeno-Associated Viral Vector

Among the various gene delivery vehicles, recombinant adeno-associated viral (rAAV) vector represents one of the most suitable means for direct introduction of therapeutic genes into the mammalian brain. The rAAV vector is the only viral vector system based on a nonpathogenic replication-defective virus. Efficient and long-term *in vivo* gene expression has been achieved without substantial toxicity or immune response (Kaplitt *et al.*, 1994; Du *et al.*, 1996; During *et al.*, 1998; Fan *et al.*, 1998; Leff *et al.*, 1999; Shen *et al.*, 2000).

AAVs are small, nonenveloped, single-stranded DNA viruses of the Parvoviridae family assigned to the genus *Dependovirus*, as productive infection requires coinfection with a helper virus such as adenovirus (Ad) or herpesvirus (Berns *et al.*, 1995). AAVs have been isolated from a variety of different species, including primates, dogs, cows, horses, sheep, and chicken, and appear to be nonpathogenic in all cases. To date, eight primate AAVs (AAV-1 to -8) have been cloned and sequenced (Srivastava *et al.*, 1983; Muramatsu *et al.*, 1996; Rutledge *et al.*, 1998; Bantel-Schaal *et al.*, 1999; Chiorini *et al.*, 1997, 1999; Xiao *et al.*, 1999 Gao *et al.*, 2002). Among them, AAV-2 has been the most extensively studied and offers a major platform for the development of gene therapy vectors. The genomic organization of AAV-2 consists of two large open reading frames (ORFs) flanked by two inverted terminal repeats (ITRs). AAV-2 ITRs are 145-nt sequences that function in *cis* as the replication origin and are required for packaging of progeny AAV DNA into virus particles. The left ORF encodes four nonstructural Rep proteins, which are essential for the replication of the AAV genome and targeted integration into the host chromosome during latent infection. The right ORF encodes three viral capsid proteins, VP1, VP2, and VP3.

The conventional method of rAAV vector production involves double-plasmid transfection followed by helper Ad infection. The vector plasmid contains a transgene expression cassette flanked by AAV ITRs, whereas the packaging plasmid expresses Rep and VP proteins in *trans* to supply required AAV helper functions. More recently, a helper-free triple-plasmid transfection method has been developed (Matsushita *et al.*, 1998). By employing a helper plasmid containing the minimum Ad genomic sequences necessary for helper functions (E1A, E1B, E2, E4ORF6, VA RNA), undesirable contamination with wild-type Ad is avoided. Although high multiplicities of infection, estimated to be between $10^3$ and $10^5$ genome copies per cell, are required for efficient gene transfer via rAAV vector transduction, this system has the ability to produce pure, high-titer rAAV vectors that can be used for clinical applications. Vectors are purified by two sequential continuous CsCl gradient, or for clinical use discontinuous iodixanol gradient, ultracentrifugations. Vector titer is determined by quantitative DNA dot-blot hybridization of DNase I-treated vector stocks, and is routinely $10^{12}$ to $10^{13}$ genome copies per milliliter.

## III. Restoration of DA Synthesis in the Striatum

### A. Biosynthetic Pathway of DA

For efficient DA synthesis, three enzymes are necessary (Fig. 1). Tyrosine hydroxylase (TH) is the rate-limiting enzyme and first converts L-tyrosine, an essential amino acid present in all types of cells, to L-DOPA. Aromatic L-amino acid decarboxylase (AADC) then converts L-DOPA to DA. Guanosine triphosphate cyclohydrolase I (GCH) is the rate-limiting enzyme for synthesis of the essential TH cofactor tetrahydrobiopterine (BH4). Because low levels of endogenous BH4

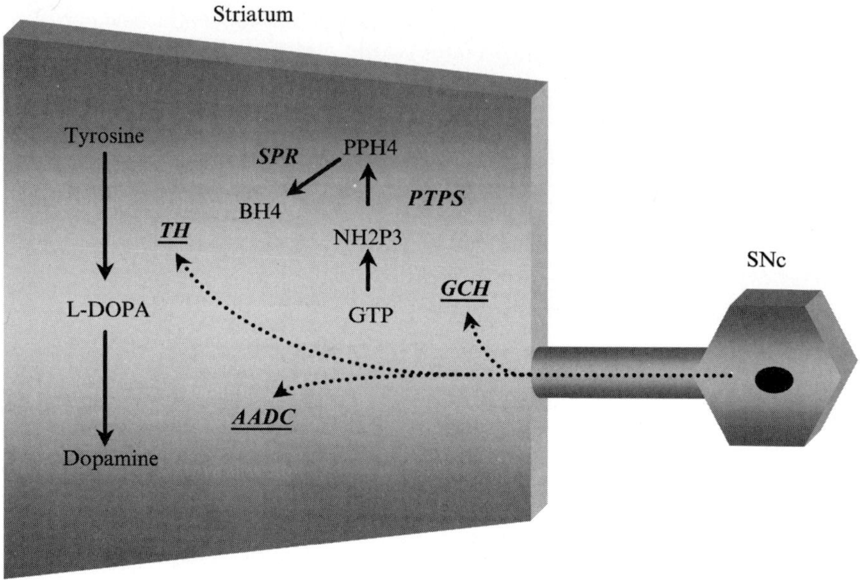


FIG. 1. Biosynthetic pathway of dopamine. Three enzymes (TH, AADC, and GCH) are necessary for efficient dopamine (DA) production. GCH is a rate-limiting enzyme for biosynthesis of BH4, an essential TH cofactor. These enzymes are transported from the substantia nigra to the striatum in an anterograde manner. TH, tyrosine hydroxylase; AADC, aromatic-L-amino-acid decarboxylase; BH4, tetrahydrobiopterin; GTP, guanosine triphosphate; GCH, GTP cyclohydrolase I; NH2P3, *d*-erythro-7,8-dihydroneopterin triphosphate; PPH4, 6-pyruvoyl tetrahydropterin; PTPS, 6-pyruvoyl tetrahydropterin synthase; SPR, sepiapterin reductase.

do not yield sufficient TH activity, GCH is considered to regulate TH activity via regulation of BH4 biosynthesis, thus indirectly controlling DA production in TH-containing DA neurons (Nagatsu *et al.*, 1997; Nagatsu and Ichinose, 1999). These enzymes are transported in an anterograde manner from the substantia nigra to the striatum. A severe loss of dopaminergic nerve terminals in advanced PD is associated with an 80–95% depletion of striatal TH and AADC activity (Zhong *et al.*, 1995; Kaddis *et al.*, 1997). AADC is present in DA-denervated striatum within nondopaminergic neurons and glial cells, but endogenous AADC activity in the striatum is considered insufficient (Nakamura *et al.*, 2000). Although L-DOPA may also function as a neurotransmitter, most L-DOPA produced in the striatum after gene transfer must be converted to DA *in situ* (Kaplitt *et al.*, 1994). GCH activity in the striatum is also decreased in PD and parallels decreases in levels of BH4, TH, and DA (Nagatsu *et al.*, 1987, 1997; Nagatsu and Ichinose, 1999). In a rat model, levels of BH4 and GCH activity in the striatum decrease to less than 27% and 32% of normal levels, respectively, after injection of 6-hydroxydopamine (6-OHDA) into the substantia nigra, selectively destroying the DA neurons (Levine *et al.*, 1981). Although BH4 can cross the blood–brain barrier, BH4 in the brain is supplied predominantly by intracellular biosynthesis and uptake of exogenous BH4 from the blood was low (Hoshiga *et al.*, 1993). GCH gene transfer into striatal cells appears to offer a more efficient method of supplying BH4 than administration of exogenous BH4.

## B. rAAV Vector Expressing DA-Synthesizing Enzyme

We produced rAAV vectors expressing TH (rAAV-TH), AADC (rAAV-AADC), GCH (rAAV-GCH), and LacZ (rAAV-LacZ) by helper-free method to investigate whether rAAV vector-mediated gene transfer of DA-synthesizing enzymes could restore local production of DA and produce behavioral recovery in PD model animals. The expression cassette of each rAAV vector contains cDNA encoding the exogenous gene with the human cytomegalovirus (CMV) immediate-early promoter, human growth hormone first intron, and simian virus 40 (SV40) polyadenylation signal sequence between two ITRs of the AAV-2 genome.

Expression of GCH via rAAV-GCH induced BH4 synthesis and enhanced DA production *in vitro*. When cells from the 293 human embryonic kidney cell line were transduced with various doses of rAAV-GCH (ranging from 0 to $5 \times 10^5$ genome copies per cell) together with fixed amounts of rAAV-TH and rAAV-AADC ($5 \times 10^5$ genome copies per cell each), GCH activity and BH4 production gradually increased in concert with gradual increases in dose of rAAV-GCH. Intracelullar L-DOPA and DA levels increased significantly with increasing amounts of rAAV-GCH (Shen *et al.*, 2000).

## C. Long-Term Behavioral Recovery in a Rat Model of PD

Hemiparkinsonian rats were made by the injection of 6-OHDA into the medial forebrain bundle. Mixtures of three rAAV vectors (rAAV-TH/-AADC/-GCH, $5 \times 10^7$ genome copies per site for each vector) were then injected into the lesioned striatum at three sites (Shen *et al.*, 2000). The rAAV vectors efficiently transduced striatal cells. A majority (>95%) of transduced cells were microtubule-associated protein (MAP2)-immunoreactive (IR) neurons and more than 90% of TH-IR cells were also positive for AADC and GCH, representing a high coexpression efficiency. This indicates that although the limited packaging capacity on the size of the DNA (<5 kb) renders the use of a single rAAV vector impossible for multiple gene transfer, cells could be simultaneously transduced using different rAAV vectors. TH-IR cells were still observed around injection sites in a wide area of the striatum, even 18 months after vector injection (Fig. 2). Rats

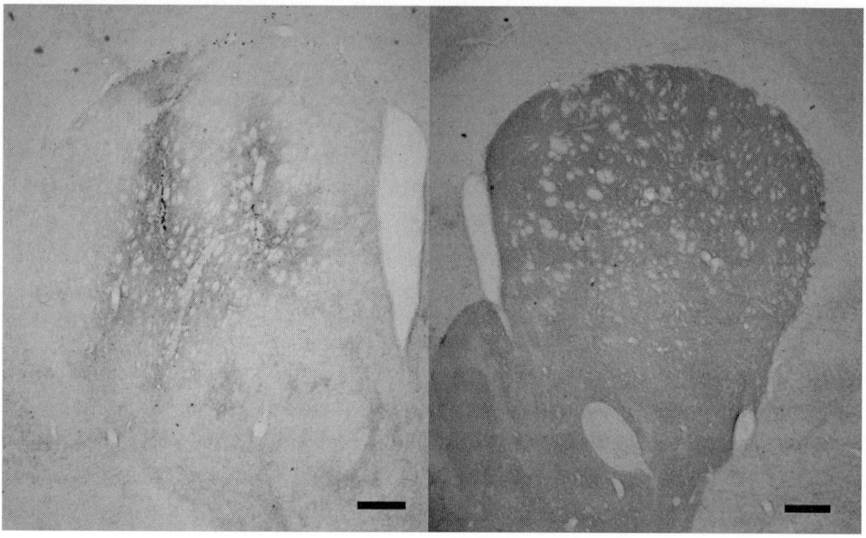


FIG. 2. TH immunostaining in the brain of 6-OHDA-lesioned rats 18 months after stereotaxic injection of rAAV-TH/-AADC/-GCH. (A) rAAV vectors were injected into the striatum of the lesioned side. Intense TH staining was observed around the two needle tracks. (B) Positive staining of the dopaminergic terminals was observed in the intact striatum. Scale bar = 2 mm.

receiving either rAAV-TH/-AADC or rAAV-TH/-AADC/-GCH demonstrated significant decreases in rates of apomorphine-induced rotation compared to pretreated levels, while the rotational behavior of rats injected with rAAV-LacZ was unaltered from baseline throughout the 18-month observation period following injection (Fig. 3). In addition, rats transduced with three rAAV vectors demonstrated a more remarkable decrease in rotation rate than those with two rAAV vectors. Levels of BH4 and DA in the lesioned striatum of rats injected with all three vectors were significantly greater than those in rats treated with rAAV-TH/-AADC or in uninjected control animals. Conversely, no significant difference in BH4 level was observed between animals injected with two vectors and control animals. Some researchers have reported that addition of AADC to systems expressing TH and GCH via genetically modified fibroblasts results in less L-DOPA production due to feedback inhibition of DA on TH (Wachtel *et al.*, 1997; Kang *et al.*, 2001). However, previous studies including our own have demonstrated that restoration of AADC and TH activity in the striatum is beneficial in 6-OHDA-lesioned rats (Fan *et al.*, 1998; Leff *et al.*, 1999), and the results presented here show that local production of striatal DA is actually enhanced by the additional expression of GCH compared with TH and AADC alone.

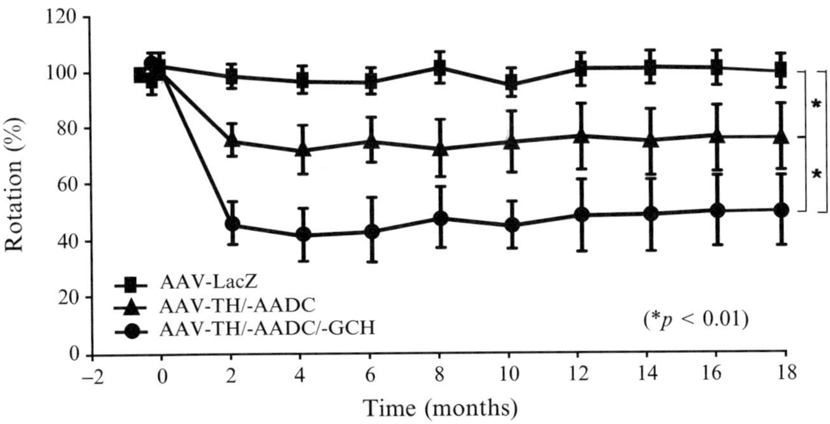


FIG. 3. Effects of intrastriatal injection of rAAV vectors on apomorphine-induced rotational behavior in 6-OHDA-lesioned rats. Rotation was evaluated every week before and after stereotaxic injection of rAAV-LacZ ($n = 5$), rAAV-TH/-AADC ($n = 8$), and rAAV-TH/-AADC/-GCH ($n = 8$).

In summary, cotransduction with three kinds of rAAV vectors results in more effective DA production and markedly increased behavioral recovery than transduction with rAAV-TH vector alone or combination of rAAV-TH and rAAV-AADC vectors. Behavioral recovery persisted at least 18 months after rAAV vector treatment.

## D. Behavioral Recovery in a Primate Model of PD

Encouraging results of rAAV vector-mediated gene transfer of DA-synthesizing enzymes in parkinsonian rats prompted us to extend preclinical explorations to a primate model of PD (Muramatsu *et al.*, 2002). Primate models are useful to evaluate motor functions more properly and to examine transduction efficiency in larger areas of striatum. Cynomolgus monkeys (*Macaca fascicularis*) had been trained to perform a fine motor task consisting of picking up four raisins or small pieces of apple with each of the two hands. Bilateral striatal lesions were made by intravenous injection of 1-methyl-4-phenyl-1,2,3,6-tetrahydropyridine (MPTP), a selective toxin of dopaminergic neurons, once a week until a stable parkinsonian syndrome was achieved. Monkeys were allowed to recover for 2 months after the last MPTP treatment to avoid the possibility that spontaneous recovery from acute toxicity of MPTP could mimic any behavioral effect of rAAV injection. Mixtures of three rAAV vectors (rAAV-TH/-AADC/-GCH) were then stereotaxically injected unilaterally into the putamen of parkinsonian monkeys. Each monkey received nine unilateral injections of rAAV vectors in three tracks in the unilateral putamen. Each injection comprised 5 $\mu$l with a 1:1:1 mixture of rAAV-TH/-AADC/-GCH ($1 \times 10^{13}$ copies of each vector genome per ml). Thus, in total, $1.5 \times 10^{11}$ genome copies of each vector were injected. As a control, 15 $\mu$l of AAV-LacZ ($1 \times 10^{13}$ vector genome copies/ml) or phosphate-buffered saline was injected into the contralateral putamen. Coexpression of enzymes in the unilateral putamen resulted in marked improvement in manual dexterity on the contralateral side to rAAV-TH/-AADC/-GCH vector injections. Monkeys picked up raisins better with the contralateral hand, in which tremor had disappeared. The ipsilateral limb remained disabled, suggesting that behavioral recovery of the contralateral limb cannot be explained by spontaneous recovery that may be expected to occur over time. Although 20% of thalamic projections from basal ganglia decussate in monkeys, unilateral transduction would primarily affect contralateral limb movement. Behavioral recovery persisted throughout the observation period (three monkeys were sacrificed for histological evaluation at 48 days, 65 days, and 50 days, respectively, and one monkey was kept alive more than 30 months after rAAV vectors injection). After rAAV vector injections, animals demonstrated apomorphine-induced dystonic postures, in which the body and face turned toward the ipsilateral side, and a circling tendency toward the ipsilateral side, indicating reduced DA receptor supersensitivity in the AAV-TH/-AADC/-GCH-injected putamen (Fig. 4). TH-IR, AADC-IR, and GCH-IR cells were present in a large region of the putamen (>90% of the putamen) (Fig. 5).

Although the convection-enhanced delivery method is reportedly efficient, a sufficient region can be transduced in primate brains by simple injection methods if high-titer rAAV vectors are available. The optimal TH:AADC:GCH ratio and the minimal volume of transduced region necessary for functional recovery are yet to be elucidated for human brains, in which the putamen is approximately 10 times larger than in monkey brains (Lange *et al.*, 1976; Lieberman *et al.*, 1995). However, a considerable therapeutic effect was obtained with a 1:1:1 ratio of rAAV vector doses and transducing a broad region of the human putamen by increasing the number of injection sites seems feasible. Except for slight infiltration of mononuclear cells and residual hemosiderin around the needle tract, hematoxylin and eosin staining revealed no signs of cytotoxicity in the rAAV vector-injected putamen. Microdialysis demonstrated that concentrations of DA in the rAAV-TH/-AADC/-GCH vector-injected putamen were increased compared with the control side. Monkeys did not show any complications related to rAAV vector injection, including dyskinesia.

In summary, rAAV vectors efficiently introduced genes for DA-synthesizing enzymes into the striatum of primates, resulting in restoration of motor function with robust transgene expression and elevated DA synthesis in the treated putamen.

## IV. Neuroprotection by GDNF Gene Delivery

An alternative approach to the treatment of PD is the protection of nigrostriatal pathways from progressive degeneration (Dunnett and Bjorklund, 1999). Glial cell line-derived neurotrophic factor (GDNF) is a small glycoprotein that provides strong trophic support for DA neurons and is a candidate for this neuroprotective strategy (Bjorklund *et al.*, 2000; Bohn, 2000). However, the applications of GDNF protein are limited because of its short duration of activity and poor ability to cross the blood–brain barrier. An attempt to treat PD patients via direct injection of GDNF protein into the ventricles was unsuccessful (Kordower *et al.*, 1999). Moreover, because PD is a progressive disorder, GDNF should be delivered continuously to support diseased DA neurons. Repeated injections of GDNF proteins into the human brain are not practical and implanting infusion devices is likely to increase patient morbidity and risks of infection over the long term. Recombinant viral vectors are powerful tools for sustained GDNF production in the brain.

Previous studies have demonstrated that GDNF gene delivered via rAAV (Mandel *et al.*, 1997, 1999; Kirik *et al.*, 2000), adenoviral (Bilang-Bleuel *et al.*, 1997; Choi-Lundberg *et al.*, 1997, 1998; Kozlowski *et al.*, 2000; Connor *et al.*, 1999, 2001), herpes simplex viral (Natsume *et al.*, 2001), or lentiviral (Kordower *et al.*, 2000) vectors protects nigral DA neurons in rodent and primate models of PD. However, GDNF gene was administered prior to or shortly after injections of neurotoxin in these studies. Because substantial numbers of DA neurons are already lost before characteristic symptoms appear in PD, a clinically more important question is whether GDNF gene delivered in a delayed manner is capable of rescuing DA neurons and improving behavioral performance in animal models with extensive nigrostriatal DA denervation already present. In our study using a rat model, 4 weeks after creation of a unilateral striatal lesion using 6-OHDA, animals received injections of rAAV vectors expressing GDNF tagged with FLAG peptide (rAAV-GDNF*flag*) or $\beta$-galactosidase (rAAV-LacZ) into the lesioned striatum (Fig. 6). At that time point, the percentage of TH-IR neurons on the lesioned side had been reduced to 35% of the intact side. At 20 weeks after rAAV vector injection, extensive loss of TH-IR cells in the substantia nigra (SN) (20% of intact side) and TH-IR fibers in the striatum (14% of intact side) was observed in rats injected with rAAV-LacZ. In contrast, injection of rAAV-GDNF*flag* significantly increased numbers of TH-IR neurons in the SN and density of TH-IR fibers in the striatum, reaching nearly 57% and 49% of the contralateral side, respectively. Levels of DA and its metabolites in the striatum were markedly higher in the rAAV-GDNF*flag* group compared with the control group. Consistent with anatomical and biochemical changes, significant behavioral recovery was observed from 4–20 weeks following rAAV-GDNF*flag*

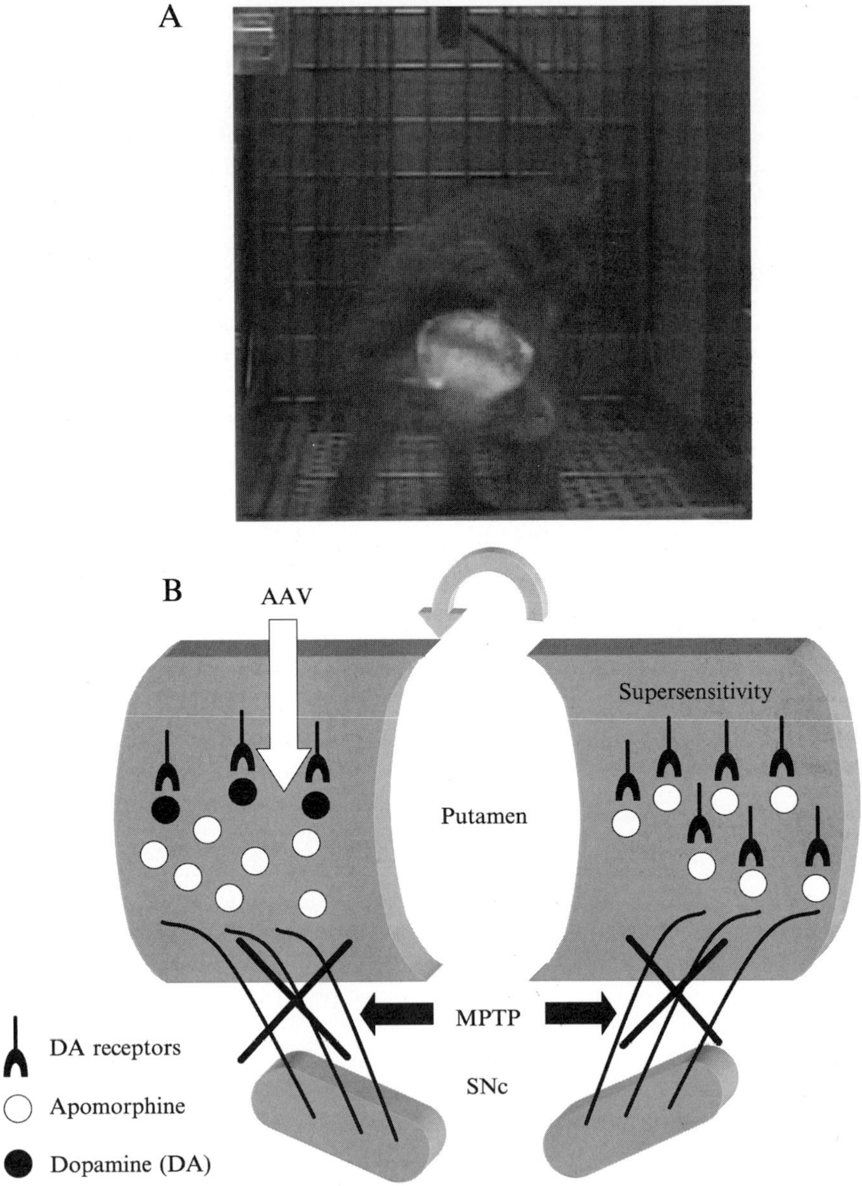


FIG. 4. Apomorphine-induced dystonic posture in an MPTP-treated monkey after gene delivery of DA-synthesizing enzymes into the unilateral putamen. (A) Administration of apomorphine, a dopamine receptor agonist, induced a dystonic posture, in which the body axis and face turned toward the left side after rAAV-TH/-AADC/-GCH injection into the left putamen. (B) Schematic representation of mechanisms of turning behavior. MPTP, a selective neurotoxin, induced

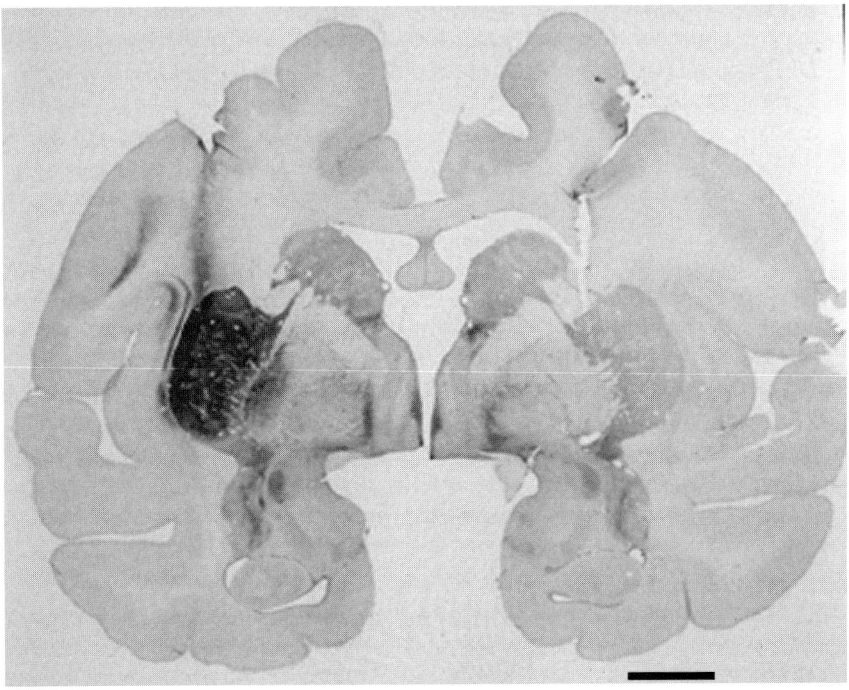

FIG. 5. Dense TH immunoreactivity in the unilateral putamen of the monkey, 65 days after rAAV-TH/-AADC/-GCH injection. Immunoreactivity in striatum on the control side was comprehensively lost.

injection. Dual-immunostaining for FLAG and TH demonstrated double positive cells in the SN. In contrast, no $\beta$-galactosidase signals could be detected in the SN of rAAV-LacZ-injected rats, although X-Gal-positive cells were detected in the striatum. Thus GDNF*flag* protein, rather than the rAAV vector per se, underwent retrograde transport to the SN, although the rAAV vector might demonstrate retrograde transport to some extent (Kaspar *et al.*, 2002). Striatal GDNF appears important for functional reinnervation. However, expression of GDNF in nigral DA neurons themselves may be detrimental, because intense local sprouting may prevent regeneration of lesioned axons toward the striatum (Grondin and Gash, 1998; Mandel *et al.*, 1999; Walton, 1999). Our results

supersensitivity of DA receptors in the striatum. Upregulation of DA receptors is shown here as simple elevation in receptor numbers, but may involve changes in downstream transduction pathways. This change was ameliorated by production of DA in the rAAV vector-treated side. Systemic administration of apomorphine caused an imbalance between the two striata, resulting in turning of the body axis toward the vector-treated side.

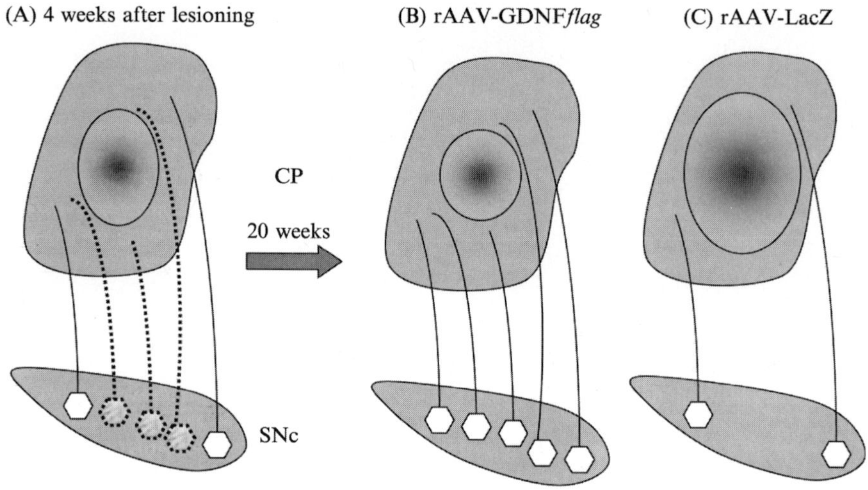


FIG. 6. Schematic drawing of the nigrostriatal pathway illustrating effects of rAAV vector-mediated delivery of GDNF gene in the striatum. At the time point of vector injection, i.e., 4 weeks after injection of 6-OHDA into the unilateral striatum, TH-IR neurons in the SN on the lesioned side had decreased to approximately 35% of the unlesioned side (A). Twenty weeks after rAAV vectors injection, TH-IR neurons in the SN and TH-IR nigrostriatal fibers in the striatum were restored to almost half the levels of the control side in rAAV-GDNF*flag* vector-injected rats (B), while further loss of TH-IR neurons and fibers was observed in rAAV-LacZ vector-injected rats (C). Some portions of the nigrostriatal connections remaining at 4 weeks postlesion might provide the substrate for sprouting and regrowth of TH-IR fibers in response to GDNF. CP, caudate-putamen; SNc, substantia nigra pars compacta.

indicate that delayed delivery of the GDNF gene using rAAV vectors halts the ongoing degeneration of nigrostriatal pathways, producing functional recovery even after substantial numbers of DA cells have been depleted.

## V. Conclusions

Until recently, clinical trials of gene therapy for neurological disorders have been largely confined to rare metabolic diseases and brain tumors at advanced stages. However, advances in gene transfer methods, and particularly in the development of improved viral vectors, have expanded the potentials of gene therapy to include the treatment of a wide range of genetic and acquired diseases. At present, the rAAV vector represents one of the most powerful vehicles for delivering therapeutic genes into mammalian brains. In rodents and nonhuman primates, a substantial number of striatal neurons can be transduced with high titer rAAV vectors through simple stereotaxic injection. Although chromosomal rearrangements in association with integration of rAAV have been indicated in a transformed cell line (Miller *et al.*, 2002), the safety and potential efficacy of rAAV vectors have been demonstrated both in animal models of several diseases and in clinical trials on patients with hemophilia (High, 2001).

Recent advances in the field of neural or embryonic stem cell research has raised the hope of cell replacement as an alternative therapy for PD (Bjorklund and Lindvall, 2000). In most protocols, dopaminergic cells are transplanted ectopically into the striatum instead of into the SN. If the primary mechanism underlying any expected functional recovery in these cell replacement therapies is restoration of dopaminergic neurotransmission, then direct gene delivery of DA-synthesizing enzymes into the striatum offers a much simpler and more straightforward approach. The significant functional recovery achieved with the triple transduction method in our primate model provides an optimistic outlook for gene therapy treatment of PD. If reconstruction of neural networks including nigrostriatal pathways is required to ameliorate more complex symptoms of PD such as dementia, the combination of gene therapy and cell transplantation appears to be the next strategy to overcome this considerable obstacle.

Although whether neurotoxin-induced PD models faithfully reflect the human disease requires further verification, neuroprotective gene therapy with neurotrophic factors holds further potential as a novel treatment of PD. Recently, neuroimaging techniques and genetic analyses for some familial cases have provided opportunities for detecting at-risk individuals before characteristic symptoms appear (Brooks, 2000; Hardy *et al.*, 2003). Earlier gene therapy with GDNF could obviously be applied to such at-risk individuals as easily as to PD patients. The elucidation of mechanisms whereby genetic mutations lead to the loss of DA neurons in familial forms of PD, in addition to the detection of factors increasing risk for PD, will supply fresh targets for gene therapy.

Developing vector constructs that allow regulation of gene expression should be necessary to avoid excess production of DA or GDNF. Studies on rAAV vectors with regulatable promoters are underway in several laboratories. For the DA replacement strategy, gene transfer of AADC alone in combination with oral administration of L-DOPA offers a shortcut toward reaching clinical trials (Bankiewicz *et al.*, 2000; Sanchez-Pernaute *et al.*, 2001). In this method, patients need to take L-DOPA for amelioration of symptoms, but DA production can be controlled by a dose of L-DOPA.

More recently, based on the observation that deep brain stimulation of the subthalamic nucleus (STN) provides antiparkinsonian effects, another palliative approach has been proposed that does not address DA replacement. This method involves rAAV vector-mediated gene transfer of glutamate decarboxylase, an enzyme for synthesis of inhibitory neurotransmitter GABA, to the STN

(During *et al.*, 2001). However, the mechanism of action of subthalamic stimulation remains to be defined, and the involvement of nonmotor territories may be deleterious. The optimal target site in the STN, where GABA synthesis is enhanced, needs to be identified to ameliorate parkinsonian symptoms while avoiding dyskinesia and cognitive side effects.

Gene therapy using rAAV vectors is now not merely an experimental strategy, but offers a novel and apparently feasible protocol for the treatment of PD.

## Acknowledgments

We would like to thank Avigen, Inc., for providing the rAAV vector production system. This study was supported by grants from the Ministry of Health, Labor, and Welfare of Japan; by Strategic Promotion System of Brain Science, in addition to Special Coordination Funds from the Japanese Ministry of Education, Culture, Sports, Science, and Technology; and by The Cell Science Research Foundation.

## References

Bankiewicz, K. S., Eberling, J. L., Kohutnicka, M., Jagust, W., Pivirotto, P., Bringas, J., Cunningham, J., Budinger, T. F., and Harvey-White, J. (2000). Convection-enhanced delivery of AAV vector in parkinsonian monkeys; in vivo detection of gene expression and restoration of dopaminergic function using pro-drug approach. *Exp. Neurol.* **164,** 2–14.

Bantel-Schaal, U., Delius, H., Schmidt, R., and zur Hausen, H. (1999). Human adeno-associated virus type 5 is only distantly related to other known primate helper-dependent parvoviruses. *J. Virol.* **73,** 939–947.

Berns, K. I., Bergoin, M., Bloom, M., Lederman, M., Muzyczka, N., Siegl, G., Tal, J., and Tattersall, P. (1995). Family Parvoviridae. *In* "Virus Taxonomy. Classification and Nomenclature of Viruses" (F. A. Murphy *et al.*, Ed.), pp. 169–178. Springer-Verlag, New York.

Bilang-Bleuel, A., Revah, F., Colin, P., Locquet, I., Robert, J. J., Mallet, J., and Horellou, P. (1997). Intrastriatal injection of an adenoviral vector expressing glial-cell-line-derived neurotrophic factor prevents dopaminergic neuron degeneration and behavioral impairment in a rat model of Parkinson disease. *Proc. Natl. Acad. Sci. USA* **94,** 8818–8823.

Bjorklund, A., and Lindvall, O. (2000). Cell replacement therapies for central nervous system disorders. *Nat. Neurosci.* **3,** 537–544.

Bjorklund, A., Kirik, D., Rosenblad, C., Georgievska, B., Lundberg, C., and Mandel, R. J. (2000). Towards a neuroprotective gene therapy for Parkinson's disease: Use of adenovirus, AAV and lentivirus vectors for gene transfer of GDNF to the nigrostriatal system in the rat Parkinson model. *Brain Res.* **886,** 82–98.

Bohn, M. C. (2000). Parkinson's disease: A neurodegenerative disease particularly amenable to gene therapy. *Mol. Ther.* **1,** 494–496.

Brooks, D. J. (2000). Morphological and functional imaging studies on the diagnosis and progression of Parkinson's disease. *J. Neurol.* **247**(Suppl. 2), II 11–18.

Carey, R. J., Pinheiro-Carrera, M., Dai, H., Tomaz, C., and Huston, J. P. (1995). L-DOPA and psychosis: Evidence for L-DOPA-induced increases in prefrontal cortex dopamine and in serum corticosterone. *Biol. Psychiatry* **38,** 669–676.

Chiorini, J. A., Yang, L., Liu, Y., Safer, B., and Kotin, R. M. (1997). Cloning of adeno-associated virus type 4 (AAV4) and generation of recombinant AAV4 particles. *J. Virol.* **71,** 6823–6833.

Chiorini, J. A., Kim, F., Yang, L., and Kotin, R. M. (1999). Cloning and characterization of adeno-associated virus type 5. *J. Virol.* **73,** 1309–1319.

Choi-Lundberg, D. L., Lin, Q., Chang, Y. N., Chiang, Y. L., Hay, C. M., Mohajeri, H., Davidson, B. L., and Bohn, M. C. (1997). Dopaminergic neurons protected from degeneration by GDNF gene therapy. *Science* **275,** 838–841.

Choi-Lundberg, D. L., Lin, Q., Schallert, T., Crippens, D., Davidson, B. L., Chang, Y. N., Chiang, Y. L., Qian, J., Bardwaj, L., and Bohn, M. C. (1998). Behavioral and cellular protection of rat dopaminergic neurons by an adenoviral vector encoding glial cell line-derived neurotrophic factor. *Exp. Neurol.* **154,** 261–275.

Connor, B., Kozlowski, D. A., Schallert, T., Tillerson, J. L., Davidson, B. L., and Bohn, M. C. (1999). Differential effects of glial cell line-derived neurotrophic factor (GDNF) in the striatum and substantia nigra of the aged Parkinsonian rat. *Gene Ther.* **6,** 1936–1951.

Connor, B., Kozlowski, D. A., Unnerstall, J. R., Elsworth, J. D., Tillerson, J. L., Schallert, T., and Bohn, M. C. (2001). Glial cell line-derived neurotrophic factor (GDNF) gene delivery protects dopaminergic terminals from degeneration. *Exp. Neurol.* **169,** 83–95.

Du, B., Wu, P., Boldt-Houle, D. M., and Terwilliger, E. F. (1996). Efficient transduction of human neurons with an adeno-associated virus vector. *Gene Ther.* **3,** 254–261.

Dunnett, S. B., and Bjorklund, A. (1999). Prospects for new restorative and neuroprotective treatments in Parkinson's disease. *Nature* **399,** A32–39.

During, M. J., Samulski, R. J., Elsworth, J. D., Kaplitt, M. G., Leone, P., Xiao, X., Li, J., Freese, A., Taylor, J. R., Roth, R. H., Sladek, J. R., Jr., O'Malley, K. L., and Redmond, D. E., Jr. (1998). In vivo expression of therapeutic human genes for dopamine production in the caudates of MPTP-treated monkeys using an AAV vector. *Gene Ther.* **5,** 820–827.

During, M. J., Kaplitt, M. G., Stern, M. B., and Eidelberg, D. (2001). Subthalamic GAD gene transfer in Parkinson disease patients who are candidates for deep brain stimulation. *Hum. Gene Ther.* **12,** 1589–1591.

Fan, D. S., Ogawa, M., Fujimoto, K. I., Ikeguchi, K., Ogasawara, Y., Urabe, M., Nishizawa, M., Nakano, I., Yoshida, M., Nagatsu, I., Ichinose, H., Nagatsu, T., Kurtzman, G. J., and Ozawa, K. (1998). Behavioral recovery in 6-hydroxydopamine-lesioned rats by cotransduction of striatum with tyrosine hydroxylase and aromatic L-amino acid decarboxylase genes using two separate adeno-associated virus vectors. *Hum. Gene Ther.* **9,** 2527–2535.

Gao, G. P., Alvira, M. R., Wang, L., Calcedo, R., Johnson, J., and Wilson, J. M. (2002). Novel adeno-associated viruses from rhesus monkeys as vectors for human gene therapy. *Proc. Nati. Acad. Sci. USA* **99,** 11854–11859.

Grondin, R., and Gash, D. M. (1998). Glial cell line-derived neurotrophic factor (GDNF): A drug candidate for the treatment of Parkinson's disease. *J. Neurol.* **245,** P35–42.

Hardy, J., Cookson, M. R., and Singleton, A. (2003). Genes and parkinsonism. *Lancet Neurol.* **2,** 221–228.

High, K. A. (2001). AAV-mediated gene transfer for hemophilia. *Ann. N.Y. Acad. Sci.* **953,** 64–74.

Hoshiga, M., Hatakeyama, K., Watanabe, M., Shimada, M., and Kagamiyama, H. (1993). Autoradiographic distribution of [14C]tetrahydrobiopterin and its developmental change in mice. *J. Pharmacol. Exp. Ther.* **267,** 971–978.

Jenner, P. (2000). Factors influencing the onset and persistence of dyskinesia in MPTP-treated primates. *Ann. Neurol.* **47,** S90–99.

Kaddis, F. G., Clarkson, E. D., Weber, M. J., Vandenbergh, D. J., Donovan, D. M., Mallet, J., Horellou, P., Uhl, G. R., and Freed, C. R. (1997). Intrastriatal grafting of Cos cells stably expressing human aromatic L-amino acid decarboxylase: Neurochemical effects. *J. Neurochem.* **68,** 1520–1526.

Kang, U. J., Lee, W. Y., and Chang, J. W. (2001). Gene therapy for Parkinson's disease: Determining the genes necessary for optimal dopamine replacement in rat models. *Hum. Cell* **14,** 39–48.

Kaplitt, M. G., Leone, P., Samulski, R. J., Xiao, X., Pfaff, D. W., O'Malley, K. L., and During, M. J. (1994). Long-term gene expression and phenotypic correction using adeno-associated virus vectors in the mammalian brain. *Nat. Genet.* **8,** 148–154.

Kaspar, B. K., Erickson, D., Schaffer, D., Hinh, L., Gage, F. H., and Peterson, D. A. (2002). Targeted retrograde gene delivery for neuronal protection. *Mol. Ther.* **5,** 50–56.

Kirik, D., Rosenblad, C., Bjorklund, A., and Mandel, R. J. (2000). Long-term rAAV-mediated gene transfer of GDNF in the rat Parkinson's model: Intrastriatal but not intranigral transduction promotes functional regeneration in the lesioned nigrostriatal system. *J. Neurosci.* **20,** 4686–4700.

Kirik, D., Georgievska, B., Burger, C., Winkler, C., Muzyczka, N., Mandel, R. J., and Bjorklund, A. (2002). Reversal of motor impairments in parkinsonian rats by continuous intrastriatal delivery of L-dopa using rAAV-mediated gene transfer. *Proc. Natl. Acad. Sci. USA* **99,** 4708–4713.

Klein, R. L., Lewis, M. H., Muzyczka, N., and Meyer, E. M. (1999). Prevention of 6-hydroxydopamine-induced rotational behavior by BDNF somatic gene transfer. *Brain Res.* **847,** 314–320.

Kordower, J. H., Palfi, S., Chen, E. Y., Ma, S. Y., Sendera, T., Cochran, E. J., Mufson, E. J., Penn, R., Goetz, C. G., and Comella, C. D. (1999). Clinicopathological findings following intraventricular glial-derived neurotrophic factor treatment in a patient with Parkinson's disease. *Ann. Neurol.* **46,** 419–424.

Kordower, J. H., Emborg, M. E., Bloch, J., Ma, S. Y., Chu, Y., Leventhal, L., McBride, J., Chen, E. Y., Palfi, S., Roitberg, B. Z., Brown, W. D., Holden, J. E., Pyzalski, R., Taylor, M. D., Carvey, P., Ling, Z., Trono, D., Hantraye, P., Deglon, N., and Aebischer, P. (2000). Neurodegeneration prevented by lentiviral vector delivery of GDNF in primate models of Parkinson's disease. *Science* **290,** 767–773.

Kozlowski, D. A., Connor, B., Tillerson, J. L., Schallert, T., and Bohn, M. C. (2000). Delivery of a GDNF gene into the substantia nigra after a progressive 6-OHDA lesion maintains functional nigrostriatal connections. *Exp. Neurol.* **166,** 1–15.

Lange, H., Thorner, G., Hopf, A., and Schroder, K. F. (1976). Morphometric studies of the neuropathological changes in choreatic diseases. *J. Neurol. Sci.* **8,** 401–425.

Langston, J. W., Quik, M., Petzinger, G., Jakowec, M., and Di Monte, D. A. (2000). Investigating levodopa-induced dyskinesias in the parkinsonian primate. *Ann. Neurol.* **47,** S79–89.

Leff, S. E., Spratt, S. K., Snyder, R. O., and Mandel, R. J. (1999). Long-term restoration of striatal L-aromatic amino acid decarboxylase activity using recombinant adeno-associated viral vector gene transfer in a rodent model of Parkinson's disease. *Neuroscience* **92,** 185–196.

Levine, R. A., Miller, L. P., and Lovenberg, W. (1981). Tetrahydrobiopetrin in striatum: localization in dopamine nerve terminals and role in catecholamine synthesis. *Science* **214,** 919–921.

Lieberman, D. M., Laske, D. W., Morrison, P. F., Bankiewicz, K. S., and Oldfield, E. H. (1995). Convection-enhanced distribution of large molecules in gray matter during interstitial drug infusion. *J. Neurosurg.* **82,** 1021–1029.

Mandel, R. J., Spratt, S. K., Snyder, R. O., and Leff, S. E. (1997). Midbrain injection of recombinant adeno-associated virus encoding rat glial cell line-derived neurotrophic factor protects nigral neurons in a progressive 6-hydroxydopamine-induced degeneration model of Parkinson's disease in rats. *Proc. Natl. Acad. Sci. USA* **94,** 14083–14088.

Mandel, R. J., Rendahl, K. G., Spratt, S. K., Snyder, R. O., Cohen, L. K., and Leff, S. E. (1998). Characterization of intrastriatal recombinant adeno-associated virus-mediated gene transfer of human tyrosine hydroxylase and human GTP-cyclohydrolase I in a rat model of Parkinson's disease. *J. Neurosci.* **18,** 4271–4284.

Mandel, R. J., Snyder, R. O., and Leff, S. E. (1999). Recombinant adeno-associated viral vector-mediated glial cell line-derived neurotrophic factor gene transfer protects nigral dopamine neurons after onset of progressive degeneration in a rat model of Parkinson's disease. *Exp. Neurol.* **160,** 205–214.

Matsushita, T., Elliger, S., Elliger, C., Podsakoff, G., Villarreal, L., Kurtzman, G. J., Iwaki, Y., and Colosi, P. (1998). Adeno-associated virus vectors can be efficiently produced without helper virus. *Gene Ther.* **5,** 938–945.

Miller, D. G., Rutledge, E. A., and Russell, D. W. (2002). Chromosomal effects of adeno-associated virus vector integration. *Nat. Genet.* **30,** 147–148.

Mochizuki, H., Hayakawa, H., Migita, M., Shibata, M., Tanaka, R., Suzuki, A., Shimo-Nakanishi, Y., Urabe, T., Yamada, M., Tamayose, K., Shimada, T., Miura, M., and Mizuno, Y. (2001). An AAV-derived Apaf-1 dominant negative inhibitor prevents MPTP toxicity as antiapoptotic gene therapy for Parkinson's disease. *Proc. Natl. Acad. Sci. USA* **98,** 10918–10923.

Muramatsu, S., Mizukami, H., Young, N. S., and Brown, K. E. (1996). Nucleotide sequencing and generation of an infectious clone of adeno-associated virus 3. *Virology* **221,** 208–217.

Muramatsu, S., Fujimoto, K., Ikeguchi, K., Shizuma, N., Kawasaki, K., Ono, F., Shen, Y., Wang, L., Mizukami, H., Kume, A., Matsumura, M., Nagatsu, I., Urano, F., Ichinose, H., Nagatsu, T., Terao, K., Nakano, I., and Ozawa, K. (2002). Behavioral recovery in a primate model of Parkinson's disease by triple transduction of striatal cells with adeno-associated viral vectors expressing dopamine-synthesizing enzymes. *Hum. Gene Ther.* **13,** 345–354.

Nagatsu, I., Arai, R., Sakai, M., Yamawaki, Y., Takeuchi, T., Karasawa, N., and Nagatsu, T. (1997). Immunohistochemical colocalization of GTP cyclohydrolase I in the nigrostriatal system with tyrosine hydroxylase. *Neurosci. Lett.* **224,** 185–188.

Nagatsu, T., and Ichinose, H. (1999). Molecular biology of catecholamine-related enzymes in relation to Parkinson's disease. *Cell. Mol. Neurobiol.* **19,** 57–66.

Nagatsu, T., Horikoshi, T., Sawada, M., Nagatsu, I., Kondo, T., Iizuka, R., and Narabayashi, H. (1987). Biosynthesis of tetrahydrobiopterin in parkinsonian human brain. *Adv. Neurol.* **45,** 223–226.

Nakamura, K., Ahmed, M., Barr, E., Leiden, J. M., and Kang, U. J. (2000). The localization and functional contribution of striatal aromatic L-amino acid decarboxylase to L-3,4-dihydroxyphenylalanine decarboxylation in rodent parkinsonian models. *Cell Transplant* **9,** 567–576.

Natsume, A., Mata, M., Goss, J., Huang, S., Wolfe, D., Oligino, T., Glorioso, J., and Fink, D. J. (2001). Bcl-2 and GDNF delivered by HSV-mediated gene transfer act additively to protect dopaminergic neurons from 6-OHDA-induced degeneration. *Exp. Neurol.* **169,** 231–238.

Rutledge, E. A., Halbert, C. L., and Russell, D. W. (1998). Infectious clones and vectors derived from adeno-associated virus (AAV) serotypes other than AAV type 2. *J. Virol.* **72,** 309–319.

Sanchez-Pernaute, R., Harvey-White, J., Cunningham, J., and Bankiewicz, K. S. (2001). Functional effect of adeno-associated virus mediated gene transfer of aromatic L-amino acid decarboxylase into the striatum of 6-OHDA-lesioned rats. *Mol. Ther.* **4,** 324–330.

Shen, Y., Muramatsu, S. I., Ikeguchi, K., Fujimoto, K. I., Fan, D. S., Ogawa, M., Mizukami, H., Urabe, M., Kume, A., Nagatsu, I., Urano, F., Suzuki, T., Ichinose, H., Nagatsu, T., Monahan, J., Nakano, I., and Ozawa, K. (2000). Triple transduction with adeno-associated virus vectors expressing tyrosine hydroxylase, aromatic-L-amino-acid decarboxylase, and GTP cyclohydrolase I for gene therapy of Parkinson's disease. *Hum. Gene Ther.* **11,** 1509–1519.

Srivastava, A., Lusby, E. W., and Berns, K. I. (1983). Nucleotide sequence and organization of the adeno-associated virus 2 genome. *J. Virol.* **45,** 555–564.

Wachtel, S. R., Bencsics, C., and Kang, U. J. (1997). Role of aromatic L-amino acid decarboxylase for dopamine replacement by genetically modified fibroblasts in a rat model of Parkinson's disease. *J. Neurochem.* **69,** 2055–2063.

Walton, K. M. (1999). GDNF: A novel factor with therapeutic potential for neurodegenerative disorders. *Mol. Neurobiol.* **19,** 43–59.

Wang, L., Muramatsu, S., Lu, Y., Ikeguchi, K., Fujimoto, K., Okada, T., Mizukami, H., Hanazono, Y., Kume, A., Urano, F., Ichinose, H., Nagatsu, T., Nakano, I., and Ozawa, K. (2002). Delayed delivery of AAV-GDNF prevents nigral neurodegeneration and promotes functional recovery in a rat model of Parkinson's disease. *Gene Ther.* **9,** 381–389.

Xiao, W., Chirmule, N., Berta, S. C., McCullough, B., Gao, G., and Wilson, J. M. (1999). Gene therapy vectors based on adeno-associated virus type 1. *J. Virol.* **73,** 3994–4003.

Zhong, X. H., Haycock, J. W., Shannak, K., Robitaille, Y., Fratkin, J., Koeppen, A. H., Hornykiewicz, O., and Kish, S. J. (1995). Striatal dihydroxyphenylalanine decarboxylase and tyrosine hydroxylase protein in idiopathic Parkinson's disease and dominantly inherited olivopontocerebellar atrophy. *Mov. Disord.* **10,** 10–17.

# HERPES SIMPLEX VIRUS VECTORS FOR PARKINSON'S DISEASE

David S. Latchman

Institute of Child Health
University College London
London WC1N 1EH, United Kingdom



## I. Introduction

Virus vectors represent a high efficiency means of gene delivery both to cultured cells *in vitro* and to the intact animal or human patient *in vivo*. As such they have two potential uses in Parkinson's disease research. First, viral vectors may ultimately be used to deliver specific genes that will be of therapeutic benefit to patients in a gene therapy procedure. Second, viral vectors can be used to deliver genes that have been identified as mutated in specific cases of Parkinson's disease. Although such strategies may ultimately lead to novel therapeutic procedures, their initial aim is to evaluate the consequences of altering the expression of wild-type or mutant forms of these genes so as to indicate their normal functional role and elucidate the manner in which their mutation results in Parkinson's disease.

These two uses will be discussed in turn.




0074-7742/03 $35.00

## II. Gene Therapy and Parkinson's Disease

Gene therapy refers to the use of a specific gene as a means of treating a particular human disease (for review see Miller, 1992; French Anderson, 1998). Initially, the potential use of such a therapy was suggested to be simple genetic diseases where the individual's disease was caused by a defective copy of a particular gene. It would then be possible to treat the disease by providing a functional copy of that gene to complement the defect.

However, it has now become clear that gene therapy could also potentially be used in diseases that do not have a genetic component or in which that genetic component has not yet been characterized. Thus, in this case, the gene used in the gene therapy procedure would be one whose protein product could relieve the symptoms of the disease. For example, in cancer one could deliver a gene encoding a factor that would render the tumor more recognizable by the immune system so promoting its rejection. Similarly, in neurodegenerative diseases one could supply the gene encoding a neurotrophic factor that could promote survival of the neurons that would otherwise die.

In this second type of gene therapy, Parkinson's disease represents a prime candidate for the application of the technique. This is because of the use of neuronal transplants from aborted fetuses to treat both animal models and human individuals with Parkinson's disease (Widner *et al.*, 1992; Lindvall *et al.*, 1996; Olanow *et al.*, 1996; Freed *et al.*, 2001). The relative success of this work, despite the lack of fetal material and ethical problems, suggests that specific interventions, for example, enhancing dopamine production, may be of value in the treatment of human Parkinson's disease (Freese *et al.*, 1996). Moreover, gene therapy would theoretically allow standardized batches of the therapeutic agent to be produced in large amounts avoiding the problems of quantity and quality encountered with fetal material resulting in material from up to 12 aborted fetuses being used to treat a single patient.

## III. Therapeutic Genes for Parkinson's Disease

### A. Tyrosine Hydroxylase

Given that Parkinson's disease results from a loss of dopaminergic neurons, one means of treating the disease would be to enhance the supply of dopamine and this effect is likely to be the primary reason for the beneficial effects noted with fetal transplants. An obvious means of enhancing dopamine production would be to enhance the levels of the tyrosine hydroxylase enzyme that is the rate-limiting step of the dopaminergic pathway. Clearly, however, tyrosine hydroxylase cannot be supplied simply as a protein as it will not be taken up by the cells that require it. Therefore it is necessary to express tyrosine hydroxylase either within genetically modified cells that produce dopamine or by direct delivery of the gene encoding this protein. Interestingly, it has been demonstrated that optimal tyrosine hydroxylase activity can be obtained only in the presence of sufficient levels of tetrahydrobiopterin and it is likely therefore to be necessary to also introduce the gene encoding the enzyme GTP-cyclohydrolase 1, which catalyzes the rate-limiting step in the tetrahydrobiopterin-synthesizing pathway (Bencsjcs *et al.*, 1996; Mandel *et al.*, 1998).

## B. Neurotrophic Factors

In addition to boosting the supply of dopamine by exogenous means, it would also be possible to produce a therapeutic effect by slowing or completely inhibiting the death of dopaminergic neurons, which is characteristic of Parkinson's disease. Thus, patients will present when a certain number of neurons have been lost but will continue to inexorably deteriorate as further neurons are lost during the progress of the disease. A means of arresting such a loss would therefore be of considerable therapeutic benefit.

Importantly, the neurotrophic factor glial-derived neurotrophic factor (GDNF) has been shown to be an important survival factor for mid-brain dopaminergic neurons that are lost in Parkinson's disease (Lin *et al.*, 1993). Moreover, it has been shown that GDNF itself can be of therapeutic benefit in both the 6-hydroxydopamine rat model of Parkinson's disease and in primate models of the disease (for review see Lapchak, 1996; Lapchak *et al.*, 1996). These studies involved the delivery of GDNF to the animal brain where it was effective since it acts by binding to a cell surface receptor and thus, unlike tyrosine hydroxylase, does not need to enter the cell to be effective. It would therefore be possible to similarly deliver GDNF to the human brain without the need for cell transplants or DNA injection. However, previous attempts to deliver neurotrophic factors in this way to humans have failed due to problems with efficacy and toxic side effects and similar attempts with GDNF in human Parkinson's disease has been unsuccessful (Kordower *et al.*, 1999). Thus, as with tyrosine hydroxylase, the most effective means of delivering GDNF is likely to be by delivering the gene encoding it.

## C. Other Potential Genes

Although most attention has focused on tyrosine hydroxylase and GDNF as potential therapeutic genes, it has been shown that sonic hedgehog, a signaling molecule that is related to transforming growth factor-$\beta$ is able to induce the production of dopaminergic neurons (Hynes *et al.*, 1995). It is possible therefore that delivery of the gene encoding sonic hedgehog or other genes with similar activity could be used to enhance the numbers of dopaminergic neurons by treating cell cultures prior to transplantation or by direct injection of DNA into the brain.

Similarly, overexpression of the antiapoptotic molecule Bcl-2 using a herpes simplex virus vector has been shown to be of benefit in an animal model of Parkinson's disease (Yamada *et al.*, 1999) and this approach could potentially be used to reduce neuronal cell death in human patients, either alone or in conjunction with GDNF.

## IV. Genetics of Parkinson's Disease

Although familial Parkinson's disease caused by a single gene mutation is relatively rare compared to nonfamilial disease, these cases are more easily analyzed and the gene responsible for Parkinson's disease has been defined in a number of such families (for reviews see Riess *et al.*, 1998; de Silva *et al.*, 2000). Most importantly, it is clear that insights gained from characterizing these genes will be of value in understanding the pathological processes that may be common to all forms of the disease. Two specific genes of this type will therefore be discussed.

### A. $\alpha$-Synuclein

In several unrelated families in which Parkinson's disease was shown to be inherited as a dominant mutation, the gene involved was identified as $\alpha$-synuclein (Polymeropoulos *et al.*, 1997; Krüger *et al.*, 1998). $\alpha$-Synuclein is a 140 amino acid protein of unknown function that is highly expressed in the brain, being particularly abundant at presynaptic terminals (for review see Clayton and George, 1999). Two distinct mutations in $\alpha$-synuclein have been identified in families with Parkinson's disease involving either an alanine-to-proline mutation at position 30 of the protein or an alanine-to-threonine change at position 53 (Polymeropoulos *et al.*, 1997; Krüger *et al.*, 1998).

Interestingly, knockout mice lacking $\alpha$-synuclein show mild defects in dopaminergic function but do not exhibit severe abnormalities resembling those found in Parkinson's disease (Abeliovich *et al.*, 2000). This indicates that the mutations do not cause the disease simply by abolishing the function of $\alpha$-synuclein. Rather, in agreement with their dominant negative nature, they are likely to act via some additional deleterious function.

It is possible that this deleterious function may also be observed with the wild-type protein since overexpression of wild-type $\alpha$-synuclein in transgenic mice resulted in the formation of inclusion bodies and dopaminergic cell loss, although the mutant proteins were not tested in this system (Masliah *et al.*, 2000). Similarly, overexpression of wild-type $\alpha$-synuclein, or the two disease-associated mutants in *Drosophila*, resulted in intracellular aggregate formation and death of dopaminergic neurons with locomotion defects also being observed in flies expressing the mutant protein (Feany and Bender, 2000).

These findings indicate that it would be of particular interest to use viral vectors to deliver either wild-type $\alpha$-synuclein or the mutant forms to cultured neuronal cells *in vitro* or to localized regions of the brain (such as the substantia nigra or the striatum) *in vivo*. The effects of the different forms can then be compared. The need to carry out such experiments is reinforced by the finding that wild-type $\alpha$-synuclein is deposited in Lewy bodies in nonfamilial Parkinson's disease indicating that it is involved in the disease process even when the normal protein is made by an unmutated gene (Spillantini *et al.*, 1997, 1998).

### B. Parkin

Although mutations in $\alpha$-synuclein are a relatively rare cause of familial Parkinson's disease, recessive mutations in the parkin gene have been shown to be the major cause of early-onset familial Parkinson's disease (Kitada *et al.*, 1998; Lücking *et al.*, 2000). Unlike mutations in $\alpha$-synuclein, parkin mutations are recessive and are likely to represent loss of function mutations since they include gene deletions and truncating mutations.

Most interestingly, the parkin protein has been shown to be a ubiquitin protein ligase that can catalyze the addition of ubiquitin to proteins targeted for degradation (Shimura *et al.*, 2000). This potential involvement of the ubiquitin protein turnover system in Parkinson's disease is supported by the finding of a single family in which Parkinson's disease is associated with a mutation in the gene encoding UCH-LI, which is a ubiquitin carboxy-terminal hydrolase (Leroy *et al.*, 1998). It is likely therefore that abnormalities in protein turnover underlie these cases of familial Parkinson's disease and that defects in this process may also be involved in nonfamilial PD.

Clearly this possibility could be tested by using viral vectors to manipulate the expression of parkin and/or UCH-LI either singly or in combination. Moreover, because the parkin mutations, unlike those in $\alpha$-synuclein, are recessive and appear to abolish the protein's function, it will be necessary also to use viruses expressing the antisense strand of the parkin gene to investigate the effect of reducing its expression.

Hence, manipulating the expression of the genes involved in familial Parkinson's disease should provide insights into the manner in which these genes cause disease and into the potential involvement of specific cellular processes in nonfamilial Parkinson's disease. Ultimately also, alteration of their expression by viral vectors may be of therapeutic benefit in Parkinson's disease providing a more direct form of gene therapy than the symptomatic gene therapy discussed above involving, for example, tyrosine hydroxylase or GDNF.

## V. Gene Delivery to the Nervous System

There therefore exists a need to develop vectors for the *in vivo* delivery of genes relevant to Parkinson's disease both to enhance our understanding of the role of specific genes and to develop therapeutic procedures. As discussed elsewhere in this book, a variety of methods have been used to deliver genes to the nervous system in animal models of Parkinson's disease and other diseases.

Thus, for example, therapeutic effects have been noted in animal models of Parkinson's disease with vectors derived from adenovirus (Horellou *et al.*, 1994; Bilang-Bleuel *et al.*, 1997), adeno-associated virus (Kaplitt *et al.*, 1994; see also Muramatsu *et al.*, this volume) or herpes simplex virus (During *et al.*, 1994; Natsume *et al.*, 2001) encoding tyrosine hydroxylase or GDNF. Obviously, however, these viral methods have potentially greater safety problems that would not be a problem with nonviral methods, although these methods are currently of much lower efficiency. This point is emphasized by the recent death of a patient in a gene therapy trial with an adenoviral vector.

Hence a major challenge is to obtain a viral vector that is highly safe but also highly effective. Although vectors derived from several different viruses may ultimately be of therapeutic benefit in human Parkinson's disease, we believe that herpes simplex virus (HSV) has a number of advantages over the other viruses.

## VI. HSV as a Vector

A number of laboratories have attempted to use HSV as a vector for gene delivery. Two basic systems for doing this have been devised. The amplicon system relies on introducing the gene of interest into a plasmid that contains an HSV origin of replication and packaging signal. This construct is introduced into cells by transfection and the cells are then infected with a helper HSV virus resulting in the amplicon becoming packaged into viral particles. This system is extensively discussed in other reviews (Fink *et al.*, 1996; Kennedy, 1997) and will not be discussed in detail here.

The alternative system that is used by our laboratory involves the introduction of the gene of interest directly into the HSV genome. This is achieved by cloning the gene into a plasmid vector in such a way that it is flanked by specific HSV viral sequences (Fig. 1). When this plasmid is transfected into cells together with HSV DNA, recombination will take place between the viral sequences in the plasmid vector and the corresponding sequences in the virus genome, resulting in the introduction of the foreign gene into the HSV genome (Fig. 1). Recombinant viruses containing the gene of interest can be detected among the nonrecombinant virus, by using a reporter gene such as $\beta$-galactosidase or green fluorescent protein (GFP). If such a gene is included in the plasmid construct together with the gene of interest, the recombinant plaques can then be recognized by their blue or green color. Alternatively, one can begin with a virus that already produces blue or green plaques and convert this to a white plaque by replacing the $\beta$-galactosidase or GFP gene with the gene of interest. This system

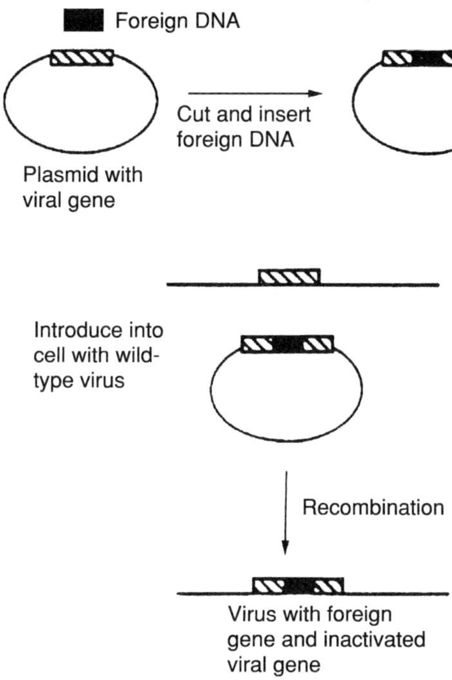


FIG. 1. Introduction of foreign DNA into the HSV-1 genome by recombination.

thus allows foreign DNA sequences to be readily introduced into the HSV genome and the resulting recombinant viruses can be purified relatively easily (for more detailed reviews of this method see Freese *et al.*, 1996; Latchman, 2001).

It is thus possible to introduce foreign genes into HSV and use it as a gene delivery vector. Such HSV-based vectors have a number of advantages over systems based on other viruses but also suffer from several disadvantages that need to be overcome before these vectors could be used for safe and efficient gene therapy of human Parkinson's disease. These advantages and disadvantages are discussed in the next sections.

## A. Advantages of HSV as a Vector

### 1. *Latent Infection of the Nervous System*

Although lentiviruses, adeno-associated virus, and adenovirus itself can all infect neuronal cells in the brain when delivered by direct injection, only HSV naturally infects human neuronal cells to produce asymptomatic latent infections (for reviews see Roizman and Sears, 1987; Latchman, 1990). Such latent infections are part of the normal life cycle of HSV in which it produces a lytic infection of peripheral skin cells and then migrates up the nerve process to establish a latent infection in sensory neurons innervating the site of the initial lesions. The periodic reactivation of the virus from these sites and its remigration to the periphery results in the characteristic pattern of recurrent cold sores or genital sores produced, respectively, by infection with HSV-1 and HSV-2. Interestingly, these infections are entirely asymptomatic in terms of the neuron itself and do not produce any damage to the nerve cell even when a reactivation occurs leading to lytic replication at the periphery. This ability to establish latent infections in neuronal cells offers an important basis for vector development, provided the other damaging effects of HSV such as its ability to replicate in nonneuronal cells can be eliminated (see below).

### 2. *Growth in Culture*

Evidently a virus may be highly efficient at incorporating foreign genes and at delivering them to neuronal cells, but it will be of no use as a vector if it cannot be grown efficiently in culture. Thus, it is necessary to grow the virus to produce sufficient stocks for experimental or therapeutic purposes. It is therefore of importance that HSV will replicate lytically in a wide range of cell types and can be grown to high titers ($10^8$–$10^9$ pfu/ml) in culture. This contrasts, for example, with the related herpes virus varicella zoster, which has proved virtually impossible to grow in culture, as well as with some of the other proposed virus vectors such as lentiviruses or adeno-associated virus, which have proved very difficult to grow to high titers in culture such that appropriate scaling up from the research laboratory setting could be effectively achieved.

3. *Genome Size*

Perhaps the greatest advantage possessed by HSV as a vector, however, is the fact that it has a very large genome that is considerably bigger than that of all the other competing virus vectors (Table I). This is of particular importance because a virus can normally package into infectious viral particles only the amount of DNA that it normally packages in its life cycle. Hence, even if all the genetic material in the smaller viruses were replaced with the desired gene, it would be impossible to introduce a very large gene or even two genes of reasonable size. In contrast, HSV contains many genes that can be removed without producing a significant effect on viral growth and hence could accommodate up to 30–40 kb of foreign genetic material.

This is of particular importance since it is likely that the effective therapy of Parkinson's disease will require the delivery of multiple genes. Thus, for example, an optimal effect may be achieved by delivering GDNF to stimulate the survival of the remaining dopaminergic neurons while also delivering tyrosine hydroxylase to stimulate further dopamine production. Indeed, as noted above, it has also been suggested that an optimal effect of tyrosine hydroxylase requires the delivery of the gene encoding GTP cyclohydrolase 1 to allow production of tetrahydrobiopterin to activate tyrosine hydroxylase activity (Bencsjcs *et al.*, 1996; Mandel *et al.*, 1998). Similarly, it has recently been shown that delivery of the genes encoding GDNF and the antiapoptotic factor Bcl-2 using HSV vectors produces enhanced neuronal survival in an animal model of Parkinson's disease compared to delivery of either gene alone (Natsume *et al.*, 2001). Evidently, as well as this potential therapeutic use, HSV could also be used to simultaneously manipulate the expression of two or more genes identified in familial Parkinson's disease so as to elucidate any interactions between them.

TABLE I
GENOME SIZE OF DIFFERENT POTENTIAL VIRUS VECTORS

| Virus | Genome size (base pairs) |
|---|---|
| Adeno-associated virus | 8,500 |
| Adenovirus | 35,000 |
| Lentivirus | 10,000 |
| Herpes simplex virus | 150,000 |

## B. Problems with the Use of HSV as a Vector

### 1. *Damaging Effects*

As noted above, in its natural life cycle HSV normally replicates in epithelial cells producing cold sores or genital sores. Hence it is a natural human pathogen. More seriously, the direct introduction of HSV into the brain as would be required in gene therapy procedures will result in a lethal encephalitis due to viral replication, which would rapidly result in the death of the individual or animal. Clearly, therefore, the virus must be disabled in some way if it is to be used as a vector. It should be noted that such disablement is required not only where the gene of interest is introduced directly into the HSV vector but also in the amplicon system where the helper will need to be disabled if damaging effects are not to occur.

In considering how to disable the virus, it is necessary to bear in mind that such disablement needs to be balanced with the need to grow the virus efficiently in culture, so that viral stocks can be prepared. Two basic methods of disabling the virus are available:

1. The simplest method of disabling the virus is to remove genes whose protein products are required for lytic replication in all cell types. Thus, viruses have been produced that lack either of the essential immediate early genes encoding ICP4 (Chiocca *et al.*, 1990; Dobson *et al.*, 1990) or ICP27 (Howard *et al.*, 1998). Removal of either of these genes renders the virus incapable of lytic replication in all cell types so that it does not produce encephalitis when directly injected into the brain. The disadvantage of this method is that to grow the virus, the gene encoding ICP4 or ICP27 must be provided in specially constructed cell lines so as to complement the defect and allow viral growth. In doing this it is essential to ensure that there is no overlap between the virus gene in the cell line and the sequences remaining in the virus itself since otherwise recombination will occur to generate a wild-type virus. It should also be noted that HSV also expresses three other immediate-early genes encoding ICP0, ICP22, and ICP47 and these will continue to be expressed in viruses lacking ICP4 and ICP27 (see below).
2. To avoid the need to grow viruses on specially constructed cell lines and the possibility of recombination to produce a wild-type virus, an alternative means of disabling the virus is to remove a gene that is necessary for growth on nondividing cells such as neurons. Thus, a virus lacking the gene encoding ICP34.5 grows perfectly well in culture on dividing cells but is incapable of replicating when introduced into neuronal cells in the brain and is therefore avirulent and does not produce encephalitis (Maclean *et al.*, 1991). Hence, the removal of this gene could produce a

virus capable of delivering genes to nondividing cells in a safe and efficient manner while being readily grown in culture on dividing cells to prepare virus stocks.

In addition to the two methods noted above that completely prevent viral replication in specific cell types, it is also possible to inactivate viral genes that are important for efficient growth of the virus but whose inactivation does not completely prevent growth of the virus. Inactivation of one or more of these genes can be used as a further safety feature in conjunction with the disabling mutations described above. An example of this type of disablement involves the viral gene encoding the Vmw65 (VP16) protein. This protein normally acts to transactivate the five viral immediate-early genes such as those encoding ICP4 and ICP27 (Goding and O'Hare, 1989). A mutation that inactivates this function of Vmw65 does not prevent viral replication but does make the virus grow more poorly, particularly at low multiplicities of infection (Ace *et al.*, 1989). We have found that this inactivation of Vmw65 together with deletion of ICP4 and ICP27 effectively prevents viral gene expression, thereby minimizing toxicity (Thomas *et al.*, 1999). It also represents an added safety feature preventing high level production of any of the immediate-early proteins that have not been specifically inactivated.

In studies in our laboratory we showed that although a virus lacking only ICP34.5 does not produce encephalitis when injected into the brain as would be required for the therapy of Parkinson's disease, it also does not produce high efficiency gene delivery when used in this way (Coffin *et al.*, 1996). However, when further disabled by removal of the gene encoding ICP27 whose protein product is required for replication in all cell types, the doubly disabled virus produces high efficiency gene delivery to the brain (Howard *et al.*, 1998).

This virus does, however, have to be grown on a cell line, which artificially expresses ICP27 since this protein is essential for lytic growth of the virus on all cell types (Sacks *et al.*, 1985). The possibility of recombination resulting in a functional ICP27 gene being regenerated within the virus is minimized, however, by completely deleting the ICP27 gene from the virus and ensuring that there is no overlap between the DNA sequences remaining in the virus genome and those contained within the cell line. Moreover, even if a rare recombination event were to occur, this would still not regenerate a wild-type virus, since the resulting virus would continue to lack ICP34.5 and would still not cause a damaging encephalitis.

Hence, this doubly disabled virus represents a relatively safe and efficient gene delivery vector, which can be used as a platform for further disablement. This has now been achieved by deleting the gene encoding the other essential immediate-early protein ICP4 and inactivating Vmw65 as described above. This results in a virus that lacks functional ICP4 or ICP27 and fails to produce significant amounts of the remaining immediate-early proteins ICP0, ICP22, or ICP47 (Thomas *et al.*,

1999). This virus has been shown to deliver genes with high efficiency and minimal toxicity to neuronal cells both *in vitro* and *in vivo* (Lilley *et al.*, 2001).

As well as growing this virus on a cell line expressing ICP4 and ICP27, it is also necessary to complement its deficiency in Vmw65 since viruses with a point mutation in this factor grow poorly (Ace *et al.*, 1989). However, unlike the genes for ICP4 or ICP27, which are completely deleted from the virus, this point mutation in the gene for Vmw65 cannot be complemented by expressing the corresponding wild-type gene in the cell line since recombination would occur between the wild-type and mutant genes to restore a wild-type Vmw65 gene. This mutation is therefore complemented by a novel strategy involving expressing the corresponding gene from equine herpes virus (EHV) in the cell line. This can functionally complement the mutation, but since the HSV and EHV genes are not homologous to one another, no recombination will occur (Thomas *et al.*, 1999).

This virus can thus be grown to high titer in culture and has four mutations preventing its lytic growth in the brain. Moreover, it delivers genes with high efficiency (Lilley *et al.*, 2001). It can therefore be used to effectively investigate the potential of gene therapy in animal models without any damaging effects due to viral infection and may ultimately be safe enough for therapeutic use in human Parkinson's disease.

2. *Long-Term Expression*

Although optimally disabled HSV can deliver a gene to a high number of neurons in the brain or the peripheral nervous system, such gene expression lasts for less than a week when most virus or cellular gene promoters are used to drive expression of the gene of interest. This is because when the virus enters latency in the neuronal cells, it shuts down expression of its own genes and similarly shuts down expression of the inserted gene. Clearly, this is not desirable if the virus is to represent an effective therapy for long-term diseases of the nervous system such as Parkinson's disease.

During viral latency, a single virus gene continues to be active. This gene produces a set of RNAs known as the latency-associated transcripts (LAT) (Fig. 2). Expression of this set of RNAs is driven by two promoters known as P1 and P2. Interestingly, when each of these promoters is placed upstream of a reporter gene within the HSV genome, P1 produces strong expression that is detectable only for short periods (Margolis *et al.*, 1992, 1993; Dobson *et al.*, 1995) whereas P2 produces very weak expression that is, however, detectable for long periods during latent infection (Ho and Mocarski, 1989; Goins *et al.*, 1994).

This suggested to us that the LAT P2 element may act as an enhancer that confers long-term expression upon the much stronger P1 promoter, allowing long-term expression during latency. We therefore tested whether LAT P2 was able to confer long-term expression upon other promoters when introduced into the virus genome. We were able to show that such an arrangement was able to

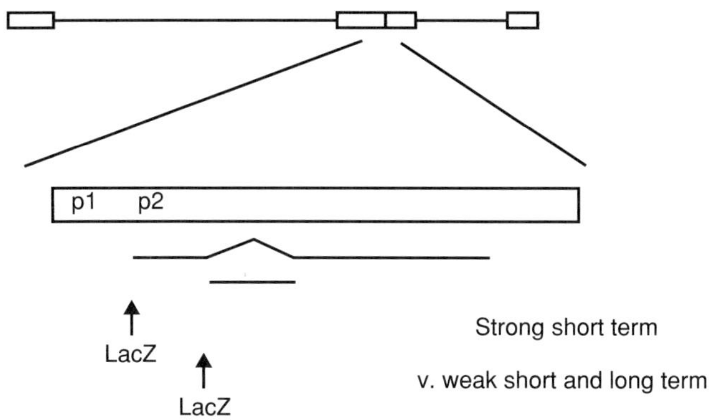


FIG. 2. Position of the Lat region (boxed) and the two promoters (p1 and p2) in the HSV genome. The two Lat RNA transcripts are illustrated. These are an 8-kb RNA that is spliced (as indicated by the triangle) to produce the 2-kb stable RNA. The effect of linking each of these promoters to a marker gene (LacZ) is also shown.

produce long-term expression of a reporter gene for up to 6 months *in vivo* (the longest period tested) and this could be observed from two promoters cloned in opposite orientations with the P2 element centrally located between them and therefore acting as a bidirectional long-term enhancer (Fig. 3) (Palmer *et al.*, 2000).

Hence this system not only overcomes the problem with long-term expression of genes inserted into the HSV genome but allows two genes to be expressed from the same construct under the control of different promoters. This may be of particular use in therapeutic applications such as Parkinson's disease where more than one gene needs to be expressed (see above).

## VII. HSV as a Vector for Parkinson's Disease

As discussed above, HSV has a number of advantages as a vector system for gene delivery to the nervous system. Moreover, considerable progress has been made with the disablement of the virus so that it represents a safe and effective

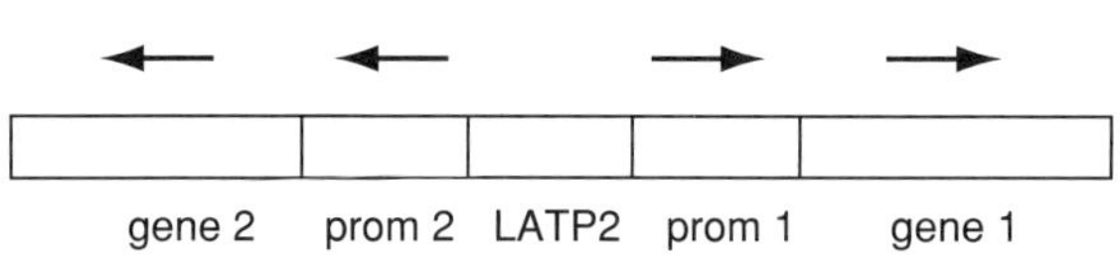


FIG. 3. Use of the Lat P2 element as an enhancer that can direct long-term gene expression from two promoters located in opposite orientations.

gene delivery system as well as with modifications that allow gene expression to continue in the long term.

We have therefore prepared a number of different derivatives of a disabled virus lacking functional genes encoding ICP27, ICP34.5, and the transactivator function of Vmw 65 (see above). These contain different combinations of genes whose protein products may be of therapeutic benefit in Parkinson's disease such as tyrosine hydroxylase (TH) or GDNF. These viruses have then been tested in the rat 6-hydroxydopamine model of Parkinson's disease in which dopaminergic neurons are destroyed on one side of the brain by a unilateral injection of 6-hydroxydopamine into the substantia nigra. The effect of striatal injection of the virus carrying the potential therapeutic gene on the rotational behavior induced by a subsequent injection of apomorphine is then assessed and compared to the effect of control virus.

Figure 4 shows a typical result in which a virus encoding TH was injected. The virus produces a strong reduction in rotational behavior for up to 20 weeks after injection compared to the effect of control virus expressing only $\beta$-galactosidase. Interestingly, improvement is observed when viruses expressing TH or GDNF are injected after the 6-hydroxydopamine lesion, as well as before, thereby mimicking the situation in the human disease where therapeutic intervention would have to occur after the onset of dopaminergic neuronal death.

Clearly, further studies are required in both the rat model and the MPTP marmoset model of Parkinson's disease to determine the optimal combination of therapeutic genes. Similarly, it will be necessary to test the effect on safety

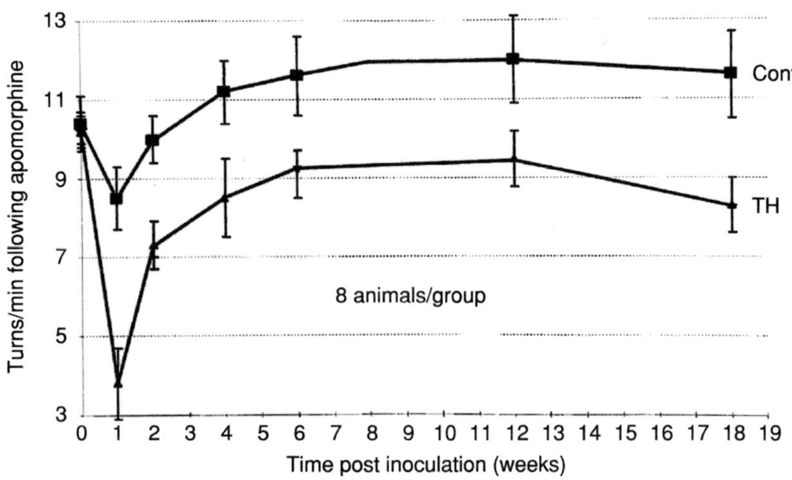


FIG. 4. Effect of control virus or virus encoding tyrosine hydroxylase on apomorphine-induced rotation in rats lesioned via injection of 6-hydroxydopamine.

and toxicity of further viral disablement such as the removal of the essential immediate-early gene encoding ICP4, which is likely to be required for human use. However, once these further studies have been completed, it is possible to envisage clinical trials with these viruses, particularly in patients in whom dopamine drug therapy has failed.

One potential problem in human use of these viruses is that the majority of humans, unlike animals, have latent wild-type HSV, predominantly in the dorsal root ganglia of the peripheral nervous system (PNS) but also in the central nervous system (CNS). Hence, injection of disabled recombinant virus could result in reactivation of potentially dangerous wild-type virus or recombination with it. However, it appears that this is unlikely to occur. Thus, studies with animals containing latent wild-type HSV in the CNS and PNS have not shown any reactivation following injection of disabled virus (Wang *et al.*, 1997). Similar results were obtained in a recent phase I clinical trial (Rampling *et al.*, 2000) in human glioma patients involving intracranial injection of a virus lacking only ICP34.5 that is expected to replicate and kill the dividing glioma cells while not affecting nondividing neurons. Although the great majority of the patients in this study had latent wild-type HSV, no reactivation of the wild-type virus was observed, even though the injected virus was only minimally disabled and could carry out lytic replication in some cell types. In addition, no other damaging effects were noted in this trial despite the fact that to achieve its effect of killing the tumor cells, the virus has only a single disabling mutation that does not prevent replication in all cell types as opposed to the four genes that would be inactivated in gene therapy vectors for human use.

## VIII. Conclusion

HSV offers unique features as a gene delivery tool for the nervous system, notably its ability to establish asymptomatic latent infections in neuronal cells and its ability to accept multiple inserted genes. The effective disablement of the virus and its modification to allow long-term expression have already allowed it to be used to test the effect of particular gene products in cultured neuronal cells (see, for example, Smith *et al.*, 1998; Wagstaff *et al.*, 1999) or in the intact animal (Smith *et al.*, 2001) as well as to demonstrate therapeutic effects in a rat model of Parkinson's disease (see Fig. 4). Hence, this virus can be used to test the effect of manipulating genes such as $\alpha$-synuclein or parkin in cultured cells or the intact animal. Moreover, the further disablement of the virus and the use of combinations of several different genes may ultimately result in a vector that can be used in patients with Parkinson's disease as a therapeutic alternative to surgery or fetal neuronal transplants.

## Acknowledgments

I thank all the members of the laboratory working in this area and in particular Rob Coffin and Zi-qun Han for the data in Fig. 4. I also thank the Parkinson's Disease Society for its generous support of my work in this area.

## References

Abeliovich, A., Schmitz, Y., Fariñas, I., Choi-Lundberg, D., Ho, W.-H., Castillo, P. E., Shinsky, N., Garcia Verdugo, J. M., Armanini, M., Ryan, A., Hynes, M., Phillips, H., Sulzer, D., and Rosenthal, A. (2000). Mice lacking $\alpha$-synuclein display functional deficits in the nigrostriatal dopamine system. *Neuron* **25,** 239–252.

Ace, C. I., McKee, T. A., Ryan, M., Cameron, J. M., and Preston, C. M. (1989). Construction and characterization of a herpes simplex virus type I mutant unable to transduce immediate-early gene expression. *J. Virol.* **63,** 2260–2269.

Bencsjcs, C., Wachtel, S. R., Milstien, S., Hatakeyama, K., Becker, J. B., and Kang, U. J. (1996). Double transduction with GTP cyclohydrolase I and tyrosine hydroxylase is necessary for spontaneous synthesis of L-DOPA by primary fibroblasts. *J. Neurosci.* **16,** 4449–4456.

Bilang-Bleuel, A., Revah, F., Colin, P., Locquet, I., Robert, J.-J., Mallet, J., and Horellou, P. (1997). Intrastriatal injection of an adenoviral vector expressing glial-cell-line-derived neurotrophic factor prevents dopaminergic neuron degeneration and behavioural impairment in a rat model of Parkinson disease. *Proc. Nat. Acad. Sci. USA* **94,** 8818–8823.

Chiocca, A. E., Choi, B. B., Cai, W., DeLuca, N., Schaffer, P. A., DiFiglia, M., Breakfield, X. O., and Martuza, R. L. (1990). Transfer and expression of the lac Z gene in rat brain neurons by herpes simplex virus mutants. *New Biol.* **266,** 739–746.

Clayton, D. F., and George, J. M. (1999). Synucleins in synaptic plasticity and neurodegenerative disorders. *J. Neurosci. Res.* **58,** 120–129.

Coffin, R. S., Maclean, A. R., Latchman, D. S., and Brown, S. M. (1996). Gene delivery to the central and peripheral nervous systems of mice using HSV1 ICP34.5 deletion mutant vectors. *Gene Ther.* **3,** 886–891.

de Silva, R.H.A., Khan, N. L., and Wood, N. W. (2000). New developments in the genetics of Parkinson's disease. *Curr. Opin. Genet. Dev.* **10,** 292–298.

Dobson, A. T., Margolis, T. P., Sedarati, F., Stevens, J. G., and Feldman, L. T. (1990). A latent nonpathogenic HSV-1 derived vector stably expresses $\beta$ galactosidase in mouse neurons. *Neuron* **5,** 353–360.

Dobson, A. T., Margolis, T. P., Gomes, W. A., and Feldman, L. T. (1995). *In vivo* deletion analysis of the herpes simplex virus type 1 latency associated transcript promoter. *J. Virol.* **69,** 2264–2270.

During, M. J., Naegele, J. R., O'Malley, K. L., and Geller, A. I. (1994). Long-term behavioural recovery in parkinsonian rats by an HSV vector expressing tyrosine hydroxylase. *Science* **266,** 1399–1403.

Feany, M. B., and Bender, W. W. (2000). A *Drosophila* model of Parkinson's disease. *Nature* **404,** 394–398.

Fink, D. J., DeLuca, N. A., Goins, W. F., and Glorioso, J. C. (1996). Gene transfer to neurons using herpes simplex virus-based vectors. *Annu. Rev. Neurosci.* **19,** 265–287.

Freed, C. R., Greene, P. E., Breeze, R. E., Tsai, W.-Y., DuMouchel, W., Kao, R., Dillon, S., Winfield, H., Culver, S., Trojanowski, J. Q., Eidelberg, D., and Fahn, S. (2001). Transplantation of embryonic dopamine neurons for severe Parkinson's disease. *N. Engl. J. Med.* **344,** 710–719.

Freese, A., Stern, M., Kaplitt, M. G., O'Connor, W. M., Abbey, M. V., O'Connor, M. J., and During, M. J. (1996). Prospects for gene therapy in Parkinson's disease. *Mov. Disord.* **11,** 469–488.

French Anderson, W. (1998). Human gene therapy. *Nature.* **392**(Suppl.), 25–30.

Goding, C. R., and O'Hare, P. (1989). Herpes simplex virus Vmw65-octamer binding protein interaction: A paradigm for combinatorial control of transcription. *Virology* **173,** 363–367.

Goins, W. F., Stenberg, L. R., Croen, K. D., Krause, P. R., Hendricks, R. L., Fink, D. J., Straus, S. E., Levine, M., and Gloriso, J. C. (1994). A novel latency-associated promoter is contained within the herpes simplex virus type 1 UL flanking repeats. *J. Virol.* **68,** 2239–2252.

Ho, D. Y., and Mocarski, E. S. (1989). Herpes simplex virus latent RNA (Lat) is not required for latent infection in the mouse. *Proc. Natl. Acad. Sci. USA* **86,** 7596–7600.

Horellou, P., Vigne, E., Castel, M. N., Barneoud, P., Colin, P., Perricaudet, M., Delare, P., and Mallet, J. (1994). Direct intracerebral gene transfer of an adenoviral vector expressing tyrosine hydroxylase in a rat model of Parkinson's disease. *NeuroReport* **6,** 49–53.

Howard, M. K., Kershaw, T., Gibb, B., Storey, W., Maclean, A. R., Zeng, B.-W., Jenner, P., Brown, S. M., Woolf, C. J., Anderson, P. N., Coffin, R. S., and Latchman, D. S. (1998). High efficiency gene transfer to the central nervous system of rodents and primates using herpes virus vectors lacking functional ICP27 and ICP34.5. *Gene Ther.* **5,** 1137–1147.

Hynes, M., Porter, J. A., Chiang, C., Chang, D., Tessier-Lavigne, M., Beachy, P. A., and Rosenthal, A. (1995). Induction of midbrain dopaminergic neurons by sonic hedgehog. *Neuron* **15,** 35–44.

Kaplitt, M. G., Leone, P., Samulski, R. J., Xiao, X., Pfaff, D. W., O'Malley, K. L., and During, M. J. (1994). Long-term gene expression and phenotypic correction using adeno-associated virus vectors in the mammalian brain. *Nat. Genet.* **8,** 148–154.

Kennedy, G. E. (1997). Potential use of herpes simplex virus (HSV) vectors for gene therapy of neurological disorders. *Brain* **120,** 1245–1259.

Kitada, T., Asakawa, S., Hattori, N., Matsumine, H., Yamamura, Y., Minoshima, S., Yokochi, M., Mizuno, Y., and Shimizu, N. (1998). Mutations in the *parkin* gene cause autosomal recessive juvenile parkinsonism. *Nature* **392,** 605–608.

Kordower, J. H., Palfi, S., Chen, E.-Y., Ma, S. Y., Sendera, T., Cochran, E. J., Mufson, E. J., Penn, R., Goetz, C. G., and Comella, C. D. (1999). Clinicopathological findings following intraventricular glial-derived neurotrophic factor treatment in a patient with Parkinson's disease. *Ann. Neurol.* **46,** 419–424.

Krüger, R., Kuhn, W., Müller, T., Woitalla, D., Graeber, M., Kösel, S., Przuntek, H., Epplen, J. T., Schöls, L., and Riess, O. (1998). Ala30P mutation in the gene encoding α-synuclein in Parkinson's disease. *Nat. Genet.* **18,** 106–108.

Lapchak, P. A. (1996). Therapeutic potentials for glial cell line-derived neurotrophic factor (GDNF) based upon pharmacological activities in the CNS. *Rev. Neurosci.* **7,** 165–176.

Lapchak, P. A., Miller, P. J., Jiao, S. S., Araujo, D. M., Hilt, D., and Collins, R. (1996). Biology of glial cell line-derived neurotrophic factor (GDNF): Implications for the use of GDNF to treat Parkinson's disease. *Neurodegeneration* **5,** 197–205.

Latchman, D. S. (1990). Molecular biology of herpes simplex virus latency. *Int. J. Exp. Pathol.* **71,** 133–141.

Latchman, D. S. (2001). Gene delivery and gene therapy with herpes simplex virus based vectors. *Gene* **264,** 1–9.

Leroy, E., Boyer, R., and Auburger, G. (1998). The ubiquitin pathway in Parkinson's disease. *Nature* **395,** 451–452.

Lilley, C. E., Groutsi, F., Han, Z. Q., Palmer, J. A., Anderson, P. N., Latchman, D. S., and Coffin, R. S. (2001). Multiple immediate-early gene-deficient herpes simplex virus vectors allowing efficient gene delivery to neurons in culture and widespread gene delivery to the central nervous system *in vivo*. *J. Virol.* **75,** 4343–4356.

Lin, L.-F.H., Doherty, D. H., Lile, J. D., Bekthesh, S., and Collins, F. (1993). A glial cell-derived neurotrophic factor for midbrain dopaminergic neurons. *Science* **260,** 1130–1132.

Lindvall, O., Brundin, P., Widner, H., Rhencrona, S., Gustavii, B., Fracowiak, R., Leenders, K. L., Sawle, G., Rothwell, J. C., Marsden, C. D., and Bjorklund, A. (1996). Grafts of fetal dopamine neurons survive and improve motor function in Parkinson's disease. *Science* **247,** 574–577.

Lücking, C. B., Dürr, A., Bonifati, V., Vaughan, J., De Michele, G., Gasser, T., Harhangi, B. S., Meco, G., Denèfle, P., Wood, N. W., Agid, Y., and Brice, A. (2000). Association between early-onset Parkinson's disease and mutations in the *Parkin* gene. *N. Engl. J. Med.* **342,** 1560–1567.

Maclean, A. R., Fareed, M. U., Robertson, L., Harland, J., and Brown, S. M. (1991). Herpes simplex virus type 1 deletion variants 1714 and 1716 pinpoint neurovirulence related sequences in Glasgow strain 17+ between immediate-early gene 1 and the 'a' sequence. *J. Gen. Virol.* **72,** 631–639.

Mandel, R. J., Rendahl, K. G., Spratt, S. K., Snyder, R. O., Cohen, L. K., and Leff, S. E. (1998). Characterization of intrastriatal recombinant adeno-associated virus-mediated gene transfer of human tyrosine hydroxylase and human GTP-cyclohydrolase I in a rat model of Parkinson's disease. *J. Neurosci.* **18,** 4271–4284.

Margolis, T. P., Sedarati, F., Dobson, A. T., Feldman, L. T., and Stevens, J. G. (1992). Pathways of viral gene expression during acute neuronal infection with HSV-1. *Virology* **189,** 150–160.

Margolis, T. P., Bloom, D. C., Dobson, A. T., Feldman, L. T., and Stevens, J. G. (1993). Decreased reporter gene expression during latent infection with HSV LAT promoter constructs. *Virology* **197,** 585–592.

Masliah, E., Rockenstein, E., Veinbergs, I., Mallory, M., Hashimoto, M., Takeda, A., Sagara, Y., Sisk, A., and Mucke, L. (2000). Dopaminergic loss and inclusion body formation in $\alpha$-synuclein mice: Implications for neurodegenerative disorders. *Science* **287,** 1265–1269.

Miller, A. D. (1992). Human gene therapy comes of age. *Nature* **357,** 455–460.

Natsume, A., Mata, M., Goss, J., Huang, S., Wolfe, D., Oligino, T., Glorioso, J., and Fink, D. J. (2001). Bcl-2 and GDNF delivered by HSV-mediated gene transfer act additively to protect dopaminergic neurons from 6-OHDA-induced degeneration. *Exp. Neurol.* **169,** 231–238.

Olanow, C. W., Kordower, J. H., and Freeman, T. B. (1996). Fetal nigral transplantation as a therapy for Parkinson's disease. *Trends Neurosci.* **19,** 102–109.

Palmer, J., Branston, R., Lilley, C. E., Robinson, M., Groutsi, F., Smith, J., Latchman, D. S., and Coffin, R. S. (2000). Development and optimisation of herpes simplex virus vectors for multiple long term gene delivery to the peripheral nervous system. *J. Virol.* **74,** 5604–5618.

Polymeropoulos, M. H., Lavedan, C., Leroy, E., Ide, S. E., Dehejia, A., Dutra, A., Pike, B., Root, H., Rubenstein, J., Boyer, R., Stenroos, E. S., Chandrasekharappa, S., Athanassiadou, A., Papapetropoulos, T., Johnson, W. G., Lazzarini, A. M., Duvoisin, R. C., Di Iorio, G., Golbe, L. I., and Nussbaum, R. L. (1997). Mutation in the $\alpha$-synuclein gene identified in families with Parkinson's disease. *Science* **276,** 2045–2047.

Rampling, R., Cruickshank, G., Papanastassiou, V., Nicoll, J., Hadley, D., Brennan, D., Petty, R., Maclean, A., Harland, J., McKie, E., Mabbs, R., and Brown, M. (2000). Toxicity evaluation of replication-competent herpes simplex virus (ICP 34.5 null mutant 1716) in patients with recurrent malignant glioma. *Gene Ther.* **7,** 859–866.

Riess, O., Jakes, R., and Krüger, R. (1998). Genetic dissection of familial Parkinson's disease. *Mol. Med. Today* **4,** 438–444.

Roizman, B., and Sears, A. E. (1987). An inquiry into the mechanisms of herpes simplex virus latency. *Annu. Rev. Microbiol.* **41,** 543–571.

Sacks, W. R., Greene, C. C., Aschmann, D. P., and Schaffer, P. A. (1985). Herpes simplex virus type 1 ICP27 is an essential regulatory protein. *J. Virol.* **55,** 796–805.

Shimura, H., Hattori, N., Kubo, S.-I., Mizuno, Y., Asakawa, S., Minoshima, S., Shimizu, N., Iwai, K., Chiba, T., Tanaka, K., and Suzuki, T. (2000). Familial Parkinson disease gene product, parkin, is a ubiquitin-protein ligase. *Nat. Genet.* **25,** 302–305.

Smith, M. D., Ensor, E. A., Coffin, R. S., Boxer, L. M., and Latchman, D. S. (1998). Bcl-2 transcription from the proximal p2 promoter is activated in neuronal cells by the Brn-3a POU family transcription factor. *J. Biol. Chem.* **273,** 16715–16722.

Smith, M. D., Melton, L. A., Ensor, E. A., Packham, G., Anderson, P., Kinloch, R. A., and Latchman, D. S. (2001). Brn-3a activates the expression of Bcl-$X_L$ and promotes neuronal survival *in vivo* as well as *in vitro*. *Mol. Cell. Neurosci.* **17,** 460–470.

Spillantini, M. G., Schmidt, M. L., Lee, V.M.Y., Trojanowski, J. Q., Jakes, R., and Goedert, M. (1997). $\alpha$-Synuclein in Lewy bodies. *Nature* **388,** 839–840.

Spillantini, M. G., Crowther, R. A., Jakes, R., Hasegawa, M., and Goedert, M. (1998). $\alpha$-Synuclein in filamentous inclusions of Lewy bodies from Parkinson's disease and dementia with Lewy bodies. *Proc. Nat. Acad. Sci. USA* **95,** 6469–6473.

Thomas, S. K., Lilley, C. E., Latchman, D. S., and Coffin, R. S. (1999). Equine herpes virus 1 gene 12 can substitute for Vmw65 in the growth of herpes simplex virus 1 (HSV1) allowing the generation of optimised cell lines for the propagation of vectors with multiple immediate-early gene defects. *J. Virol.* **79,** 7399–7409.

Wagstaff, M. J. D., Collaco-Moraes, Y., Smith, J., de Belleroche, J., Coffin, R. S., and Latchman, D. S. (1999). Protection of neuronal cells from apoptosis by hsp27 delivered with a herpes simplex virus-based vector. *J. Biol. Chem.* **274,** 5061–5069.

Wang, Q., Guo, J., and Jia, W. (1997). Intracerebral recombinant HSV-1 vector does not reactivate latent HSV-1. *Gene Ther.* **4,** 1300–1304.

Widner, H., Tetrud, J., and Rehncrona, S. (1992). Bilateral fetal mesencephalic grafting in two patients with parkinsonism induced by 1-methyl-4-phenyl-1,2,3,6-tetrahydropyridine (MPTP). *N. Engl. J. Med.* **327,** 1556–1563.

Yamada, M., Oligino, T., Mata, M., Goss, J. R., Glorioso, J. C., and Fink, D. J. (1999). Herpes simplex virus vector-mediated expression of Bcl-2 prevents 6-hydroxydopamine-induced degeneration of neurons in the substantia nigra *in vivo*. *Proc. Nat. Acad. Sci. USA* **96,** 4078–4083.

# GENE THERAPY FOR STROKE

K. Abe and W. R. Zhang

Department of Neurology
Graduate School of Medicine and Dentistry
Okayama University
Okayama 700–8558, Japan



## I. Introduction

In the central nervous system, neurons are particulary sensitive to various injuries such as ischemia, hypoxia, hypoglycemia, infection, and trauma. These vulnerabilities of neurons make it difficult to cure patients suffering from the above injuries in clinical situations. The vulnerability is different even within the neuronal populations (Abe *et al.*, 1991, 1993, 1995). Normal differentiation of neuronal and glial cells and their maintenance are under control of neurotrophic factors (NTFs).[1]In addition, excessive administration of NTFs greatly

[1]Abbreviations used: Ad-GDNF, adenoviral vector containing glial cell line-derived neurotrophic factor gene; Ad-LacZ, adenoviral vector containing the *E. coli lacZ* gene; BDNF, brain-derived neurotrophic factor; bFGF, basic fibroblast growth factor; BMP, bone morphogenetic protein; CBF, cerebral blood flow; CNTF, ciliary neurotrophic factor; DG, dentate gyrus; EGF, epidermal growth factor; FGF, fibroblast growth factor; GDNF, glial cell line-derived neurotrophic factor; GFP, green fluorescent protein; HE, hematoxylin and eosin; HSV, herpes simplex virus; MCA, middle cerebral artery; MCAO, middle cerebral artery occlusion; NAIP, neuronal apoptosis inhibitory protein; NGF, nerve growth factor; NMDA, *N*-methyl-D-aspartate; NTF, neurotrophic factor; PH, pleckstrin homology; SH2, Src homology 2; SVZ, subventricular zone; TGF-$\beta$, transforming growth factor-$\beta$; TNF, tumor necrosis factor; TTC, 2,3,5-triphenyltetrazolium chloride; TUNEL, terminal deoxynucleotidyl transferase-mediated dUTP-biotin *in situ* nick end labeling; VEGF, vascular endothelial growth fator.




0074-7742/03 $35.00

protects sensitive brain tissue from the injury. During the past several years, adenoviral gene transfer techniques have achieved increasing interest in many fields of ischemic brain injury. Adenovirus-mediated (Ad–) *Escherichia coli lacZ* gene was successfully transferred into normal and ischemic rodent brains (Abe *et al.*, 1997b; Barth *et al.*, 1996; Kitagawa *et al.*, 1998c), and Ad-interleukin-1 receptor antagonist gene expression reduced brain ischemic injury in rodents (Betz *et al.*, 1995; Yang *et al.*, 1997, 1998). Therefore, gene therapy for cerebrovascular disease could potentially become a better therapy in the near future (Heistad and Faraci, 1996; Abe, 2000).

On the other hand, brain cells have long been believed not to differentiate again once they matured in normal development. Recent advancements in molecular cell biology discovered the presence of neural stem cells even in mature adult mammalian brain. It is now expected that neural stem cells have the potential to compensate for and recover neural functions that were lost due to ischemic stroke. These stem cells are also under the control of NTFs to differentiate into a certain species of neural cells. Therefore, understanding the possible therapeutic potential of the role of NTFs in the differentiation of neural stem cells should be essential in developing treatment for human stroke patients.

This chapter, therefore, first summarizes the present understanding of the role of NTFs in normal and ischemic brain and the therapeutic potential of NTF protein itself and the gene therapy. For example, exogenous glial cell line-derived neurotrophic factor (GDNF) gene transfer successfully reduced the infarct size in a time-dependent manner, but the therapeutic time window was shorter than the effect of GDNF protein itself. The possible therapeutic potential of neural stem cells by modifying gene therapy is also discussed in terms of ischemic brain injury.

## II. Neurotrophic Factors

### A. Physiological Role

A number of protein peptides are present that affect target cells in different ways (Abe, 2000). Endocrine hormones have a remote effect on their target cells via the systemic circulation. In contrast, NTFs affect target cells by paracrine, autocrine, or juxtacrine (Abe, 2000). These suggest a local role of NTFs for cell proliferation, migration, differentiation, and production of extracellular matrix in the nervous system. A study of epidermal growth factor (EGF) suggests that paracrine stimulation transiently activates intracellular signals and primarily affects proliferation, whereas juxtacrine stimulation continuously activates intracellular signals and primarily affects differentiation and survival (Higashiyama *et al.*,

1995). NTFs usually promote cell proliferation, but sometimes inhibit such a proliferation under certain circumstances or for a certain cell species. For example, fibroblast growth factor (FGF) inhibits proliferation of chondroblast, and a constant activation of FGF with FGF receptor mutation causes achondroplasia (Deng *et al.*, 1996; Naski *et al.*, 1996). Transforming Growth factor (TGF)-$\beta$ was originally found to promote transformation and growth of fibroblasts. However, a recent study suggests that TGF-$\beta$ usually inhibits cell growth for most epidermal cells (Royuela *et al.*, 1998).

Based on their receptors, NTFs are usually classified into three or four subclasses (Table I). NTFs that work with tyrosine kinase receptors are a major group consisting of EGF, FGF, HGF, IGF-1, neurotrophins (NGF, BDNF, NT3, NT4/5), PDGF, and VEGF. Once these proteins bind the receptors as monomer (FGF) or dimer (neurotrophins), the receptors form a dimer and the intracellular tyrosine kinase domains are activated, which then transphosphorylate intracellular signal proteins that possess the SH2 (Src homology 2), SH3, or PH (pleckstrin homology) domain (Table I). GDNF and associated proteins such as neurturin, artemin, and persephin bind a receptor complex of GFR$\alpha$ and c-ret as a monomer, then the intracellular tyrosine kinase domain is activated. Cytokines such as CNTF, LIF, and IL-2 bind dimeric or trimeric receptors, and activate intracellular tyrosine kinase proteins such as JAK (Table I). The TGF-$\beta$ superfamily consists of the activin family (activin-$\beta$A and -$\beta$B), BMP family (BMP2$\sim$8, OP1$\sim$2, GDF5$\sim$7, etc.), and TGF-$\beta$ family (TGF-$\beta$ 1$\sim$3). TGF-$\beta$ ligands bind to types 1 and 2 receptors, and the receptors form tetrameric functional complex. Then type 2 receptors phosphorylate the serine-threonine kinase domain of type 1 receptors, so that the type 1 receptors activate subsequent signal systems such as Smad (Table I).

Many different classes of neurons, glial cells, and other cell types develop from multipotent neural crest cells. Specific growth factors control the fate of certain neural crest cells. Bone morphogenetic proteins (BMPs) 2 and 4 promote the differentiaiton of sympathetic neurons, GGF represses neurogenesis and promotes Schwann cell differentiation, and TGF-$\beta$ promotes smooth muscle cell differentiation. In the optic nerve, PDGF secreted by astrocytes maintains the proliferation of oligodendrocyte progenitor cells. Postnatally, these progenitor cells begin to differentiate into oligodendrocytes or astrocytes with or without CNTF secreted by astrocytes (Lillien and Raff, 1990). In sympathoadrenal lineage of the neural crest, sympathoadrenal precursor cells differentiate into adrenal chromaffin cells, noradrenergic and cholinergic sympathetic neurons with glucocorticoids, FGF/NGF, or LIF/CNTF, respectively. Thus, NTFs are actively involved in many steps of normal cell development and differentiation (Jessell and Sanes, 2000).

Even after differentiation, NTFs are also involved in the normal maintenance and survival of neuronal cells. Constant secretion of NTFs keeps cell survival signals activated and death signals inactive. For example, the presence of nerve

TABLE I
CLASSIFICATION OF NEUROTROPHIC FACTORS BASED ON THEIR RECEPTORS

| Type of receptors and neurotrophic factor family | Neurotrophic factors as ligands | Intracellular active domain | Downstream signal proteins |
|---|---|---|---|
| Tyrosine kinase receptors | | Tyrosine kinase domain | SH2, SH3, PH |
| EGF receptor family | EGF, HB-EGF, TGF-$\alpha$ | | |
| Eph family | Ephrin (LERK) | | |
| FGF receptor family | FGF 1 ~9 | | |
| HGF receptor family | HGF, MSP | | |
| Insulin receptor family | Insulin, IGF-1 | | |
| NGF receptor family | Neurotrophins (NGF, BDNF, NT3, NT4/5) | | |
| PDGF receptor family | PDGF-A and B, CSF-1, steel factor | | |
| Ret receptor family | GDNF, neurturin, artemin, persephin | | |
| Tie family | Angiopoietin-1 and -2 | | |
| VEGF receptor family | VEGF-A, B, and C, PIGF | | |
| Cytokine receptors | CNTF, LIF, IL-2, IL-6 | Intracellular portion (gp130/LIFR$\beta$) | JAK |
| Serine/threonine kinase receptor | | Serine-threonine kinase domain | Smad 1~8 |
| TGF-$\beta$ superfamily | BMP family: BMP2 ~8, OP1 and 2 GDF5 ~7<br>Activin family: Activin $\beta$A and $\beta$B<br>TGF-$\beta$ family: TGF-$\beta$1~3 | | |

growth factor (NGF) maintains cell survival by activating cell survival Bcl-2 and inhibiting death signals such as Apaf-1 and caspases (Abe, 2000; Jessell and Sanes, 2000). However, once NTFs are lost, constant activation of cell survival signals is lost, which permits activation of death signals that finally result in the neuronal cell death. Many other stimuli such as reactive oxygen species (ROS), FAS ligand, and tumor necrosis factor (TNF) activate a distinct biochemical pathway that triggers a family of caspases and leads to cell death.

### B. Induction after Ischemia

NTFs such as BDNF, bFGF, GDNF, NGF, and TGF-$\beta$1 were induced in brain cells under pathological conditions such as ischemia and kinate-induced seizure (Kato *et al.*, 1992; Logan *et al.*, 1992; Comelli *et al.*, 1993; Humpel *et al.*, 1994; Lee *et al.*, 1994). However, each NTF species shows a different temporal and cellular distribution. GDNF was originally shown to promote survival and neurite outgrowth of dopaminergic and motor neurons as well as peripheral sensory and sympathetic neurons (Henderson *et al.*, 1994; Li *et al.*, 1995; Tomac *et al.*, 1995). Although immunoreactive GDNF was not detectably present in sham control brain (Abe and Hayashi, 1997), it began to be expressed in neuronal cytoplasm of the cerebral cortex at 1 h of reperfusion after 90 min of transient middle cerebral artery occlusion (MCAO). The induction reached a maximum at 3 h with a gradual decrease by 1 day. Brain cells positive for GDNF-like immunoreactivity were mainly large pyramidal neurons in layers 4–6 of cerebral cortex, and no glial cell was stained in the brain sections (Abe and Hayashi, 1997).

Although initially identified as a neurotrophic molecule for parasympathetic ciliary ganglion neurons which is abundantly expressed in astroglial and Schwann cells, ciliary neurotrophic factor (CNTF) also affects the survival of other neuronal cells (Barbin *et al.*, 1984; Arakawa *et al.*, 1990; Ip *et al.*, 1991). CNTF levels were significantly higher in the frontal, parietotemporal, and occipital lobes at 2 days after 120 min of transient MCAO (Hu *et al.*, 1997). With subsequent immunohistochemical analysis, astrocytes were predominantly stained in the marginal region of the infarct from 1 day to 7 days of reperfusion after 90 min of transient MCAO (Abe, 2000).

Also known as vascular permeability factor, vascular endothelial growth factor (VEGF) is mitogenic for endothelial cells. By alternative splicing, four different isoforms exist *in vivo*: VEGF206, VEGF189, VEGF165, and VEGF121. VEGF165 and VEGF121 act as diffusible factors. A recent report (Hayashi *et al.*, 1997) showed that immunoreactive VEGF was normally present in the ependymal cells, but was not in neurons, glial cells, vascular endothelial cells, or pial cells in control brains. However, it became positive in neurons of the ischemic cerebral cortex after 1 h to 1 day of reperfusion after 90 min of transient MCAO with a peak at 3 h. Pial cells of the MCA area became positive for VEGF from 1 h with a peak at 3 h, which was sustained until 3–7 days of reperfusion.

GDNF was originally isolated from cultured glial cells, but the gene was expressed only in neurons at an early stage of ischemia (Abe and Hayashi, 1997). Because the induction of CNTF was predominantly observed in the astroglia of the marginal region (Abe, 2000), injured neurons may stimulate the synthesis of CNTF in astroglia, which may in turn support survival and regeneration of neurons in the injured regions. Because the VEGF gene was also induced in other brain cells under hypoxic conditions or in continuous MCAO (Fischer *et al.*, 1995; Ijichi *et al.*, 1995; Nomura *et al.*, 1995; Kovacs *et al.*, 1996), sensitivity for inducing the VEGF gene may be different in neural cells under conditions of hypoxia/ischemia or continuous ischemia/reperfusion. Neovascularization usually begins 2–5 days after the ischemic insult and continues for months (Krupinski *et al.*, 1994). Therefore, the sustained expression of the VEGF gene in the pia mater after ischemia (Hayashi *et al.*, 1997) suggests its role in new vessel formation in the affected area. On the other hand, the rapid and transient induction of VEGF in neurons (Hayashi *et al.*, 1997) may be related to the protection of the vascular system against ischemic insult.

As pointed out above, NTFs are induced in different neural cell types (neuronal, astroglial, and pial) with different temporal profiles, suggesting a different role of each NTF for each neural cell species at a particular stage after ischemia (Table II).

TABLE II
TEMPORAL AND CELLULAR PROFILES OF NEUROTROPHIC FACTOR EXPRESSION AFTER 90 MIN OF TRANSIENT MCA OCCLUSION IN RAT BRAIN

| Neurotrophic factors and brain cell | Control | Time after reperfusion | | | | |
|---|---|---|---|---|---|---|
| | | 1 h | 3 h | 1 day | 3 days | 7 days |
| GDNF | | | | | | |
| Neuron only | – | + | +++ | ± | – | – |
| CNTF | | | | | | |
| Astrocyte only | – | – | – | + | +++ | ++ |
| VEGF | | | | | | |
| Neuron | – | + | ++ | ± | – | – |
| Pia mater | – | + | ++ | + | + | ± |

## III. Application of Neurotrophic Factors

### A. Neurotrophic Factors Protect Neural Cells against Ischemic Injury

Constant support by NTFs is essential for survival and differentiation of normally developing neurons, but is also important in the protection and recovery of mature neurons under pathological conditions (Lin *et al.*, 1993). Galectin-1 regulates initial axonal growth in peripheral nerves after axotomy (Horie *et al.*, 1999). Protective effects of GDNF for various injuries to central and peripheral nervous tissue have been reported (Schaar *et al.*, 1993; Krieglstein *et al.*, 1995). CNTF infusion prevented learning disabilities and neuronal loss induced by ischemia in gerbils (Wen *et al.*, 1995).

Topical administration of NTFs such as GDNF, NT-3, VEGF, IGF-1, and BDNF to the ischemic brain surface has recently been tried in brain ischemia models to examine a possible direct protective effect of NTF proteins against ischemic brain damage. Abe *et al.* (1997a) first reported amelioration of brain edema by topical application of GDNF in reperfused rat brain (Fig. 1). Kitagawa *et al.* (1998b) reported that topical application of GDNF greatly reduced the infarct size (48%) and brain edema (30%) at 24 h of continuous MCAO in rats. The reduction of infarct size was not related to a change of cerebral blood flow (CBF), but was accompanied by marked reduction of positive cells for TUNEL and caspases in the affected area. Thus, GDNF showed a direct protective effect against ischemic brain damage, but not secondary by improving CBF. Inducton of TUNEL was mainly located in the ischemic core, and that of caspases-1–3 was mainly in the penumbra, suggesting that the main mechanism of ischemic neuronal cell death is necrosis in the core and apoptosis in the penumbra. Although the infarct size became smaller by GDNF treatment, the dissociative spatial distribution among TUNEL and caspases-1–3-positive cells was similar to that with vehicle treatment. These data suggest that treatment with GDNF was protective due not only to the reduction of apoptotic process, but also to the reduction of necrotic process.

Although similar results were obtained with different NTFs and animal models, ischemic injury was less in the short period of transient focal ischemia than in continuous ischemia. In fact, reduction of brain edema by GDNF was greater in transient MCAO (Abe *et al.*, 1997a) than continuous MCAO (Kitagawa *et al.*, 1998a) in rats. In transient MCAO models, treatment of ischemic rats with topical application of NT-3, VEGF, and IGF-1 ameliorated infarct sizes associated with reduction of the number of positive cells for TUNEL and caspases (Hayashi *et al.*, 1998; Zhang *et al.*, 1999; Wang *et al.*, 2000). However, the pharmacological potency was strongest for GDNF, and GDNF > NT-3 > VEGF > IGF-1 in this order. Yamashita *et al.* (1997) delivered BDNF into the

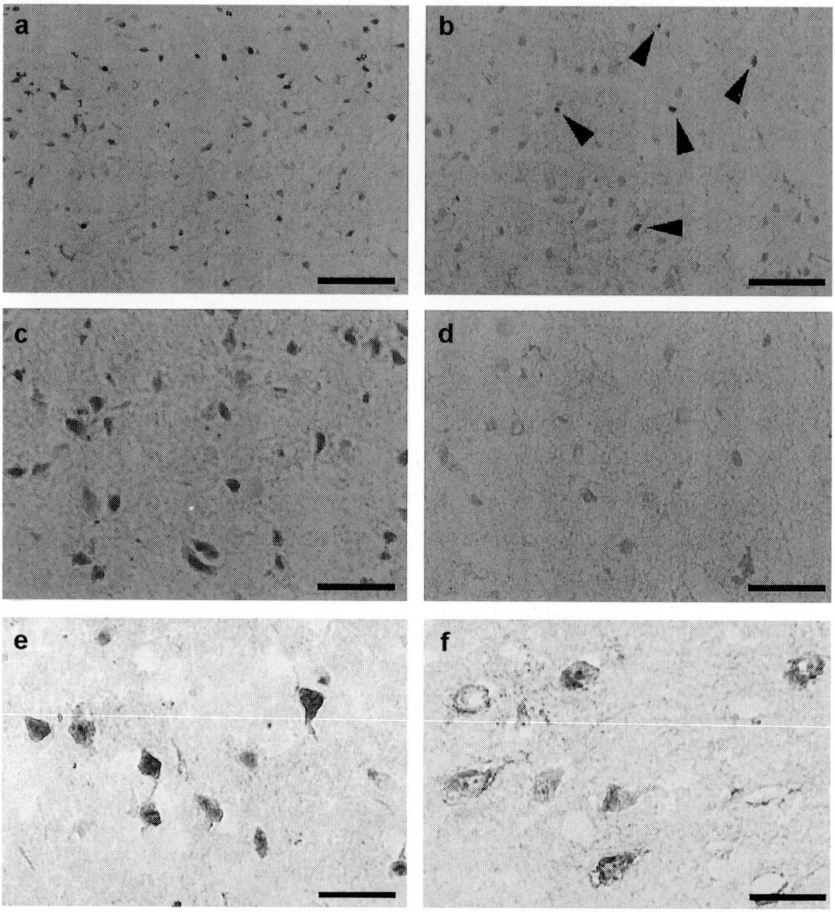


FIG. 1. Representative stainings of TUNEL (a–d) and HSP70 (e and f). Panels represent cerebral cortical sections of the occluded MCA area at 1 (a and b) and 2 (c–f) days of reperfusion. Left columns (a, c, and e) were from rats treated with vehicle, and right columns (b, d, and f) were with GDNF. Note many TUNEL-positive cells in the vehicle group (a and c) in contrast to only several (b, arrowheads) or no (d) TUNEL-positive cells in the GDNF-treated group. HSP70 staining is only slightly reduced in the GDNF (f) than vehicle (e) group.

territory of MCA via an osmotic minipump (1 $\mu$/h), and found a 33% reduction in total infarct volume.

The protective mechanism of NTFs on ischemic brain injury is still to be determined, but exogenously added excessive amount of NTFs could partially inhibit activation of apoptotic signal cascade including caspases, and could in turn potentiate survival signal pathways (Abe, 2000). However, GDNF was also protective in the cases of continuous ischemia where the necrotic mechanism is more involved than apoptotic mechanism (Kitagawa *et al.*, 1998b). Thus treatment with NTFs may have the potential to be protective against both apoptotic and necrotic mechanisms of cell death. These strong effects of NTFs against ischemic brain injury could be applicable to improving the clinical outcome for human stroke, or even to prolonging the therapeutic time window for another effective therapy.

## B. Therapeutic Time Window of Neurotrophic Factors

In a study with transient MCAO, topical application of GDNF significantly reduced infarct size in a time-dependent manner (Zhang *et al.*, 2001). The extent of penetration of GDNF was roughly two-thirds of the affected hemisphere that covers the MCAO territory. Amelioration of the infarct size by application of GDNF was strongest at 0 h, still effective at 1 h, but became insignificant at 3 h (Fig. 2). This effect was closely related to the reduction of the numbers of TUNEL, active caspase-3 and -9-positive cells (Fig. 3). Because there was no significant difference between the vehicle and GDNF groups in the rCBF before, during, and after MCAO, the protective effect was not secondary to rCBF improvement but primarily to reducing apoptotic pathways by GDNF. DNA fragmentation detected by the TUNEL method can be found in both apoptotic and necrotic neurons, although they can be discriminated morphologically (Charriaut-Marlangue *et al.*, 1996). The larger number of TUNEL-positive cells than active caspase-3 or caspase-9-positive cells (Fig. 3) may be due to colocalization of necrotic and apoptotic cells in the boundary region of infarction. Caspase-3 is activated through caspase-9 or another pathway (Li *et al.*, 1997;

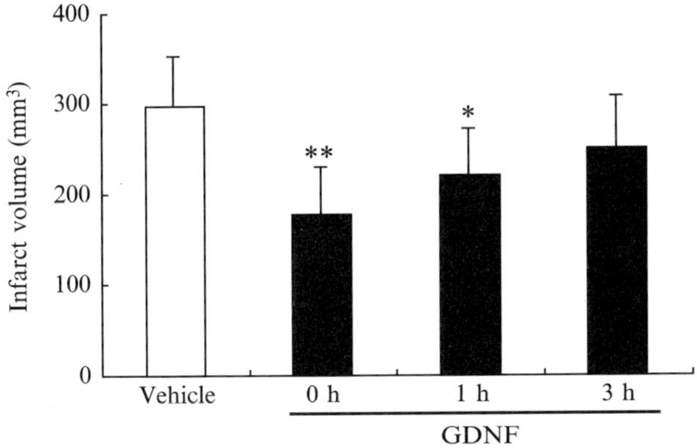


FIG. 2. Effects of GDNF (GDNF: 0, 1, and 3 h) on infarct volume at 24 h after 90 min of transient MCAO. Note significant reduction with GDNF treatment at 0 and 1 h (*$p < 0.05$, **$p < 0.01$).

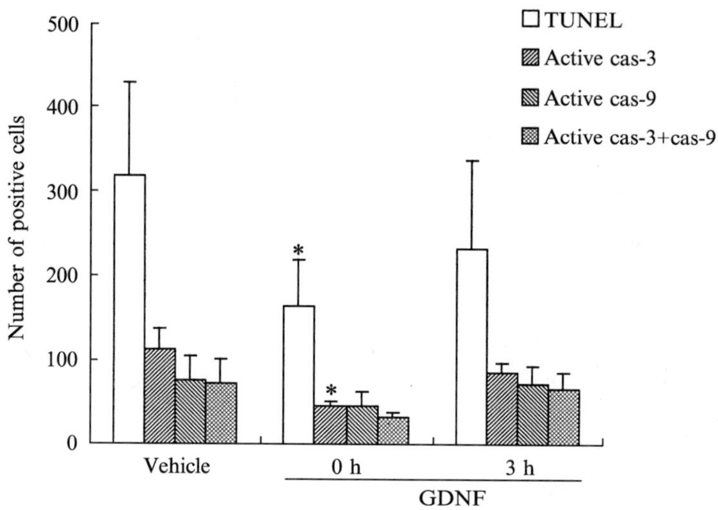


FIG. 3. Number of TUNEL and active caspase-3 and -9-positive neurons with or without GDNF at 12 h after transient MCAO. Note significant reduction in the number of both TUNEL and active caspase-3-positive cells in the border of infarcts at 0 h of treatment (*$p < 0.05$).

Kitagawa *et al.*, 1998b; Porter and Janicke, 1999). The presence of active caspase-9-negative but active caspase-3-positive cells suggests an activation of the alternative pathway in the ischemic brain (Porter and Janicke, 1999). Double-positive cells suggest the serial activation of caspase-9 and caspase-3 in the cells. Treatment with GDNF reduced the number of TUNEL and active caspase-3-positive cells, while the reduction was weak in active caspase-9- or double-positive cells (Fig. 3). Thus, the effect of GDNF was stronger in suppressing active caspase-3 than active caspase-9, or the effect of GDNF may be weak in the pathway of cytochrome *c*, caspase-9, and caspase-3.

Therapeutic time windows of many drugs have been reported after cerebral ischemia. Combinations of an NMDA receptor antagonist (MK-801) and a free radical scavenger (α-phenyl-*tert*-butylnitrone, PBN) increased the therapeutic window up to 3 h (Barth *et al.*, 1996). Combinations of a GABA agonist and glutamate antagonist showed a synergistic neuroprotective effect up to 4 h following the onset of cerebral ischemia (Lyden *et al.*, 2000). However, such a therapeutic time window has not been previously reported for neurotrophic factors such as GDNF or others. Our study demonstrated a time-dependent ameliorative effect of GDNF after transient cerebral ischemia, whereas the time window was shorter than previously reported with other drugs mentioned above. Because the effect of GDNF seemed to be weaker in the cytochrome *c* and caspase-9 pathway, potentiation of this pathway with other drugs may further prolong the time window in the future.

## IV. Gene Therapy

### A. Gene Therapy of Stroke by Neurotrophic Factor Gene Transfer

Using virus vectors (Table III), gene delivery systems have been reported in many fields including genetic diseases as well as some acquired diseases such as cerebrovascular diseases and cancer (Heistad and Faraci, 1996; Karpati *et al.*, 1996; Abe *et al.*, 1997b). Initial experimental gene transfer into the brain was performed with replication-defective herpes simplex virus (HSV) type 1 (Pallela *et al.*, 1989; Fink *et al.*, 1992; Wolfe *et al.*, 1992). Although HSV vectors can carry a relatively large gene up to 36 kb, cytopathogenicity such as neuronal death may be a major disadvantage (Breakfield, 1993; Culotta, 1993; Mulligan, 1993). The advantages and disadvantages of each vector are shown in Table III. Retroviral vectors also have a potential to deliver gene into the brain (Price *et al.*, 1987; Walsh and Cepko, 1988). However, retroviral vectors do not infect nondividing cells such as neurons, and *in vivo* risks such as mutagenesis and tumorogenesis are also indicated (Mulligan, 1993; Karpati *et al.*, 1996). Lentivirus can transduce a gene to nondividing cells and has long-term expression, but has a potential insertional mutagenesis (Kafri *et al.*, 1997). Direct injection of plasmid vector with or without a conjugation of liposome also carries genes into the brain and muscle (Ono *et al.*, 1990; Danko *et al.*, 1993). However, the efficacy of gene expression is low (Ono *et al.*, 1990; Culotta, 1993).

With replication-defective adenoviral vector, the *E. coli lacZ* gene was already expressed at 8 h after inoculation in control brain. The cell number of expression and staining gradually increased with a peak at 7 days (Fig. 4; Abe *et al.*, 1997b). In brains exposed to continuous MCAO, the *lacZ* gene was expressed at 8 h of ischemia in the cerebral cortex around the injection needle (Fig. 5a, needle route as arrows). After injection of adenovirus-mediated (Ad-) *lacZ* gene into ischemic rat brain after transient 90 min of MCAO, no or only a few cells expressed the *lacZ* gene at 8 h (Fig. 5e) and 2 days of reperfusion. However, many brain cells expressed the gene at 7 days (Fig. 5f and g). Cells expressing the *lacZ* gene were also found in the contralateral cerebral cortex (Fig. 5h). At 21 days of reperfusion, the majority of brain cells were no longer stained for $\beta$-gal blue color. Double staining showed only a part (5–10%) of the cells was double positive for GFAP and $\beta$-gal (Fig. 5d, filled arrowheads). There was only minimum traumatic injury and leukocyte infiltration (Fig. 5b, arrow). Transient expression of adenoviral vector-mediated gene may be advantageous as an aspect of the disappearance of a foreign gene from the tissue after the it is no longer required.

Based on the strong protective effect of NTF protein against ischemic brain damage (Abe *et al.*, 1997a; Kitagawa *et al.*, 1998b) and the considerable transfer of a foreign gene into the ischemic brain (Fig. 5), an adenovirus vector containing

TABLE III

CHARACTERISTICS OF GENE THERAPY VECTORS FOR ISCHEMIC STROKE

| Characteristics | Plasmid | HSV | Retrovirus | Lentivirus | Adenovirus | AAV |
|---|---|---|---|---|---|---|
| Insert size up to | Unlimited | 30 kb or more | 7 kb | 7 kb | 30 kb | 4 kb |
| Route of delivery | *Ex/in vivo* | *Ex/in vivo* | *Ex vivo* | *Ex/in vivo* | *Ex/in vivo* | *Ex/in vivo* |
| Infection to neurons | Few | Yes | No | Yes | Yes | Yes |
| Genomic integration | Very poor | No | Yes | Yes | No | Yes |
| Expression level | Very low | High | Low | Low | High | Moderate |
| Duration of expression | Short | Long | Short[a] | Long | Short[a] | Long |
| Preparation | Easy | Moderate | Moderate | Moderate | Moderate | Difficult |
| Immunological and safety problem | None | High inflammation cell toxicity | Few mutagenesis? | Few mutagenesis? | Moderate inflammation cell toxicity? | None |

[a]Usually considered as a disadvantage in gene therapy, but as an advantage for acute stroke.

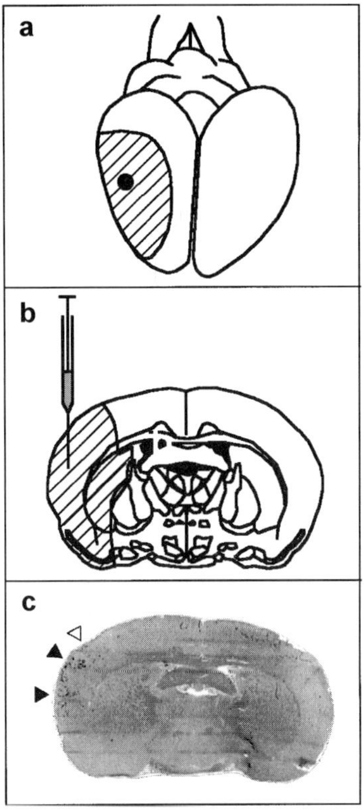


FIG. 4. Position of needle injection for gene delivery from top (a) and frontal (b) views into MCA area (shaded). At 7 days (c), *E. coli LacZ* gene is expressed in the rat cerebral cortex (arrowheads).

the GDNF gene (Ad-GDNF) was prepared, and a possible protective effect of the Ad-GDNF transfer was examined after transient MCAO in rat. Pretreatment of animals with Ad-GDNF 24 h before the subsequent 90 min of transient MCAO effectively reduced infract volume (Fig. 6) without affecting regional CBF compared to the vehicle or Ad-LacZ animal groups (Kitagawa *et al.*, 1999). With immunohistological analyses, treatment with Ad-GDNF greatly reduced the numbers of positive cells for TUNEL, caspase-3, and cytochrome *c*. There was no leukocyte infiltration in Ad-GDNF-treated rats. Inhibition of cytosolic release of cytochrome *c* by Ad-GDNF suggests a target of protection in an apoptotic pathway via cytochrome *c* and caspase-3 in the penumbra (Fujimura *et al.*, 1998).

Many other gene therapeutic studies for cereral ischemia have been sucessfully tried in animal models. Adenovirus-mediated neuronal apoptosis inhibitory protein (NAIP) and an interleukin-1 receptor antagonist had protective effects against alchemic brain injury (Betz *et al.* 1995; Xu *et al.* 1997). HSV mediated

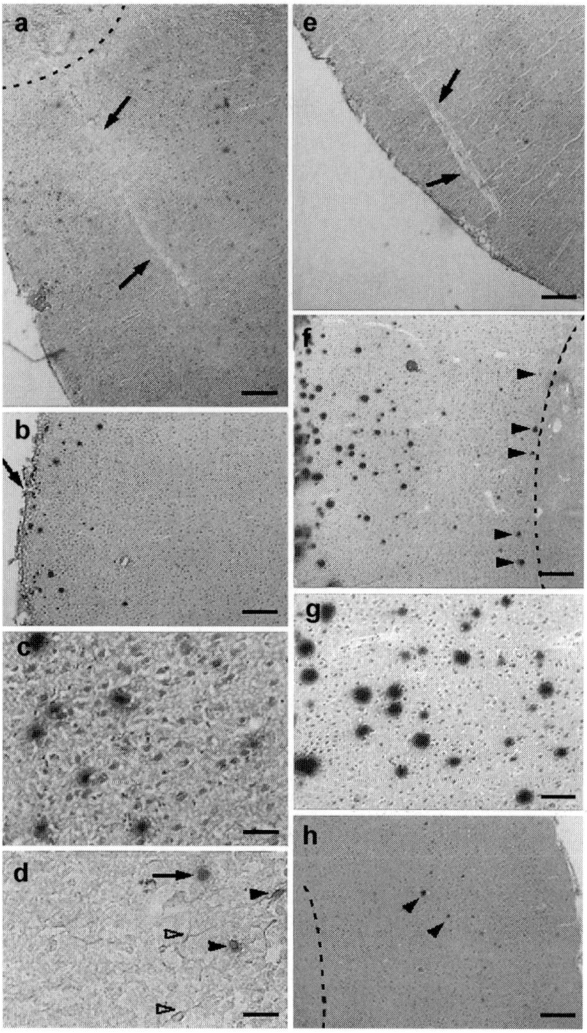


FIG. 5. Expression of the *lacZ* gene in rat brain sections from continuous MCAO (a–d) and reperfusion after 90 min of transient MCAO (e–h). Note considerable amount of expression of *lacZ* gene in the continuous ischemia from 8 h (a) to 2 days (b and c) with a few cells positive for both GFAP and X-Gal (d, filled arrowheads). Also note no staining at 8 h of reperfusion (e), in contrast to strong staining at 7 days (f and g) with some positive cells in the contralateral hemisphere (h, arrowheads).

HSP72 or Bcl-2 gene transfer improved brain damage in focal ischemia (Linnik *et al.*, 1995; Lawrence *et al.*, 1997; Yang *et al.*, 1997; Yenari *et al.*, 1998). Each vector has some disadvantages such as toxicity or efficacy of gene expression, and each experiment including our data has other problems to be resolved such

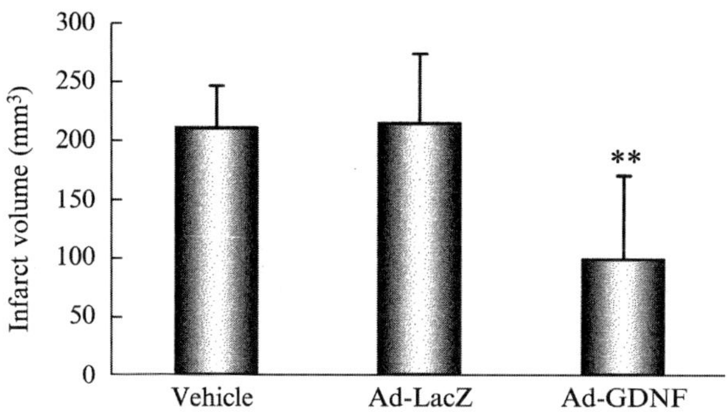


FIG. 6. Effect of adenovirus-mediated gene therapy with GDNF on infarct volume. Note reduction of infarct volume with Ad-GDNF treatment (** $p < 0.01$).

as timing of gene transfer (pretreatment), spatial limitation of the exogenous gene expression around the injection site or periventricular ependymal cells, or the gene delivered. Development of a more effective and safer vector may be necessary for future practical gene therapy (Ferrari *et al.*, 1997; Verma and Somia, 1997). Finally, the route of administration may be another important aspect for gene therapy of stroke. Direct injection of the vector into brain parenchyma or ventricle may not be practical as a future clinical application. Development of ischemic tissue-specific targeted gene transfer by intraarterial or venous administration may be expected (Abe *et al.*, 1997b). A few additional approaches of gene therapy for stroke could be possible (Heistad and Faraci, 1996; Abe *et al.*, 1997b; Abe, 2000). Stabilization of atherosclerotic plaque and restenosis (Morishita *et al.*, 1997, 1998) or a stimulation of collateral vessels may also be feasible. When these problems are successfully resolved, gene therapy could offer great hope for stroke therapy in the future.

## B. Therapeutic Time Window of Neurotrophic Factor Gene Therapy

Previous studies showed Ad–GDNF treatment significantly reduced TUNEL- or active caspase-3-positive cells ( Hermann *et al.*, 2001), and decreased the size of brain infarction when administered before middle cerebral artery occlusion (MCAO) (Kitagawa *et al.*, 1999). However, the therapeutic time window is not still clear, especially when Ad-GDNF should be administered after MCAO. Our recent study demonstrated that topical application of Ad-GDNF significantly reduced infarct size in a time-dependent manner. Amelioration of the infarct size by GDNF application was effective at 0 h, but became insignificant at 1 h (Fig. 7). This effect was closely related to the reduction of the numbers

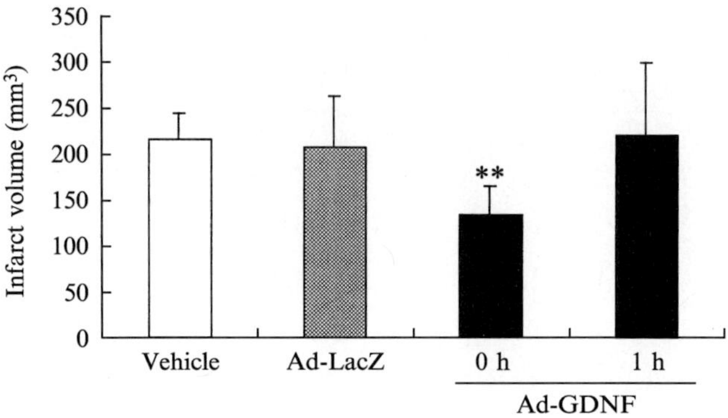


FIG. 7. Effects of Ad-GDNF (GDNF 0 h and 1 h) on infarct volume at 24 h after 90 min of transient MCAO. Note significant reduction of the Ad-GDNF treated only at 0 h (**$p < 0.01$).

of TUNEL, active caspase-3, but not to caspase-9-positive cells (Table IV). Because there was no significant difference between the vehicle, Ad-LacZ, and Ad-GDNF groups in the rCBF before, during, and after MCAO, the protective effect was not secondary to rCBF improvement. Our group demonstrated that Ad-GDNF used in this study had the ability to infect 293 cells and produce 20-kDa GDNF *in vitro* (Kitagawa *et al.*, 1999). Furthermore, in the immunohistochemical analysis, GDNF staining of the Ad-GDNF-treated group at 8 or 24 h after reperfusion was greater than in the vehicle- and Ad-LacZ-treated groups (Table IV). Thus, GDNF gene was successfully transferred into the brain, and the protein was produced after 8 h of reperfusion. DNA fragmentation detected by TUNEL can be found in both apoptotic and necrotic neurons, although they can be discriminated morphologically (Charriaut-Marlangue *et al.*, 1996). The larger number of TUNEL-positive cells than active caspase-3 or -9-positive cells (Table IV) at 24 h after reperfusion may be due to colocalization of necrotic and apoptotic cells in the boundary region of infarction.

Caspase-3 is activated through caspase-9 or another pathway (Li *et al.*, 1995; Kitagawa *et al.*, 1998b; Porter and Janicke, 1999). In the previous study, our group demonstrated that GDNF protein reduced the number of active caspase-3 but not active caspase-9 at 12 h of reperfusion (Zhang *et al.*, 2001). In this study, the number of active caspase-3-positive cells was reduced with Ad-GDNF treatment at 24 h but not at 8 h (Table IV), and the number of active caspase-9 was not affected by Ad-GDNF treatment. It may be too early to reduce the active caspase-3-positive cells at 8 h to produce enough GDNF protein from the virus vector.

The therapeutic time window of many drugs has been reported after cerebral ischemia. Combinations of an NMDA receptor antagonist (MK-801) and a free radical scavenger (PBN) increased the therapeutic window for up to 3 h (Barth

TABLE IV
EFFECT OF GENE THERAPY WITH Ad-GDNF ON STAINING FOR ACTIVE CASPASE-9, ACTIVE CASPASE-3, TUNEL, AND GDNF IN THE ISCHEMIC HEMISPHERE AFTER TRANSIENT MCA OCCLUSION[a]

| | Sham | Vehicle | Ad-LacZ | Ad-GDNF |
|---|---|---|---|---|
| ac cas-9 | | | | |
| 8 h | – | 3+ | 3+ | 3+ |
| 24 h | – | 2+ | 2+ | 2+ |
| ac cas-3 | | | | |
| 8 h | – | 2+ | 2+ | 2+ |
| 24 h | – | 2+ | 2+ | + |
| TUNEL | | | | |
| 8 h | – | + | ± | ± |
| 24 h | – | 4+ | 4+ | 3+ |
| GDNF | | | | |
| 8 h | ± | + | + | 2+ |
| 24 h | ± | 2+ | 2+ | 3+ |

[a]Staining was performed at 8 h or 24 h after transient MCAO, and was counted into the following six grades: no staining, a few (2–50), small (51–150), moderate (151–300), many (301–500), and large (501–800) number of stained cells into (–), (±), (+), (2+), (3+), and (4+), respectively ($n = 4$, in each time point).

*et al.*, 1996). Combinations of a GABA agonist and glutamate antagonist showed a synergistic neuroprotective effect up to 4 h following the onset of cerebral ischemia (Lyden *et al.*, 2000). Our group also reported the GDNF protein had a neuroprotective effect up to 1 h (Barth *et al.*, 1996). However, such a therapeutic time window has not been reported in adenovirus-mediated gene delivery such as Ad-GDNF or others. This study first demonstrated a time-dependent effect of Ad-GDNF after transient cerebral ischemia, while the time window was shorter than the GDNF protein itself and other drugs mentioned above.

## V. Stem Cells

### A. Physiological Differentiation and Ischemic Activation of Neural Stem Cells

Stem cells possess two potentials—self-renewal and multidirectional differentiation. Development and differentiation of the nervous system depend on the expression of particular genes at particular places and times. This spatial and temporal pattern of gene expression is regulated by both internal molecular programs and external epigenetic environments. Internal influences include (1) transcription factors that act at a DNA level to control gene expression, and (2) cell surface and secreted molecules that control the fate of neighboring cells. External factors include secreted factors, nutrients, sensory stimuli, and social experience that mediate patterned changes in the activity of nerve cells. From these neuronal or glial progenitor cells, mature neurons or glial cells are finally produced (Abe, 2000). Many gene factors and NTFs play pivotal role in the differentiation of neural crest cells into neurons, glia, or muscle, and further differentiation (Takahashi *et al.*, 1999; Abe, 2000). As for the cerebral cortex, the neurons are generated from an epithelial layer of progenitor cells that lines the lateral ventricles. Early in normal development, the progenitor cells expand the population of neuronal precursor cells by rapid proliferation, then give rise both to neurons and to additional progenitor cells, and finally generate only neurons (Jessell and Sanes, 2000). Stem cells are present in the hippocampal dentate gyrus (DG) and subventricular zone (SVZ) of striatum of adult mammalian brain (Gould *et al.*, 1999; Kempermann and Gage, 1999). Three DG layers, the hilus, subgranular layer, and granular cell layer, generate stem cells for the hippocampus. To the olfactory bulb, new neurons are provided from the SVZ of the striatum. Although the exact origin of neuronal progenitor cells is still debated as to subventricular cells, ependymal cells, or subventricular astrocytes (Johansson *et al.*, 1999; Doetsch *et al.*, 1999), it is interesting that differentiated cells such as ependymal cells or astrocytes could be a neuronal progenitor. These data suggest the possibility of making brain cells dedifferentiate and redifferentiate into neuronal progenitor cell, and, therefore, further suggest a potential use of these cells to recover and hopefully rearrange the neuronal network after destruction of the brain by ischemia.

Dividing cells can be detected after intraperitoneal injection and incorporation of BrdU. With double staining for neuronal marker NeuN and calbindin-D28k, neurogenesis and age-related decrease were found in hippocampal DG of adult rats (Kuhn *et al.*, 1996). These data explain a part of the decreased plasticity of aged brain. Based on such a study in normal brain, a possible activation of neural stem cells began to be examined in ischemic brains. Proliferation of neuronal precursor cells in DG was accelerated after transient forebrain ischemia in mice and gerbils (Takagi *et al.*, 1999; Iwai *et al.*, 2001a). It was found that the number of BrdU-labeled cells increased in the DG and paraventricle region at 3, 7, and 10 days after reperfusion. Liu and Shrap (1999) found a 12-fold increase of BrdU incorporation at the subgranular zone of DG between 1 and 2 weeks after 10 min of global ischemia in gerbils. A 2 or 4 min occlusion time was the threshold for ischemia-induced stem cell proliferation in the DG. Confocal microscopy confirmed that 60% of the BrdU-positive cells in the granule cell layer expressed neuronal markers NeuN, Calbindin, and

MAP-2 at 1 month after the BrdU labeling, and that many cells expressed glial marker GFAP in the dentate hilus between 2 weeks and 1 month after BrdU injection (Liu *et al.*, 1998). In addition to ischemia, activation of neurogenesis was also reported in a systemic chemoconvulsant model (Parent *et al.*, 1997). It was found that hippocampal network plasticity associated with epileptogenesis may arise from aberrant connections formed by newly born dentate granule cells.

Such recent work suggests that neural stem cells in DG respond to ischemic injury by increasing the rate of proliferation in the subgranular zone both in mice and gerbils. The neuronal and glial progeny derived from the stem cells migrated into different anatomical locations such as the granule cell layer and the dentate hilus. Enhanced neurogenesis in the DG may be a compensatory adaptive response to ischemic injury and could promote functional recovery after ischemia. Expression of highly polysialylated neural cell adhesion molecule (PSA-NCAM) in mossy fiber of granule cells in adult rat DG could be involved in migration of stem cells (Seki and Arai, 1999; Iwai *et al.*, 2001b).

## B. Gene Therapy of Stroke by Modifying Neural Stem Cells

Transplantation of brain cells into ischemic brain has been performed. For example, fetal rat hippocampal neurons were stereotaxically transplanted into 5-day-old ischemic hippocampal CA1 lesions of adult rat (Aoki *et al.*, 1993). In the recipient brain, the transplanted cells survived well and formed clusters in the host CA1 subfield at 14 or 100 days after transplantation. Of the animals, 67% exhibited supragranular mossy fiber sprouting in DG. Hodges *et al.* (1997) reported that fetal CA1 grafts promoted recovery from cognitive deficits in marmosets induced by excitotoxic damage to the CA1 field.

With various stem cells and mordern techniques, novel challenges have been tried. The hematopoietic myeloid progenitor cells ($1 \times 10^5/\mu l$) cultured from adult mouse bone marrow were implanted into the penumbra of the ischemic striatum at 4 or 24 h after 3 h of transient MCAO in mice (Li *et al.*, 1999). In another experiment, fresh bone marrow cells ($1 \times 10^6/10\ \mu l$) including the supportive tissue (stroma) obtained directly from adult rats were grafted into the penumbra of the ischemic striatum at 4 or 24 h after 2 h of transient MCAO in rats. Data demonstrate that adult hematopoietic bone marrow cells that were labeled with green fluorescent protein (GFP) survive in the adult mouse and rat brains after ischemia. Migration of SVZ cells was more prominent in the rat model than in the mouse, suggesting that the grafting microenvironment such as stroma increased the potential of stem cell migration. Colocalization of mitotic and apoptotic cells in the graft suggests that both ischemic injured brain and the graft cells may revert to an early stage of development to promote repair.

A neuroepithelial stem cell line (MHP36) is genetically engineered to replicate at 33°C, but cease replication and differentiate at body temperature of 37°C into neurons or glia. MHP36 were transplanted (75,000 cells/3 $\mu$l) at 2 weeks after 60 min of transient MCAO into eight sites of the striatum and cortex (Veizovic *et al.*, 1999). The grafted animals recovered from the sensorimotor deficit to a level similar sham controls throughout 18 weeks of testing. With this cell line (MHP36), Sinden *et al.* (1997) showed a recovery from functional deficit in adult rats following 15 min of global ischemia, and found an extensive repopulation of the grafted cells into neuronal and glial phenotypes in the hippocampus. Further study showed that MHP36 stem cells restored cognitive function in marmosets after NMDA-induced excitotoxic hippocampal CA1 lesions as effectively as fetal homografts (Hodges *et al.*, 1999; Virley *et al.*, 1999).

Two different stroke therapies could be applicable using stem cells. One is to activate intrinsic stem cells located in the hippocampal DG or SVZ of the caudate (Abe, 2000). Craig *et al.* (1996) showed that infusion of EGF into adult

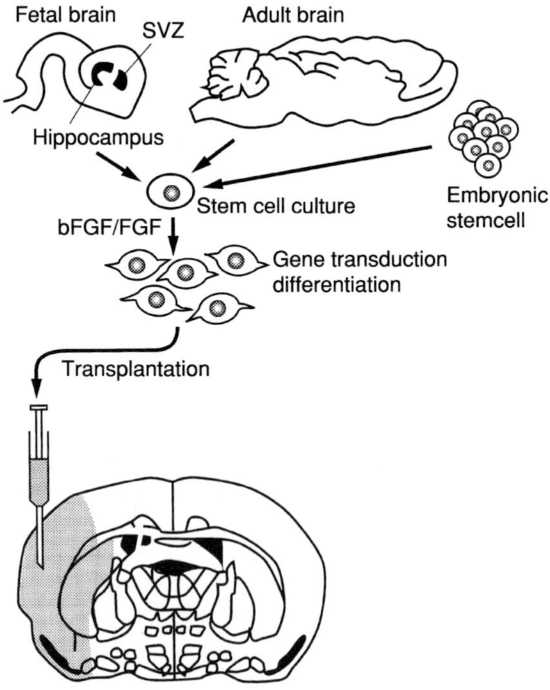


FIG. 8. Strategy of gene therapy with stem cell modification. Stem cells from some sources can be transplanted after specific genes are transfected to induce particular cell types.

mouse ventricle induced migration of subependymal cells away from the lateral ventricle walls into adjacent parenchyma. Kuhn *et al.* (1997) found a neurogenesis in the olfactory bulb after injection of bFGF or EGF into the ventricle. As mentioned above, these stem cells are activated after ischemic injury. Addition of NTFs could further potentiate such a spontaneous activation of stem cells after ischemia. Another approach is to transplant manipulated stem cells originating from fetal or adult brain, or even embryonic stem cells (Flax *et al.*, 1998; Vescovi *et al.*, 1999) into the injured lesion of the brain (Fig. 8). However, we are not able to make these stem cells divide, differentiate, and migrate to fulfill the space lost by ischemia and to form a functionally normal neuron network at present. To develop future therapy for human stroke, a greater understanding of and experimental studies on the role of each NTF, on genetic markers at each stage of stem cell differentiation, and on genetic direction to a certain species of brain cells are essential. Even if it takes time, it is promising.

## Acknowledgments

This work was partly supported by Grants-in-Aid for Scientific Research (B) 12470141 and (Hoga) 12877211 from the Ministry of Education, Science, Culture and Sports of Japan, and by Grants (K. Tashiro, Y. Itoyama) and Comprehensive Research on Aging and Health (H11-Choju-010, No. 207, A. Koizumi) from the Ministry of Health and Welfare of Japan, and by the Cytoscientific Foundation.

## References

Abe, K. (2000). Therapeutic potential of neurotrophic factors and neural stem cells against ischemic brain injury. *J. Cereb. Blood Flow Metab.* **20,** 1393–1408.

Abe, K., and Hayashi, T. (1997). Expression of the GDNF gene in rat brain after transient MCA occlusion. *Brain Res.* **776,** 230–234.

Abe, K., and Kogure, K. (1993). Selective gene expression after brain ischemia. *Prog. Brain Res.* **96,** 221–236.

Abe, K., Tanzi, R. E., and Kogure, K. (1991). Induction of HSP70 mRNA after transient ischemia in gerbil brain. *Neurosci. Lett.* **125,** 166–168.

Abe, K., Aoki, M., Kawagoe, J., Yoshida, T., Hattori, A., Kogure, K., and Itoyama, Y. (1995). Ischemic delayed neuronal death: A mitochondrial hypothesis. *Stroke* **26,** 1478–1489.

Abe, K., Hayashi, T., and Itoyama, Y. (1997a). Amelioration of brain edema by topical application of GDNF in reperfused rat brain. *Neurosci. Lett.* **231,** 37–40.

Abe, K., Setoguchi, Y., Hayashi, T., and Itoyama, Y. (1997b). In vivo adenovirus-mediated gene transfer and the expression in ischemic and reperfused rat brain. *Brain Res.* **763,** 191–202.

Aoki, H., Onodera, H., Yae, T., Jian, Z., and Kogure, K. (1993). Neural grafting to ischemic CA1 lesions in the rat hippocampus: An autoradiographic study. *Neuroscience* **56,** 345–354.

Arakawa, Y., Sendtner, M., and Thoenen, H. (1990). Survival effect of ciliary neurotrophic factor (CNTF) on chick embryonic motoneurons in culture: Comparison with other neurotrophic factors and cytokines. *J. Neurosci.* **10,** 3507–3515.

Barbin, G., Manthorpe, M., and Varon, S. (1984). Purification of the chick eye ciliary neurotrophic factor. *J. Neurochem.* **43,** 1468–1478.

Barth, A., Barth, L., and Newell, D. W. (1996). Combination therapy with MK-801 and alpha-phenyl-tert-butyl-nitrone enhances protection against ischemic neuronal damage in organotypic hippocampal slice cultures. *Exp. Neurol.* **141,** 330–336.

Betz, A. L., Yang, G. Y., and Davidson, B. L. (1995). Attenuation of stroke size in rats using an adenoviral vector to induce overexpression of interleukin-1 receptor antagonist in brain. *J. Cereb. Blood Flow Metab.* **15,** 547–551.

Breakfield, X. O. (1993). Gene delivery into the brain using virus vectors. *Nat. Genet.* **3,** 187–189.

Charriaut-Marlangue, C., Margaill, I., Represa, A., Popovici, T., Plotkine, M., and Ben-Ari, Y. (1996). Apoptosis and necrosis after reversible focal ischemia: An in situ DNA fragmentation analysis. *J. Cereb. Blood Flow Metab.* **16,** 186–194.

Comelli, M. C., Guidolin, D., Seren, M. S., Zanoni, R., Canella, R., Rubini, R., and Manev, H. (1993). Time course, localization and pharmacological modulation of immediate early inducible genes, BDNF and trkB mRNAs in the rat brain following photochemical stroke. *Neuroscience* **55,** 473–490.

Craig, C. G., Tropepe, V., Morshead, C. M., Reynolds, B. A., Weiss, S., and van der Kooy, D. (1996). In vivo growth factor expansion of endogenous subependymal neural precursor cell populations in the adult mouse brain. *J. Neurosci.* **16,** 2649–2658.

Culotta, E. (1993). New startups move in as gene therapy goes commercial. *Science* **260,** 914–915.

Danko, I., Fritz, J. D., Latendresse, J. S., Herweijer, H., Schltz, E., and Wolf, J. A. (1993). Dystrophin expression improves myofiber survival in mdx muscle following intramuscular plasmid DNA injection. *Hum. Mol. Genet.* **2,** 2055–2061.

Deng, C., Wynshaw-Boris, A., Zhou, F., Kuo, A., and Leder, P. (1996). Fibroblast growth factor receptor 3 is a negative regulator of bone growth. *Cell* **84,** 911–921.

Doetsch, F., Doetsch, F., Caille, I., Lim, D. A., Garcia-Verdugo, J. M., and Alvarez-Buylla, A. (1999). Subventricular zone astrocytes are neural stem cells in the adult mammalian brain. *Cell* **97,** 703–716.

Eriksson, P. S., Perfilieva, E., Bjork-Eriksson, T., Alborn, A. M., Nordborg, C., Peterson, D. A., and Gage, F. H. (1998). Neurogenesis in the adult human hippocampus. *Nat. Med.* **4,** 1313–1317.

Ferrari, F. K., Xiao, X., McCarty, D., and Samulski, R. J. (1997). New developments in the generation of Ad-free, high-titer rAAV gene therapy vectors. *Nat. Med.* **3,** 1295–1297.

Fink, D. J., Sternberg, L. R., Weber, P. C., Mata, M., Goins, W. F., and Glorioso, J. C. (1992). In vivo expression of b-galactosidase in hippocampal neurons by HSV-mediated gene transfer. *Hum. Gene Ther.* **3,** 11–19.

Fischer, S., Sharma, H. S., Karliczek, G. F., and Schaper, W. (1995). Expression of vascular permeability factor/vascular endothelial growth factor in pig cerebral microvascular endothelial cells and its upregulation by adenosine. *Mol. Brain Res.* **28,** 141–148.

Flax, J. D., Aurora, S., Yang, C., Simonin, C., Wills, A. M., Billinghurst, L. L., Jendoubi, M., Sidman, R. L., Wolfe, J. H., Kim, S. U., and Snyder, E. Y. (1998). Engraftable human neural stem cells respond to developmental cues, replace neurons, and express foreign genes. *Nat. Biotechnol.* **16,** 1033–1039.

Fujimura, M., Morita-Fujimura, Y., Murakami, K., Kawase, M., and Chan, P. H. (1998). Cytosolic redistribution of cytochrome c after transient focal cerebral ischemia in rats. *J. Cereb. Blood Flow Metab.* **18,** 1239–1247.

Gould, E., Reeves, A. J., Graziano, M. S., and Gross, C. G. (1999). Neurogenesis in the neocortex of adult primates. *Science* **286,** 548–552.

Hayashi, T., Abe, K., Suzuki, H., and Itoyama, Y. (1997). Rapid induction of VEGF gene after transient MCA occlusion in rat. *Stroke* **28,** 2039–2044.

Hayashi, T., Abe, K., and Itoyama, Y. (1998). Reduction of ischemic damage by application of VEGF in rat brain after transient ischemia. *J. Cereb. Blood Flow Metab.* **18,** 887–895.

Heistad, D. D., and Faraci, F. M. (1996). Gene therapy for cerebral vascular disease. *Stroke* **27,** 1688–1693.

Henderson, C. E., Phillips, H. S., Pollock, R. A., Davies, A. M., Lemeulle, C., Armanini, M., Simpson, L. C., Moffet, B., Vandlen, R. A., Koliatsos, V. E., and Rosenthal, A. (1994). GDNF: A potent survival factor for motoneurons present in peripheral nerve and muscle. *Science* **266,** 1062–1064.

Herman, D. M., Kilic, E., Kugler, S., Isenmann, S., and Bahr, M. (2001). adino virus-mediated glial cell line-derived neurotrophic factor (GDNF) expression protects against subsequent cortical cold injury in rats. *Neurobiol Dis.* **8,** 964–973.

Higashiyama, S., Iwamoto, R., Goishi, K., Raab, G., Taniguchi, N., Klagsbrun, M., and Mekada, E. (1995). The membrane protein CD9/DRAP 27 potentiates the juxtacrine growth factor activity of the membrane-anchored heparin-binding EGF-like growth factor. *J. Cell Biol.* **128,** 929–938.

Hodges, H., Virley, D., Williams, C., Ridley, R. M., Sinden, J. D., Kershaw, T. R., Harland, S., Gray, J. A., and Lantos, P. L. (1997). Faetal CA1 grafts promote recovery from cognitive deficits induced by excitotoxic damage to the CA1 field in marmosets. *J. Cereb. Blood Flow Metab.* **17**(Suppl. 1), S451.

Hodges, H., Virley, D., Ridley, R. M., Sinden, J. D., Kershaw, T. R., French, S., Harland, S., Gray, J. A., and Lantos, P. L. (1999). Conditionally immortal MHP36 cells restore cognitive function in marmosets after excitotoxic hippocampal CA1. *J. Cereb. Blood Flow Metab.* **19**(Suppl. 1), S617.

Horie, H., Inagaki, Y., Sohma, Y., Nozawa, R., Okawa, K., Hasegawa, M., Muramatsu, N., Kawano, H., Horie, M., Koyama, H., Sakai, I., Takeshita, K., Kowada, Y., Takano, M., and Kadoya, T. (1999). Galectin-1 regulates initial axonal growth in peripheral nerves after axotomy. *J. Neurosci.* **19,** 9964–9974.

Hu, J., Saito, T., Abe, K., and Deguchi, T. (1997). Increase of ciliary neurotrophic factor (CNTF) in the ischemic rat brain as determined by a sensitive enzyme-linked immunoassay. *Neurol. Res.* **19,** 593–598.

Humpel, C., Hoffer, B., Stronberg, I., Bektesh, S., Collins, F., and Olson, L. (1994). Neurons of the hippocampal formation express glial cell line-derived neurotrophic factor messenger RNA in response to kainate-induced excitation. *Neuroscience* **59,** 791–795.

Ijichi, A., Sakuma, S., and Tofilon, P. J. (1995). Hypoda-induced vascular endothelial growth factor expression in normal rat astrocyte cultures. *Glia* **14,** 87–93.

Ip, N., Yi, Y. P., van de Stadt, I., Panayotatos, N., Alderson, R. F., and Lindsay, R. M. (1991). Ciliary neurotrophic factor enhances neuronal survival in embryonic rat hippocampal cultures. *J. Neurosci.* **11,** 3124–3134.

Iwai, M., Abe, K., Kitagawa, H., and Hayashi, T. (2001a). Gene therapy with adenovirus-mediated GDNF and neural stem cells activation after ischemic brain injury. *Hum. Cell* **14,** 27–38.

Iwai, M., Hayashi, T., Seki, T., Zhang, W. R., Sato, K., Manabe, Y., and Abe, K. (2001). Induction of PSA-NCAM in postischemic gerbil hippocampus mainly dissociated with neural stem cell proliferation. *Brain Res.* **902,** 288–293.

Jessell, T. M., and Sanes, J. R. (2000). The generation and survival of nerve cells. *In* "Principles of Neural Science," 4th ed. (E. R. Kandel, J. H. Schwartz, and T. M. Jessell, Eds.), pp. 1041–1062. MacGraw-Hill, New York.

Johansson, C. B., Momma, S., Clarke, D. L., Risling, M., Lendahl, U., and Frisen, J. (1999). Identification of a neural stem cell in the adult mammalian central nervous system. *Cell* **96,** 25–34.

Kafri, T., Blomer, U., Peterson, D. A., Gage, F. H., and Verma, I. M. (1997). Sustained expression of genes delivered directly into liver and muscle by lentiviral vectors. *Nat. Genet.* **17,** 314–317.

Karpati, G., Lochmuller, H., Nalbantoglu, J., and Durham, H. (1996). The principles of gene therapy for the nervous system. *Trends Neurosci.* **19,** 49–54.

Kato, T., Nakano, S., Kogure, K., Sasaki, H., Koiwai, K., Yamasaki, Y., Katagiri, T., and Sasaki, H. (1992). The binding of basic fibroblast growth factor to ischemic neurons in the rat. *Neuropathol. Appl. Neurobiol.* **18,** 282–290.

Kempermann, G., and Gage, F. H. (1999). New nerve cells for the adult brain. *Sci. Am.* **280,** 48–53.

Kitagawa, H., Abe, K., Hayashi, T., Mitsumoto, H., Koga, N., and Itoyama, Y. (1998a). Ameliorative effect of GDNF on brain edema after permanent MCAO in rats. *Neurol. Res.* **20,** 333–336.

Kitagawa, H., Hayashi, T., Mitsumoto, Y., Koga, N., Itoyama, Y., and Abe, K. (1998b). Reduction of ischemic brain injury by topical application of GDNF after permanent MCAO in rats. *Stroke* **29,** 1417–1422.

Kitagawa, H., Setoguchi, Y., Fukuchi, Y., Mitsumoto, Y., Mori, T., and Abe, K. (1998c). Induction of DNA fragmentation and HSP 72 immunoreactivity by adenovirus-mediated gene transfer in normal gerbil hippocampus and ventricle. *J. Neurosci. Res.* **54,** 38–45.

Kitagawa, H., Sasaki, C., Sakai, K., Mori, A., Mitsumoto, Y., Mori, T., Fukuchi, Y., Setoguchi, Y., and Abe, K. (1999). Adenovirus-mediated gene transfer of GDNF prevents ischemic brain injury after transient MCA occulusion in rats. *J. Cereb. Blood Flow Metab.* **19,** 1336–1344.

Kovacs, Z., Ikezaki, K., Samoto, K., Inamura, T., and Fukui, M. (1996). VEGF and flt: Expression time kinetics in rat brain infarct. *Stroke* **27,** 1865–1873.

Krieglstein, K., Suter-Crazzolara, C., Fisher, W. H., and Unsicker, K. (1995). TGF-$\beta$ superfamily promote survival of midbrain dopaminergic neurons and protect them against MPP + toxicity. *EMBO J.* **14,** 736–742.

Krupinski, J., Kaluza, J., Kumar, P., Kumar, S., and Wang, J. M. (1994). Role of angiogenesis in patients with cerebral ischemic stroke. *Stroke* **25,** 1794–1798.

Kuhn, H. G., Dickinson-Anson, H., and Gage, F. H. (1996). Neurogenesis in the dentate gyrus of the adult rat: Age-related decrease of neuronal progenitor proliferation. *J. Neurosci.* **16,** 2027–2033.

Kuhn, H. G., Winkler, J., Kempermann, G., Thal, L. J., and Gage, F. H. (1997). Epidermal growth factor and fibroblast growth factor-2 have different effects on neural progenitors in the adult rat brain. *J. Neurosci.* **17,** 5820–5829.

Lawrence, M. S., McLaughlin, J. R., Sun, G.- H., Ho, D. Y., McIntosh, L., Kunis, D. M., Sapolsky, R. M., and Steinberg, G. K. (1997). Herpes simplex viral vectors expressing bcl-2 are neuroprotective when delivered after a stroke. *J. Cereb. Blood Flow Metab.* **17,** 740–744.

Lee, T. H., Abe, K., Aoki, M., Kogure, K., and Itoyama, Y. (1994). The protective effect of L-threo-3, 4-dihydroxyphenylserine on ischemic hippocampal neuronal death in gerbils. *Stroke* **25,** 1425–1432.

Li, L., Wu, W., Lin, L.F.H., Lei, M., Oppenheim, R. W., and Houenou, L. J. (1995). Rescue of adult mouse motoneurons from injury-induced cell death by glial cell line-derived neurotrophic factor. *Proc. Natl. Acad. Sci. USA* **92,** 9771–9775.

Li, P., Nijhawan, D., Budihardjo, I., Srinivasula, S. M., Ahmad, M., Alnemri, E. S., and Wang, X. (1997). Cytochrome c and dATP-dependent formation of apaf-1/caspase-9 complex initiates an apoptotic protease cascade. *Cell* **91,** 479–489.

Li, Y., Chopp, M., Chen, J., Gautam, S. C., Dou, D., Zhang, L., Wang, L., Xu, Y., Powers, C., Zhang, X., and Jiang, F. (1999). Intracerebral grafting of adult bone marrow cells after stroke in adult mice and rats. *J. Cereb. Blood Flow Metab.* **19**(Suppl. 1), S615.

Lillien, L. E., and Raff, M. C. (1990). Differentiation signals in the CNS: Type-2 astrocyte development in vitro as a model system. *Neuron* **5,** 111–119.

Lin, L.F.H., Doherty, D. H., Lile, J. D., Bektesh, S., and Collins, F. (1993). GDNF: A glial cell line-derived neurotrophic factor for midbrain dopaminergic neurons. *Science* **260,** 1130–1132.

Linnik, M. D., Zahos, P., Geschwind, M. D., and Federoff, H. J. (1995). Expression of bcl-2 from a defective herpes simplex virus-1 vector limits neuronal death in focal cerebral ischemia. *Stroke* **26,** 1670–1674.

Liu, J., and Sharp, F. R. (1999). Ischemia induced neurogenesis in the dentate gyrus: An injury-dependent neuroplasticity. *J. Cereb. Blood Flow Metab.* **19**(Suppl. 1), S614.

Liu, J., Solway, K., Messing, R. O., and Sharp, F. R. (1998). Increased neurogenesis in the dentate gyrus after transient global ischemia in gerbils. *J. Neurosci.* **18,** 7768–7778.

Logan, A., Frautschy, S. A., Gonzalez, A. M., Sporn, M. B., and Baird, A. (1992). Enhanced expression of transforming growth factor beta 1 in the rat brain after a localized cerebral injury. *Brain Res.* **587,** 216–225.

Lyden, P. D., Jackson-Friedman, C., Shin, C., and Hassid, S. (2000). Synergistic combinatorial stroke therapy: A quantal bioassay of a GABA agonist and a glutamate antagonist. *Exp. Neurol.* **163,** 477–489.

Morishita, R., Sugimoto, T., Aoki, M., Kida, I., Tomita, N., Moriguchi, A., Maeda, K., Sawa, Y., Kaneda, Y., Higaki, J., and Ogihara, T. (1997). In vivo transfection of cis element "decoy" against nuclear factor-kappaB binding site prevents myocardial infarction. *Nat. Med.* **3,** 894–899.

Morishita, R., Yamada, S., Yamamoto, K., Tomita, N., Kida, I., Sakurabayashi, I., Kikuchi, A., Kaneda, Y., Lawn, R., Higaki, J., and Ogihara, T. (1998). Novel therapeutic strategy for atherosclerosis: Ribozyme oligonucleotides against apolipoprotein(a) selectively inhibit apolipoprotein(a) but not plasminogen gene expression. *Circulation* **98,** 1898–1904.

Mulligan, R. C. (1993). The basic science of gene therapy. *Science* **260,** 926–932.

Nakafuku, M., and Nakamura, S. (1995). Establishment and characterization of a multipotential neural cell line that can conditionally generate neurons, astrocytes, and oligodendrocytes in vitro. *J. Neurosci. Res.* **41,** 153–168.

Nakagawa, Y., Kaneko, T., Ogura, T., Suzuki, T., Torii, M., Kaibuchi, K., Arai, K., Nakamura, S., and Nakafuku, M. (1996). Roles of cell-autonomous mechanisms for differential expression of region-specific transcription factors in neuroepithelial cells. *Development* **122,** 2449–2464.

Naski, M. C., Wang, Q., Xu, J., and Ornitz, D. M. (1996). Graded activation of fibroblast growth factor receptor 3 by mutations causing achondroplasia and thanatophoric dysplasia. *Nat. Genet.* **13,** 233–237.

Nomura, M., Yamagishi, S., Harada, S., Hayashi, Y., Yamashima, T., Yamashita, J., and Yamamoto, H. (1995). Possible participation of autocrine and paracrine vascular endothelial growth factors in hypoxia-induced proliferation of endothelial cells and pericytes. *J. Biol. Chem.* **270,** 28316–28324.

Ono, T., Fujino, Y., Tsuchiya, T., and Tsuda, M. (1990). Plasmid DNAs directly injected into mouse brain with lipofectin can be incorporated and expressed by brain cells. *Neurosci. Lett.* **117,** 259–263.

Pallela, T. D., Hidaka, Y., Silverman, L. J., Levine, M., Glorioso, J., and Kelley, W. N. (1989). Expression of human HPRT mRNA in brains of mice infected with a recombinant herpes simplex virus-1 vector. *Gene* **80,** 137–144.

Parent, J. M., Yu, T. W., Leibowitz, R. T., Geschwind, D. H., Sloviter, R. S., and Lowenstein, D. H. (1997). Dentate granule cell neurogenesis is increased by seizures and contributes to aberrant network reorganization in the adult rat hippocampus. *J. Neurosci.* **17,** 3727–3738.

Porter, A. G., and Janicke, R. U. (1999). Emerging roles of caspase-3 in apoptosis. *Cell Death Differ.* **6,** 99–104.

Price, J., Turner, D., and Cepko, C. (1987). Lineage analysis in the vertebrate nervous system by retrovirus-mediated gene transfer. *Proc. Natl. Acad. Sci. USA* **84,** 156–160.

Royuela, M., De Miguel, M. P., Bethencourt, F. R., Sanchez-Chapado, M., Fraile, B., and Paniagua, R. (1998). Transforming growth factor beta 1 and its receptor types I and II. Comparison in human normal prostate, benign prostatic hyperplasia, and prostatic carcinoma. *Growth Factors* **16,** 101–110.

Sakakibara, S., and Okano, H. (1997). Expression of neural RNA-binding proteins in the postnatal CNS: Implications of their roles in neuronal and glial cell development. *J. Neurosci.* **17,** 8300–8312.

Schaar, D. G., Sieber, B. A., Dreyfus, C. F., and Black, I. (1993). Regional and cell-specific expression of GDNF in rat brain. *Exp. Neurol.* **124,** 368–371.

Seki, T., and Arai, Y. (1999). Temporal and spatial relationships between PSA-NCAM-expressing, newly generated granule cells, and radial glia-like cells in the adult dentate gyrus. *J. Comp. Neurol.* **410,** 503–513.

Shimazaki, T., Arsenijevic, Y., Ryan, A. K., Rosenfeld, M. G., and Weiss, S. (1999). A role for the POU-III transcription factor Brn-4 in the regulation of striatal neuron precursor differentiation. *EMBO J.* **18,** 444–456.

Sinden, J. D., Rashid-Doubell, F., Kershaw, T. R., Nelson, A., Chadwick, A., Jat, P. S., Noble, M. D., Hodges, H., and Gray, J. A. (1997). Recovery of spatial learning by grafts of a conditionally immortalized hippocampal neuroepithelial cell line into the ischaemia-lesioned hippocampus. *Neuroscience* **81,** 599–608.

Takagi, Y., Nozaki, K., Takahashi, J., Yodoi, J., Ishikawa, M., and Hashimoto, N. (1999). Proliferation of neuronal precursor cells in the dentate gyrus is accelerated after transient forebrain ischemia in mice. *Brain Res.* **831,** 283–287.

Takahashi, J., Palmer, T. D., and Gage, F. H. (1999). Retinoic acid and neurotrophins collaborate to regulate neurogenesis in adult-derived neural stem cell cultures. *J. Neurobiol.* **38,** 65–81.

Tomac, A., Lindqvist, E., Lin, L.F.H., Ogren, S. O., Young, D., Hoffer, B. J., and Olson, L. (1995). Protection and repair of the nigrostriatal dopaminergic system by GDNF in vivo. *Nature* **373,** 335–339.

Torii, M., Matsuzaki, F., Osumi, N., Kaibuchi, K., Nakamura, S., Casarosa, S., Guillemot, F., and Nakafuku, M. (1997). Transcription factors Mash-1 and Prox-1 delineate early steps in differentiation of neural stem cells in the developing central nervous system. *Development* **126,** 443–456.

Veizovic, T., Stroemer, P., Beech, J., Sowinski, P., and Hodges, H. (1999). Stem cell grafts resolve sensory dysfunction following MCAo in the rat. *J. Cereb. Blood Flow Metab.* **19**(Suppl. 1), S616.

Verma, I. M., and Somia, N. (1997). Gene therapy-promises, problems, and prospects. *Nature* **389,** 239–242.

Vescovi, A. L., Parati, E. A., Gritti, A., Poulin, P., Ferrario, M., Wanke, E., Frolichsthal-Schoeller, P., Cova, L., Arcellana-Panlilio, M., Colombo, A., and Galli, R. (1999). Isolation and cloning of multipotential stem cells from the embryonic human CNS and establishment of transplantable human neural stem cell lines by epigenetic stimulation. *Exp. Neurol.* **156,** 71–83.

Virley, D., Ridley, R. M., Sinden, J. D., Kershaw, T. R., Harland, S., Rashid, T., French, S., Sowinski, P., Gray, J. A., Lantos, P. L., and Hodges, H. (1999). Primary CA1 and conditionally immortal MHP36 cell grafts restore conditional discrimination learning and recall in marmosets after excitotoxic lesions of the hippocampal CA1 field. *Brain* **122,** 2321–2335.

Walsh, C., and Cepko, C. L. (1988). Clonally related cortical cells show several migration patterns. *Science* **241,** 1342–1345.

Wang, J. M., Hayashi, T., Zang, W. R., Sakai, K., Shiro, Y., and Abe, K. (2000). Reduction of ischemic brain injury by topical application of IGF-1 after transient MCA occlusion in rats. *Brain Res.* **859,** 381–385.

Wen, T. C., Matsuda, S., Yoshimura, T., Kawabe, T., and Sakanaka, M. (1995). Ciliary neurotrophic factor prevents ischemia-induced learning disability and neuronal loss in gerbils. *Neurosci. Lett.* **191,** 55–58.

Wolfe, J. H., Deshmann, S. L., and Fraser, N. W. (1992). Herpes virus vector gene transfer and expression of b-glucuronidase in the central nervous system of MPS VII mice. *Nat. Genet.* **1,** 379–384.

Xu, D. G., Crocker, S. J., Doucet, J.-P., St-Jean, M., Tamai, K., Hakim, A. M., Ikeda, J.-E., Liston, P., Thompson, C. S., Korneluk, R. G., MacKenzie, A., and Robertson, G. S. (1997). Elevation of neuronal expression of NAIP reduces ischemic damage in the rat hippocampus. *Nat. Med.* **3,** 997–1004.

Yamashita, K., Wiessner, C., Lindholm, D., Thoenen, H., and Hossmann, K. A. (1997). Post-occlusion treatment with BDNF reduces infarct size in a model of permanent occlusion of the middle cerebral artery in rat. *Metab. Brain Dis.* **12,** 271–280.

Yang, G. Y., Zhao, Y. J., Davidson, B. L., and Betz, A. L. (1997). Overexpression of interleukin-1 receptor antagonist in the mouse brain reduces ischemic brain injury. *Brain Res.* **751,** 181–188.

Yang, G.-Y., Liu, X.-H., Kadoya, C., Mao, Y.-J., Davidson, B. L., and Betz, A. L. (1998). Attenuation of ischemic inflammatory response in mouse brain using an adenoviral vector to induce overexpression of interleukin-1 receptor antagonist. *J. Cereb. Blood Flow Metab.* **18,** 840–847.

Yenari, M. A., Fink, S. L., Sun, G. H., Chang, L. K., Patel, M. K., Kunis, D. M., Onley, D., Ho, D. Y., Sapolsky, R. M., and Steinberg, G. K. (1998). Gene therapy with HSP72 is neuroprotective in rat models of stroke and epilepsy. *Ann. Neurol.* **44,** 584–591.

Zhang, W. R., Kitagawa, H., Hayashi, T., Sasaki, C., Sakai, K., Warita, H., Shiro, Y., Suenaga, H., Ohmae, H., Tsuji, S., Itoh, T., Nishimura, O., Nagasaki, H., and Abe, K. (1999). Topical application of neurotrophin-3 attenuates ischemic brain injury after transient middle cerebral artery occlusion in rats. *Brain Res.* **842,** 211–214.

Zhang, W. R., Hayashi, T., Iwai, M., Nagano, I., Sato, K., Manabe, Y., and Abe, K. (2001). Time dependent amelioration against ischemic brain damage by glial cell line-derived neurotrohpic factor after transient MCAO in rat. *Brain Res.* **903,** 253–256.

# GENE THERAPY FOR MUCOPOLYSACCHARIDOSIS

Assumpció Bosch

Departament de Bioquímica i Biologia Molecular
Facultat de Veterinària
Centre de Biotecnologia Animal i Teràpia Gènica (CBATEG)
Universitat Autònoma de Barcelona
08193 Bellaterra, Barcelona, Spain

Jean Michel Heard

Unité Rétrovirus et Transfert Génétique
Département de Neuroscience
Institut Pasteur 75725, Paris, France



## I. Definition and General Considerations on Lysosomal Storage Diseases

Lysosomes contain a variety of degradation enzymes, mostly acid hydrolases, involved in the turnover of extracellular, membrane-bound, and secreted cell metabolites, including molecules with lipids, saccharides, and protein components. Genetic deficiency in any of the lysosomal enzymes causes cell dysfunction as a consequence of the accumulation of insoluble partial catabolites in lysosomes. Clogging of undigested substrate results in lysosome enlargement that is readily observable on tissue sections, leading to the typical image of lysosomal storage that characterizes those diseases (Fig. 1A).

Though lysosomal enlargement likely alters cell functions and viability, the precise cascade of events linking histopathological features to clinical symptoms is partially understood for most lysosomal storage diseases (LSD). More than 30




0074-7742/03 $35.00

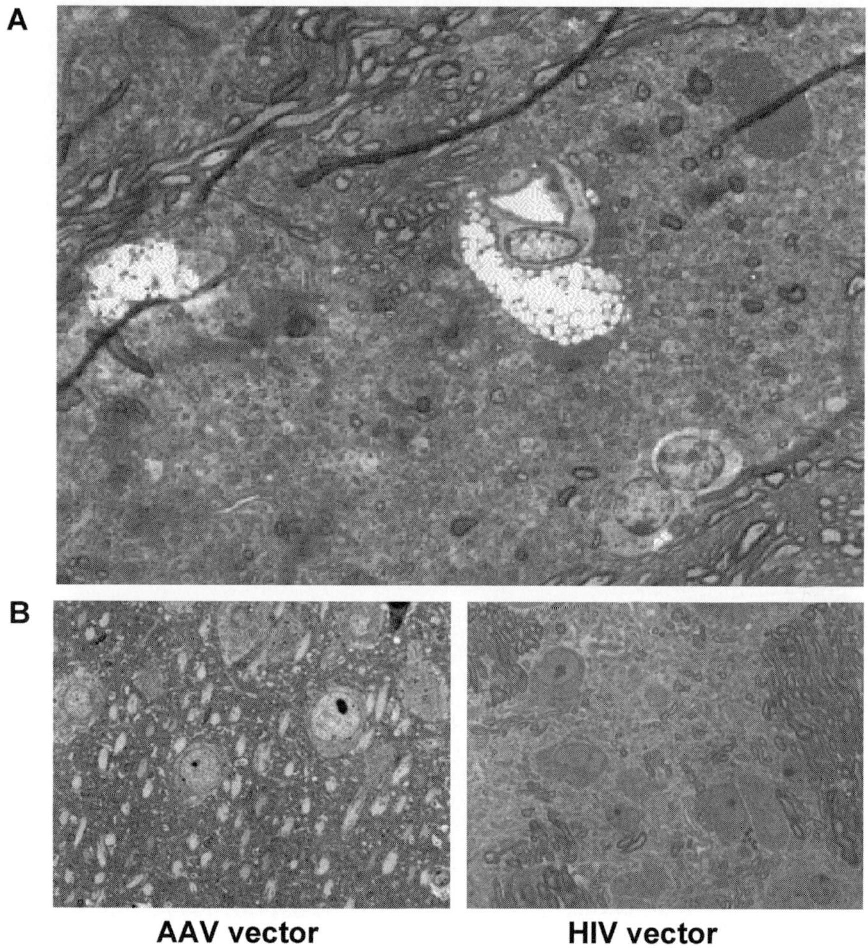


Fig. 1. Lysosomal storage lesions in brain tissue of a MPS VII mouse (A) and correction after stereotactic injection of viral vectors coding for β-glucuronidase (B).

different diseases with patent lysosomal storage lesions in tissues have been recognized (Table I). Every tissue may be affected, and symptomatology is highly polymorphic depending on which enzyme is deficient, but also within patients suffering deficiency of the same enzyme. Clinical diagnosis is therefore often uneasy and may be delayed for patients exhibiting atypical manifestations. Consequently, it is likely that psychomotor retardations of currently unknown etiology will be further recognized as LSDs.

The prevalence of LSD as a whole is equivalent to that of cystic fibrosis (1:2500), although each form of LSD is a rare disease. The identification of genetic abnormalities associated with LSDs tremendously improved diagnosis and characterization of most of these disorders. Reliable prenatal diagnosis based on biochemical or genetic assays is now available for most of these diseases. As the majority of LSD-associated genetic abnormalities are recessive traits, families with a first-born affected child may anticipate the outcome of the following pregnancies. However, though it is technically achievable, with respect to the low prevalence of those diseases, to the cost of the biochemical or genetic diagnosis, and to the absence of curative treatment for most of them, a systematic screening for LSDs in neonates cannot currently be considered as a reasonable option.

## II. Addressing Lysosomal Enzymes to the Lysosomes: Enzyme Replacement Therapy

Though LSDs are clinically and genetically heterogeneous, they share common biochemical features that have important implications for the design of potential therapies.

The observation that cultured fibroblasts from patients with Hurler disease secreted a factor that induced the regression of lysosomal storage lesions in cultured fibroblasts from patients with Hunter disease was a hallmark in the study of lysosomal enzymes (Fratantoni *et al.*, 1968). It suggested that LSDs could be amenable to treatment. One of the corrective factors was soon identified as $\alpha$-L-iduronidase (IDUA) (Sando and Neufeld, 1977), a lysosomal enzyme whose deficiency causes Hurler disease (MPS I). A few years later, the observation that not all the active enzyme was taken up by deficient cells led to the identification of the mannose-6-phosphate receptor (M6PR) as the protein mediating IDUA uptake (Kaplan *et al.*, 1977). The subcellular localization of the receptor, which is ubiquitously expressed in different cellular structures, such as the Golgi apparatus, the endoplasmic reticulum, and the lysosomal and cytoplasmic membranes, elucidated a biosysthesis and secretion pathway that is shared by most lysosomal enzymes (Fig. 2). Lysosomal enzymes that are synthesized in the endoplasmic reticulum are further glycosylated and the mannose residues are phosphorylated in the Golgi apparatus, where they become the ligand of the M6PR. The interaction between the lysosomal enzyme and the M6PR receptor allows the enzyme to traffic through endosomal compartments, lysosomes, and the plasma membrane. A fraction of the cargo, which varies according to enzymes and cell types, is secreted in the extracellular medium (Pfeffer, 1991). A similar trafficking of M6PRs in cells that are genetically deficient for the synthesis of one of the lysosomal enzyme allows capturing of the missing molecule at the cell surface and internalization into endosomal and lysosomal compartments. This mechanism supports the feasibility of a cross-correction of enzyme-deficient cells by extracellular enzyme supply. It is the rationale for a potential treatment of LSDs by enzyme replacement therapy (ERT).

TABLE I
TYPES OF MPS

| Type of MPS | Eponym | Enzyme deficiency | CNS symptoms | Systemic implications | Animal models | References for animal models |
|---|---|---|---|---|---|---|
| MPS I | Hurler | α-L-Iduronidase | Corneal clouding | Dysostosis multiplex | Cat, mouse | Haskins *et al.* (1983); Clarke *et al.* (1997) |
| | | | Mental retardation | Organomegaly<br>Heart disease<br>Death in childhood | | |
| | Scheie | α-L-Iduronidase | Corneal clouding | Stiff joints<br>Normal life span | Dog | Spellacy *et al.* (1983) |
| MPS II | Hunter (severe) | Iduronate sulfatase | Mental retardation | Dysosotosis multiplex | Dog | Wilkerson *et al.* (1998) |
| | | | | Organomegaly<br>Death before 15 years | | |
| | Hunter (mild) | Iduronate sulfatase | | Short stature<br>Survival between 20 and 60s | | |
| MPS III A | Sanfilippo A | Heparan *N*-sulfatase | Profound mental retardation<br>Hyperactivity | | Mouse, dog | Fischer *et al.* (1998); Bhaumik *et al.* (1999) |
| MPS III B | Sanfilippo B | α-*N*-Actylglucosaminidase | Profound mental retardation<br>Hyperactivity | | Mouse, bird | Li *et al.* (1999); Aronovich *et al.* (2001) |
| MPS III C | Sanfilippo C | Acetyl-CoA:α-glucosaminide acetyltransferase | Profound mental retardation<br>Hyperactivity | | | |

| | | | | | | |
|---|---|---|---|---|---|---|
| MPS III D | Sanfilippo D | *N*-Acetylglucosamine 6-sulfatase | Profound mental retardation<br>Hyperactivity | | Goat | Thompson *et al.* (1992) |
| MPS IV A | Morquio A<br>Also milder forms | Galactose 6-sulfatase | Corneal clouding | Skeletal abnormalities | | |
| MPS IV B | Morquio B | $\beta$-Galactosidase | Corneal clouding | Skeletal abnormalities | | |
| MPS VI | Maroteaux-Lamy<br>Also milder forms | Arylsulfatase B | Corneal clouding | Dysostosis multiplex<br><br>Death before 20s | Cat, rat, mouse | Jezyk *et al.* (1977);<br>Yoshida *et al.* (1993);<br>Evers *et al.* (1996) |
| MPS VII | Sly | $\beta$-Glucoronidase | Mental retardation | Dysostosis multiplex<br>Organomegaly<br>Death before 20s | Dog, mouse, cat | Haskins *et al.* (1984);<br>Birkenmeier *et al.* (1989);<br>Fyfe *et al.* (1999) |

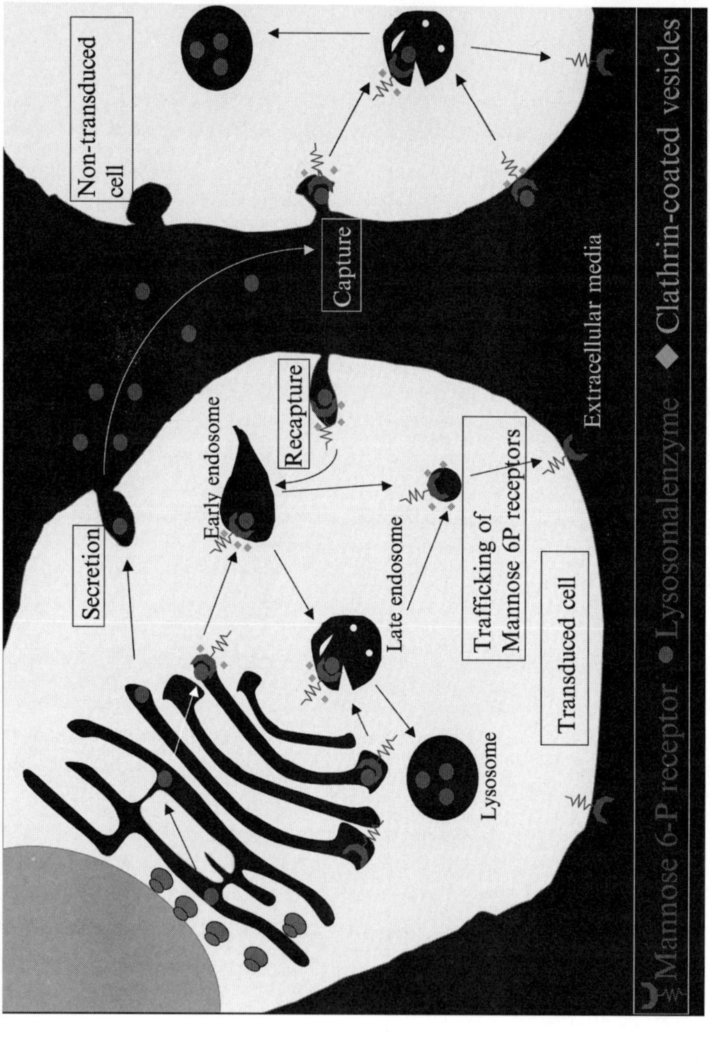


FIG. 2. Schematic representation of the biosynthesis and transport pathway of lysosomal enzymes.

Other receptors involved in the trafficking of lysosomal enzyme have been characterized later on, including the galactose receptor of hepatocytes and the mannose receptor of macrophages (Robbins *et al.*, 1981). Enzyme replacement therapy through the mannose receptor is used with success for treating Gaucher type I patients with a modified glucocerebrosidase enzyme preparation targeting macrophages (Barton *et al.*, 1991; Grabowski *et al.*, 1998). In addition, enzyme replacement via the M6PR has been recently proposed for patients with Fabry disease (Schiffmann *et al.*, 2000), Pompe disease (Van den Hout *et al.*, 2000), mild forms of MPS I (Scheie disease), and Maroteaux-Lamy disease (MPS VI).

In spite of the tremendous hope brought by ERT, this treatment suffers serious limitations. Notwithstanding the cost and the sophistication of enzyme preparations that are infused to patients, which restricts their use to rich countries, the main drawback of ERT is the impermeability of the blood–brain barrier, which avoids enzyme delivery to the central nervous system (CNS). Thus, no benefit can be expected on brain damage. This is actually a serious limitation, as most LSDs affect the CNS. As a consequence, ERT is currently considered only for patients suffering an LSD form in which brain damage is not anticipated. Extension of this treatment to others patients would imply that additional methods are available for enzyme delivery to the CNS. Candidate strategies rely on cell therapy and gene therapy. They are currently explored in animal models, and more actively in the context of a subgroup of LSDs affecting the degradation of mucopolysaccharides, the mucopolysaccharidosis.

## III. Mucopolysaccharidosis

Mucopolysaccharidosis (MPS) form a subgroup of LSDs caused by a defect in the degradation pathway of glycosaminoglycans (GAG or mucopolysaccharides), including keratan, dermatan, and heparan sulfates. Although deficiencies affect every tissue, clinical expression varies depending on the missing enzyme (Table I). Neurological symptoms are often predominant. They include severe and progressive motor impairments and mental retardation. Histopathology of the CNS reveals characteristic vacuolizations in neurons, glia, and perivascular cells, without known predominance in specific locations. Other frequent symptoms include hepatomegaly, skeletal abnormalities, corneal clouding, and progressive respiratory, cardiac, and renal dysfunctions leading to premature death.

In the white population, the most frequent MPSs are Hurler disease (MPS I), Hunter disease (MPS II), and Sanfilippo disease (MPS III). These severe forms are the most common and are always associated with serious neurological symptoms. Rarer MPSs, such as Scheie disease (the mild form of MPS I), Morquio disease (MPS IV), and Maroteaux-lamy disease (MPS VI) usually do not affect the CNS. Sly disease (MPS VII) is a very rare from of MPS with major CNS degradation.

Patients affected with mild forms of MPS have incomplete enzyme deficiency. Low enzyme activity appears sufficient to maintain a normal quality of life and to reach adulthood. Levels equivalent to less than 1% of normal values are sufficient to ensure efficient correction. Enzyme activity as low as 0.23% of normal levels was measured in fibroblasts from one adult patient with Scheie disease, whose clinical symptoms were limited to stiff joints and corneal clouding (Ashton *et al.*, 1992). Consistent observations were made in MPS VI patients (Brooks *et al.*, 1991).

The dramatic outcome of MPSs in the absence of treatment, the feasibility of a cross-correction of the enzyme deficiency, our relatively precise knowledge of the pathophysiology, and the observation that limited enzyme delivery might be beneficial explain why these diseases are considered as promising for innovative therapies based on cell transplantation and genetic correction. However, research toward that goal depends on experimentation with valuable animal models.

## IV. Animal Models for Mucopolysaccharidosis

Animal models relevant to the various human MPS are available. They were either created in laboratory mice by chemical mutagenesis or by the selective disruption (knockout, K.O.) of the gene relevant to the human disease, or recognized in domestic animals for which veterinarians further carefully maintained breeders as a colony.

Models of MPS I consist of gene K.O. mice (Clarke *et al.*, 1997), dog (Spellacy *et al.*, 1983), and cat (Haskins *et al.*, 1983) colonies. There is no animal model available for MPS II. A K.O. mouse model (Bhaumik *et al.*, 1999; Li *et al.*, 1999) and a dog model (Fischer *et al.*, 1998) exist for MPS IIIa. A model for MPS IIId exists in the goat (Thompson *et al.*, 1992). There is no animal model available for MPS IVa or IVb. Knockout mice are available for MPS VI (Evers *et al.*, 1996), as well as a model in the rat (Yoshida *et al.*, 1993) and in the cat (Jezyk *et al.*, 1977). Animal models for MPS VII have been heavily investigated. The $\beta$-glucuronidase-deficient mouse (MPS VII), which was isolated after chemical mutagenesis, reproduces the clinical features of severe human MPSs, including abnormal skeletal development, severe neurological damages, corneal clouding, and deafness (Birkenmeier *et al.*, 1989). Considerable lysosomal storage occurs in every tissue, including the neocortex and the hippocampus. Animals die around 6 months of age, apparently from both progressive neurological degradation and locomotor disability. Excellent models of MPS VII are also available in the dog (Haskins *et al.*, 1984) and the cat (Fyfe *et al.*, 1999). Recently, a genetically engineered mouse exclusively expressing a nonactive form of human $\beta$-glucuronidase has been generated to obtain a mouse model for MPS VII that is tolerant to the human enzyme (Sly *et al.*, 2001).

These multiple models are of crucial importance for pathophysiology studies. Indeed, the natural outcome and the cascade of pathological events that lead to severe pathology remain imperfectly understood for the various MPSs. Animal models are also highly valuable for designing therapeutic strategies. Candidate strategies are first explored in mouse models. The availability of large animal models allows preclinical evaluation of efficacy and tolerance, before clinical trials are considered in children. However, though most of these large animal models are valuable equivalents of the human disease, the high cost and the need for special facilities for maintaining animal colonies often limit their use.

## V. Treatment Strategies for Mucopolysaccharidosis

### A. Enzyme Replacement Therapy

Enzyme replacement therapy (ERT) is undoubtedly the most promising therapeutic approach for MPSs, as well as for some other forms of LSDs, such as Gaucher disease (Grabowski *et al.*, 1998), Fabry disease (Eng *et al.*, 2001), and Pompe disease (Van den Hout *et al.*, 2000), in which remarkable clinical benefits are currently obtained. Repeated infusions of a recombinant enzyme that targets M6P-Rs are presently performed with some success in MPS patients with mild forms of MPS I (Scheie disease) (Kakkis *et al.*, 2001a) and in patients with MPS VI. ERT is also currently considered for MPS II patients.

Though enzyme production costs and limited markets have been considered as serious limitations to the development of ERT, technical advances in the biotechnology industry as well as incentive regulations favoring investments for curing orphan diseases stimulate the commitment of biotechnology companies in the field. Nevertheless, ERT remains an extremely expensive therapeutic approach, which restricts treatment to patients living in developed countries. However, for those countries, the cost of ERT, which allows patients to be cured, compete with that of palliative care.

The efficacy of ERT in MPS was initially demonstrated in the MPS I canine model, which developed a disease very similar to severe forms of human Hurler disease (Shull *et al.*, 1994; Kakkis *et al.*, 1996). Animals received intravenous infusions of recombinant human IDUA for months at a dose from 0.1 to 0.5 mg/kg/week. Experiments were also performed in MPS I cats (Kakkis *et al.*, 2001b). Clinical improvement was obvious, although enzyme levels and histological correction were not obtained in brain, cartilage, and heart valves. The main drawback of the treatment was the development of an antibody response against human IDUA. A similar response developed when a canine enzyme was infused. Thus, immune complex deposition and complement activation appeared as potential adverse effects when treating humans. Phase I human trials in Scheie patients confirmed clinical benefits, whereas immunization appeared as a rare event in those patients, the frequency of which was comparable to that previously observed in Gaucher patients receiving recombinant glucocerebrosidase (Kakkis *et al.*, 2001a). These results seem to be currently confirmed in larger patient cohorts. It is, however, noticeable that patients with Scheie disease have partial enzyme deficiency, and may therefore be more tolerant to the drug than patients with severe MPS I.

ERT was recently explored in other MPS. A macrophage-targeted form of recombinant NAGLU was administrated to MPS IIIb mice, leading to substantial reduction of storage in peripheral tissues (Yu *et al.*, 2000). Though some clinical benefits were observed after enzyme infusion in MPS VI cats, tissue distribution was suboptimal and the development of high-titer antibody appeared as a potential detrimental effect on the efficacy of ERT (Brooks *et al.*, 1998). A phase I trial has been initiated in human patients, the conclusion of which is not yet available. ERT has been extensively investigated in the MPS VII mouse model. When treatment was started in adult animals, a reversion of storage lesions was observed in fixed macrophages of peripheral organs (Vogler *et al.*, 1996, 1999). However, treatment failed to correct CNS lesions when initiated later than 14 days after birth. In contrast, initiation at birth demonstrated that learning, memory, and hearing deficits can be prevented by ERT in newborn MPS VII mice (O'Connor *et al.*, 1998).

These later experiments, as well as the observations made in other models (Downs-Kelly *et al.*, 2000; Kakkis *et al.*, 2001b), point out the major limitation of ERT for the treatment of MPS, which is the inaccessibility of brain tissue to the therapeutic enzyme that is a consequence of the blood–brain barrier. In humans, the blood–brain barrier is already functional at birth and MPSs are rarely diagnosed before several months of life. Thus, recombinant enzymes are not able to reach the CNS and ERT is not effective on the progression of the neurological pathology. This limitation explains why ERT trials have currently been limited to few patients in whom the absence of brain damage could be anticipated, as in Scheie disease and MPS VI. In the more frequent severe forms of MPS I, in MPS II and in MPS III, ERT is not an option so far, although benefit could be expected in terms of patient comfort. Therefore, delivering enzyme to the CNS is currently the most critical issue with respect to curing MPSs. It is likely that this goal can be achieved by combining ERT with other therapeutic approaches, such as cell-based therapy and gene therapy.

## B. Cell-Based Therapy

### 1. *Allogeneic Bone Marrow Transplantation*

Therapies for LSD in general, and for MPS in particular, have initially focused on bone marrow transplantation (BMT). For MPS IH there is a clinical history of more than 20 years and more than 200 patients treated. Though benefit was initially reported in a few patients, data remained controversial for several years, essentially because of the heterogeneity of patient conditions and of the transplantation protocols. Scarcity of donors, poor engraftment, and high morbidity and mortality were considered major drawbacks. Benefits were undoubted at the level of visceral tissues, corneal clouding was improved, and life span was prolonged. However, skeletal deformities were usually not corrected and, most importantly, the neurodegenerative process was not prevented.

After thorough follow-up of treated patients for years, meta-analysis of trial data, and confrontation of clinical records with genetic diagnosis of mutations, a consensus has been reached with respect to which patients should be candidates for allogeneic BMT. A reasonable chance of success implies that a donor with less than 2 HLA mismatches, and who is not heterozygous for the mutation, is available within the family. Children must be transplanted early in life, in the absence of any neurological symptom and with normal intelligence performance (Peters *et al.*, 1996, 1998). Most children who satisfy these conditions are currently transplanted. However, this represents a rare option. In this context, it is admitted that BMT leads to improved skeletal development and can stabilize or slow down the neurodegenerative process in severe forms of MPS I (Whitley *et al.*, 1993; Hoogerbrugge *et al.*, 1995; Krivit *et al.*, 1995).

Allogeneic BMT is rarely considered in other MPSs, though a few patients with MPS II, MPS IV, and MPS VI have been transplanted. Results are still ambiguous. It is currently acknowledged that BMT is ineffective in MPS III.

### 2. *Syngeneic Bone Marrow Transplantation*

Syngeneic BMT has been extensively explored in the MPS VII mouse model. It consists in transplanting bone marrow cells from a +/+ animal to a −/− littermate. Several groups documented therapeutic enzyme levels in peripheral tissues of transplanted animals (Slavin and Yatziv, 1980; Yatziv *et al.*, 1982; Hoogerbrugge *et al.*, 1987; Poorthuis *et al.*, 1994). After high-dose conditioning irradiation of adult animals, Birkenmeier (1991) described complete correction of spleen, liver, cornea, and glomerular mesangial cells. A partial correction of meninges and perivascular cells in the brain and in renal tubular epithelial cells was also reported, as well as increased life span to near normal. CNS pathology and bone malformations did not revert. Transplantation at birth showed improvement in mobility, life span, and facial dysmorphism (Sands *et al.*, 1993).

Histologically, lysosomal storage was reduced in visceral and bone tissue as well in the brain. The auditory function was partially preserved, involving a decreased thickening of the tympanic membrane and middle ear mucosa (Sands *et al.*, 1995). Despite morphological corrections, animal's performance was equal to or even worse than nontreated animals on behavioral tests, probably as a detrimental consequence of irradiation in the neonatal period (Bastedo *et al.*, 1994). Nonablative BMT at birth was recently shown to improve bone morphology as well as visceral glycosaminoglycan storage and reproductive capacity, although CNS was not corrected (Soper *et al.*, 2001).

As an alternative to BMT, syngeneic mononuclear phagocytes, either from peripheral or bone marrow origin, were intravenously infused into adult or newborn MPS VII mice (Freeman *et al.*, 1999). Cells were able to home into the reticuloendothelial system and persisted for at least 30 days. Liver and spleen pathology was corrected but not the CNS, even in mice transplanted as newborns and in which some $\beta$-glucuronidase positive cells were detected in meninges and brain parenchyma.

3. *Syngeneic BMT with Genetically Modified Cells*

The development of viral vectors able to efficiently transduce cells in culture led to experiments in which autologous deficient cells were administrated to deficient animals after the genetic defect had been corrected through gene transfer. These experiments were most often performed using retrovirus-derived gene transfer vectors in the MPS VII mouse model. Initial experiments where genetically modified cells were reintroduced *in vivo* showed that not only cells reexpressing the deficient lysosomal enzyme were corrected, but also surrounding nontransduced cells that could internalize the secreted enzyme via mannose-6-phosphate receptors (Wolfe *et al.*, 1990; Peters *et al.*, 1991; Anson *et al.*, 1992). Bone marrow cells previously exposed to gene transfer vectors were then used for syngeneic BMT (Wolfe *et al.*, 1992; Maréchal *et al.*, 1993). Genetically modified hematopoietic cells were still present 21 weeks after treatment. Although $\beta$-glucuronidase activity was detected in many tissues including brain, only liver and spleen showed significant correction of lysosomal storage disease.

Clinical trials have been performed in MPS I patients, in whom autologous bone marrow cells or peripheral lymphocytes that had been previously exposed to a retrovirus vector encoding IDUA were transplanted (Fairbairn *et al.*, 1996; Krause *et al.*, 2001). Because retroviral vectors can infect only dividing cells, hematopoietic stem cells were induced to proliferate *ex vivo* using cytokines and growth factors. The low gene transfer efficiency, the deleterious consequences of the *in vitro* treatments of transplanted cells, and the absence of selective pressure for corrected cells probably explained poor successes. It may be expected that lentiviral vectors, which are more efficient and able to transduce resting cells, would achieve better results (Chen *et al.*, 2000; Stein *et al.*, 2001).

4. *Systemic Enzyme Delivery Using Skin Fibroblasts*

Skin fibroblasts were used as a source of enzyme in MPS patients. Attempts initially performed by the transplantation of allogeneic normal skin fibroblasts in MPS II allowed some long-term biochemical improvement, but had no effect on the course of the disease and were not investigated on a large scale (Dean *et al.*, 1976). Skin fibroblasts offer the possibility of autologous transplantation as they can be easily obtained from a skin biopsy and genetically modified using retroviral vectors. Secretion of human $\beta$-glucuronidase or IDUA was obtained from neoorgans that were formed as a combination of genetically modified cells with collagen, bFGF, and polytetrafluoroethylene fibers. When implanted in the peritoneal cavity of adult mice or dogs, neoorgans persisted for several months (Moullier *et al.*, 1993a,b, 1995; Salvetti *et al.*, 1995). In all cases correction of liver and spleen pathology was obtained, while very little enzyme activity was also detected in brain, probably due to macrophage/monocyte cells that captured the enzyme in the periphery and migrated toward the brain (Moullier *et al.*, 1993a). A similar approach has recently been used in MPS VII dogs with similar results (Wolfe *et al.*, 2000).

5. *Enzyme Delivery in the Brain through Transplanted Cells*

Immortalized fibroblasts that had been genetically modified with a retrovirus coding for $\beta$-glucuronidase were implanted in the neocortex of MPS VII mice (Taylor and Wolfe, 1997). Enzyme activity was detected above background levels for up to 2 months in neocortex and thalamus but decreased over time. Correction was detected 2 mm from the graft but storage material was evident at 4 mm from transplanted cells. Similarly, amniotic epithelial cells transduced with an adenovirus-$\beta$-glucuronidase were implanted into the corpus striatum (Kosuga *et al.*, 2001). Nine weeks after implantation evidence for cell dissemination through the brain parenchyma was evident. $\beta$-Glucuronidase expression was detected in the ipsilateral cortex and in the meninges of the contralateral ventricles but not in the cortex of the contralateral hemisphere. At 9 weeks, complete correction of lysosomal storage was found in the injected hemisphere but not in the contralateral where neurons and glia showed some vacuoles. The authors did not show long-term experiments, thus we are not able to assess the therapeutic potential of this approach. The same group had used a similar strategy for systemic gene therapy of MPS VII with short-term correction of spleen and liver (Kosuga *et al.*, 2000).

The recent characterization of neural progenitor cells suggested that they could be used as an intracerebral source of enzyme for treating MPS. Neuronal progenitor cells isolated from neonatal normal mouse cerebellum were immortalized, genetically modified with retrovirus vector, and implanted in the lateral ventricle of newborn MPS VII mice (Snyder *et al.*, 1995). Most treated mice showed diffused engraftment and significant $\beta$-glucuronidase expression in brain tissues. The accumulation of lysosomal storage material was reduced in neurons and glial cells located near the graft. As the potential drawback of immortalized cells is uncontrolled proliferation and tumor formation, attempts were made to repeat these experiments using primary cells. This may be achieved using lentivirus vectors that efficiently transduce a variety of resting cells. They are more efficient than classic retrovirus vectors and less toxic than adenovirus vectors. Neural progenitor cells could be efficiently transduced with lentivirus vector without losing their ability to give rise to multiple cell types (Hughes *et al.*, 2002). In contrast, the same cells differentiate to glia when exposed to adenovirus and were poorly transduced when exposed to adeno-associated virus vectors.

Primary cultures of human astrocytes can be obtained from human cerebral cortex biopsies. Such cells were genetically modified using adenoviral vectors and implanted in the striatum of nude mice (Serguera *et al.*, 2001). Although secretion of $\beta$-glucuronidase was detected, engrafted cells were rapidly eliminated, probably due to adenoviral toxicity.

### 6. *Cell Encapsulation*

Encapsulation protects cells from immune destruction and thus offers the possibility of engrafting nonautologous material. Encapsulation consists in embedding individual cells (microencapsulation) or a cell population (macroencapsulation) into inert polymer made of a porous membrane that allows entry of nutrients and secretion of therapeutic peptides or small proteins. Pores are too small to permit cells trafficking, impeding access to the cellular effectors of the immune system. A variety of cell types can be efficiently encapsulated, including baby hamster kidney cells, fibroblasts, and $C_2C_{12}$ cells (Déglon *et al.*, 1996; Bachoud-Levi *et al.*, 2000; Ross *et al.*, 1999, 2000a). Cells can be genetically modified prior to encapsulation in order to overexpress a therapeutic protein. High producer cell clones can be isolated and amplified.

Immortalized mouse fibroblasts were genetically engineered to overexpress $\beta$-glucuronidase and microencapsulated in alginate before implantation either subcutaneously for systemic delivery (Ross *et al.*, 2000a) or into the lateral cerebral ventricles for intracranial delivery (Ross *et al.*, 2000b). Secretion was demonstrated in both locations. However, cell viability decreased overtime. The subcutaneous implantation led to disease correction in liver and spleen, but a strong immune response against the lysosomal enzyme was raised. Injection in the cerebral hemisphere allowed enzyme detection in the whole brain at 3 weeks with levels higher than 7,000% of normal activity at the injection sites. However, secretion had decreased to 65% of normal levels at the implantation sites after 8 weeks. Histological sections showed correction of storage vacuoles in the whole brain at early stages. Injections were performed 3 mm from the suprachiasmatic nucleus of the hypothalamus, where neurons that generate circadian rhythms are located. Behavioral tests, measured by circadian rhythm, indicated some behavioral improvement.

As persistent proliferation inside the capsule decreases cell viability, myoblastic cells that can differentiate into resting myotubes have been proposed as efficient vehicles for cell encapsulation (Déglon *et al.*, 1996). Using this approach in MPS VII mice we could observe long-term survival of the $\beta$-glucuronidase-secreting encapsulated cells. However, an intense antibody response against the protein sequestrated it from the circulation and impeded reaching target tissues (unpublished data).

### C. Gene Therapy for Central Nervous System Lesions

Combination of ERT and intracranial enzyme delivery for systemic and CNS correction, respectively, would provide a cure for MPSs. Enzyme delivery in the CNS of deficient organisms could be achieved by inducing secretion from a population of residing cells through direct intracranial gene transfer. As enzyme activity is presumably required throughout the CNS for efficient treatment, the issue is then to determine how far from the transduced cell population the enzyme will be transported.

Direct intracranial injections of gene transfer vectors were performed by stereotaxic methods to induce enzyme overexpression and secretion from resident neurons and glial cells. Experiments were mostly performed in the MPS VII mouse model. Mice received a functional $\beta$-glucuronidase cDNA by the mean of a gene transfer vector, which could be herpes virus, adenovirus vectors, adeno-associated vectors, or lentivirus vectors. Consistent results provided evidence that enzyme expression was not restricted to the area in which the vector had been injected (Ghodsi *et al.*, 1998; Bosch *et al.*, 2000a). Activity could be demonstrated in far remote locations, including the contralateral hemisphere when injection was unilateral. These data indicated that brain cells were able to take up enzyme from the extracellular environment and, more importantly, suggested that $\beta$-glucuronidase could be transported over long distances in the brain by axonal transport. Distribution of the lysosomal enzyme through the cerebrospinal fluid was also considered (Stein *et al.*, 1999). These studies demonstrated that gene therapy could not only prevent the development of new lesions but also reverse preexisting damages.

The feasibility of preventing the development of lesions was demonstrated in newborn MPS VII mice and dogs. Experiments in mice were performed using a recombinant adeno-associated vector (AAV) encoding $\beta$-glucuronidase, which was injected intravenously at birth, a time when the blood–brain barrier is still permeable to viral vector (Daly *et al.*, 1999b; Elliger *et al.*, 1999). Therapeutic enzyme activity was achieved at 1 week in visceral organs, but also in brain and retina. At 4 months, animals showed little lysosomal storage in meninges, neurons, and glia. At 1 year, visual and auditory functions were preserved (Daly *et al.*, 2001). Bone length, body weight, and life span were nearly normal. $\beta$-Glucuronidase levels were maintained in all tissues, although $\beta$-glucuronidase cDNA was no longer present in the brain indicating that AAV vectors may not be effective for persistently correcting MPS disease in this context. A very similar approach has been explored in MPS VII dogs that received an intravenous bolus of a retrovirus vector encoding $\beta$-glucuronidase at birth (Xu *et al.*, 2002). It is noticeable that these vectors integrate and therefore persist in transduced cells.

Newborn mice were also injected with AAV vector intracranially. Results showed normal levels of lysosomal enzyme for 18 weeks and reduced lysosomal distention in most of the brain (Frisella *et al.*, 2001). Ganglioside GM2 and GM3 accumulation in MPS VII mice was recently described and prevented by vector administration, suggesting overall improvement of lysosomal functions. Morris Water Maze tests demonstrated some improvement of spatial learning in young treated mice compared to control age-matched MPS VII mice.

The reversion of preexisting lesions was established by treating adult animals. Transient correction was reported after the *in situ* injection of an adenovirus vector (Ghodsi *et al.*, 1998; Stein *et al.*, 1999). Sustained correction after the *in situ* injection of an AAV-2 vector has been shown (Skorupa *et al.*, 1999; Bosch *et al.*, 2000a). Some authors reported loss of transgene expression over time (Sferra *et al.*, 2000). Using an HIV-based vector we have documented enzyme delivery and reversal of pathology in the entire brain of MPS VII mice (Bosch *et al.*, 2000b). Multiple injections of VSV-pseudotyped lentivirus vectors coding for $\beta$-glucuronidase were performed into both hemispheres and one in the cerebellum. Enzyme activity was detected in 20% of the total brain, which was enough to restore normal histology of neurons, glia, and perivascular cells in 90% of the brain. Whereas previous studies noticed that cerebellum was not corrected after intrastriatal vector delivery (Sands *et al.*, 1994; O'Connor *et al.*, 1998; Frisella *et al.*, 2001), we observed that cerebellar cells could be corrected. Both AAV and HIV vectors allowed stable expression of $\beta$-glucuronidase for at least 4 months (Fig. 3). Activity actually accumulated over time, and storage vacuoles progressively disappeared from the entire brain (Fig. 1B). No toxicity related to local overexpression of the lysosomal enzyme or infiltration of inflammatory cells was observed in brains transduced with AAV or lentiviral vectors. Recently, MPS VII mice intracranially injected with feline-derived lentivirus vectors (FIV) encoding $\beta$-glucuronidase showed significant cognitive recovery 4 weeks after treatment (Brooks *et al.*, 2002). Moreover, behavioral improvement was associated with expression of genes involved in neuronal plasticity.

Herpes simplex viruses (HSV) have also been tested in the brain of the MPS VII mouse model (Wolfe *et al.*, 1996; Zhu *et al.*, 2000). Expression driven from the LAT promoter has been demonstrated that implies expression from latently

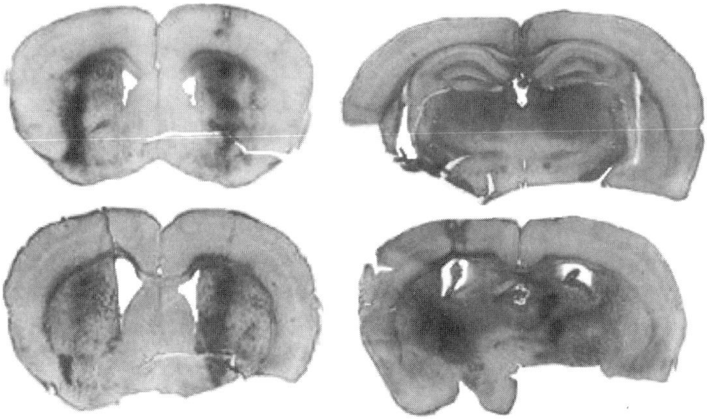

FIG. 3. Detection of $\beta$-glucuronidase activity (see arrows) on coronal brain sections of an MPS VII mouse treated with stereotactic injections of an HIV-based vector.

infected neuronal cells. This resulted in long-term activity of the therapeutic protein and expression to areas in the brain away from the injection site since HSV is able to move transneuronally. Despite the neuronal tropism of HSV vectors, the use of these vectors for CNS gene therapy is limited by the toxicity of the virus and the immunogenicity of viral proteins.

Alternative administration routes other than direct injection in the brain parenchyma have been explored. AAV and adenovirus vectors were delivered into the cerebrospinal fluid by intraventricular or intrathecal administration (Ghodsi *et al.*, 1998; Elliger *et al.*, 1999). Injection in the cisterna magna resulted in moderate levels of enzyme in the CNS but also in other tissues such as liver, heart, and muscle both in 3-day-old and adult MPS VII mice (Elliger *et al.*, 1999). On the other hand, intraventricular injection transduced the ependymal cells lining the ventricles and resulted in some penetration into the underlying brain tissue (Ghodsi *et al.*, 1998). Hyperosmolality induced by mannitol treatment, which reversed the natural flow of cerebral fluid, enhanced the penetration of adenoviral particles and the secreted lysosomal enzyme into the parenchyma (Ghodsi *et al.*, 1999). This increased the overall activity of $\beta$-glucuronidase in the whole brain and consequently the extent of correction of lysosomal storage material. However, continuous reversion of the brain fluid flow could have detrimental effects on brain physiology.

Correction of ocular pathology in MPS has also been achieved by direct injection of adenoviral vectors (Li and Davidson, 1995; Kamata *et al.*, 2001). Intravitreous or subretinal injections of adenovirus $\beta$-glucuronidase allowed the enzyme to reach retinal pigment epithelium and correct storage vacuoles. However, corneal clouding was corrected only when injections were performed in the anterior chamber of the eyes or, even better, inside the corneal stroma. Widespread $\beta$-glucuronidase activity in the eye was obtained due to the secretion and recapture mechanism. Moreover, readministration of an adenoviral vector was effective, which is consistent with the immune privilege of the eye (Anand *et al.*, 2002). Recently, gene therapy for eye disease has been reported successfully in a canine model of childhood blindness (Acland *et al.*, 2001) using AAV-2 vectors. Studies with AAV vectors for targeting retinal abnormalities in MPS animal models are currently being performed.

In adult animals, the systemic administration of viral vectors did not result in significant enzyme activity in the brain, as discussed above for BMT or ERT. Intravenous injection of adenovirus, AAV, and FIV showed correction in visceral tissues, mostly in the liver and rarely in the spleen (Watson *et al.*, 1998; Stein *et al.*, 1999, 2001). On the other hand, the intramuscular injection of AAV-2 vectors in an adult MPS VII mouse model confirmed high expression in skeletal muscles but little delivery to other tissues (Watson *et al.*, 1998), as previously shown in newborn mice (Daly *et al.*, 1999a).

## VI. Perspectives

The ability of cells to internalize lysosomal enzyme present in the extracellular environment provides a strong rationale for developing therapeutic approaches for lysosomal storage diseases. Methods and materials developed over the past two decades suggest that treatment should be available for some of these terrible diseases in the near future.

As a reliable prenatal diagnosis is presently available for most of these diseases, genetic counseling may prevent the birth of affected children in exposed families. However, although it will be technically feasible in the near future, the systematic screening for MPS at birth will not be implemented until a safe and efficient therapy is available. Consequently, the birth of a first-born affected child cannot be prevented so far in families that are not aware that they may transmit the recessive genetic abnormality. The diagnosis of MPS is usually not made at birth but rather after a few months or years of life.

Treatments must therefore be conceived for application to young children. They will likely involve a combination of ERT with gene or cell transfer into the brain. Main issues currently reside in transferring results obtained in mice to large animal models. This is especially crucial with respect to enzyme delivery into the CNS. Intracranial injection in the brain parenchyma or in the cisternae of either cells or gene transfer vectors may provide a clinically relevant approach only if enzyme spreading throughout the brain can be obtained through a limited number of injections. This raises questions about the capacity of the injected cells, or of the injected vectors, as well as that of the secreted enzyme, to be transported to remote locations, such that not more than two to four injections will allow spreading all over the brain.

Efficient enzyme spreading in the brain supposes the development of new tools. These may consist of new cell types that will exhibit efficient migration in brain tissues. They may also be derived from more efficient gene transfer vectors, such as vectors derived from AAV serotypes other than serotype 2 that has been used so far (Davidson *et al.*, 2000). Pseudotyping HIV vectors with more neurotropic envelopes (Desmaris *et al.*, 2001) or using newly characterized neurotropic vectors such as canine adenovirus (Peltékian *et al.*, 2002) could also enhance vector spreading in the brain. Modifications made in the enzyme itself may also facilitate spreading through diffusion or axonal transport (Xia *et al.*, 2001; Miana-Mena *et al.*, 2002).

Ideally, the enzyme would be delivered through the bloodstream. This supposes the characterization of sequences capable of allowing passage through the blood–brain barrier that could be fused to lysosomal enzyme in order to confer this property, as previously suggested (Schwarze *et al.*, 1999).

## References


Acland, G. M., Aguirre, G. D., Ray, J., Zhang, Q., Aleman, T. S., Cideciyan, A. V., Pearce-Kelling, S. E., Anand, V., Zeng, Y., Maguire, A. M., Jacobson, S. G., Hauswirth, W. W., and Bennett, J. (2001). Gene therapy restores vision in a canine model of childhood blindness. *Nat. Genet.* **28,** 92–95.

Anand, V., Duffy, B., Yang, Z., Dejneka, N. S., Maguire, A. M., and Bennett, J. (2002). A deviant immune response to viral proteins and transgene product is generated on subretinal administration of adenovirus and adeno-associated virus. *Mol. Ther.* **5,** 125–132.

Anson, D. S., Bielicki, J., and Hopwood, J. J. (1992). Correction of mucopolysaccharidosis type I fibroblasts by retroviral-mediated transfer of the human alpha-L-iduronidase gene. *Hum. Gene Ther.* **3,** 371–379.

Aronovich, E. L., Johnston, J. M., Wang, P., Giger, U., and Whitley, C. B. (2001). Molecular basis of mucopolysaccharidosis type III B in emu (Dromaius novaenollandiae): an avian model of Sanfilippo syndrome type B. *Genomics* **74,** 299–305.

Ashton, L. J., Brooks, D. A., McCourt, P. A., Muller, V. J., Clements, P. R., and Hopwood, J. J. (1992). Immunoquantification and enzyme kinetics of alpha-L-iduronidase in cultured fibroblasts from normal controls and mucopolysaccharidosis type I patients. *Am. J. Hum. Genet.* **50,** 787–794.

Bachoud-Levi, A. C., Deglon, N., Nguyen, J. P., Bloch, J., Bourdet, C., Winkel, L., Remy, P., Goddard, M., Lefaucheur, J. P., Brugieres, P., Baudic, S., Cesaro, P., Peschanski, M., and Aebischer, P. (2000). Neuroprotective gene therapy for Huntington's disease using a polymer encapsulated BHK cell line engineered to secrete human CNTF. *Hum. Gene Ther.* **11,** 1723–1729.

Barton, N. W., Brady, R. O., Dambrosia, J. M., Di Bisceglie, A. M., Doppelt, S. H., Hill, S. C., Mankin, H. J., Murray, G. J., Parker, R. I., Argoff, C. E., *et al.* (1991). Replacement therapy for inherited enzyme deficiency—macrophage-targeted glucocerebrosidase for Gaucher's disease. *N. Engl. J. Med.* **324,** 1464–1470.

Bastedo, L., Sands, M. S., Lambert, D. T., Pisa, M. A., Birkenmeier, E., and Chang, P. L. (1994). Behavioral consequences of bone marrow transplantation in the treatment of murine mucopolysaccharidosis type VII. *J. Clin. Invest.* **94,** 1180–1186.

Bhaumik, M., Muller, V. J., Rozaklis, T., Johnson, L., Dobrenis, K., Bhattacharyya, R., Wurzelmann, S., Finamore, P., Hopwood, J. J., Walkley, S. U., and Stanley, P. (1999). A mouse model for mucopolysaccharidosis type III A (Sanfilippo syndrome). *Glycobiology* **9,** 1389–1396.

Birkenmeier, E. H. (1991). Correction of murine mucopolysaccharidosis type VII (MPS VII) by bone marrow transplantation and gene transfer therapy. *Hum. Gene Ther.* **2,** 113.

Birkenmeier, E. H., Davisson, M. T., Beamer, W. G., Ganschow, R. E., Vogler, C. A., Gwynn, B., Lyford, K. A., Maltais, L. M., and Wawrzyniak, C. J. (1989). Murine mucopolysaccharidosis type VII. Characterization of a mouse with beta-glucuronidase deficiency. *J. Clin. Invest.* **83,** 1258–1266.

Bosch, A., Perret, E., Desmaris, N., and Heard, J. M. (2000a). Long-term and significant correction of brain lesions in adult mucopolysaccharidosis type VII mice using recombinant AAV vectors. *Mol. Ther.* **1,** 63–70.

Bosch, A., Perret, E., Desmaris, N., Trono, D., and Heard, J. M. (2000b). Reversal of pathology in the entire brain of mucopolysaccharidosis type VII mice after lentivirus-mediated gene transfer. *Hum. Gene Ther.* **11,** 1139–1150.

Brooks, A. I., Stein, C. S., Hughes, S. M., Heth, J., McCray, P. M., Jr., Sauter, S. L., Johnston, J. C., Cory-Slechta, D. A., Federoff, H. J., and Davidson, B. L. (2002). Functional correction of established central nervous system deficits in an animal model of lysosomal storage disease with feline immunodeficiency virus-based vectors. *Proc. Natl. Acad. Sci. USA* **99,** 6216–6221.

Brooks, D. A., McCourt, P. A., Gibson, G. J., Ashton, L. J., Shutter, M., and Hopwood, J. J. (1991). Analysis of N-acetylgalactosamine-4-sulfatase protein and kinetics in mucopolysaccharidosis type VI patients. *Am. J. Hum. Genet.* **48,** 710–719.

Brooks, D. A., Hopwood, J. J., and King, B. M. (1998). Immune response to enzyme replacement therapy: Clinical signs of hypersensitivity reactions and altered enzyme distribution in a high titre rat model. *Biochim. Biophys. Acta* **1407,** 163–172.

Chen, W., Wu, X., Levasseur, D. N., Liu, H., Lai, L., Kappes, J. C., and Townes, T. M. (2000). Lentiviral vector transduction of hematopoietic stem cells that mediate long-term reconstitution of lethally irradiated mice. *Stem Cells* **18,** 352–359.

Clarke, L. A., Russell, C. S., Pownall, S., Warrington, C. L., Borowski, A., Dimmick, J. E., Toone, J., and Jirik, F. R. (1997). Murine mucopolysaccharidosis type I: Targeted disruption of the murine alpha-L-iduronidase gene. *Hum. Mol. Genet.* **6,** 503–511.

Daly, T. M., Okuyama, T., Vogler, C., Haskins, M. E., Muzyczka, N., and Sands, M. S. (1999a). Neonatal intramuscular injection with recombinant adeno-associated virus results in prolonged beta-glucuronidase expression in situ and correction of liver pathology in mucopolysaccharidosis type VII mice. *Hum. Gene Ther.* **10,** 85–94.

Daly, T. M., Vogler, C., Levy, B., Haskins, M. E., and Sands, M. S. (1999b). Neonatal gene transfer leads to widespread correction of pathology in a murine model of lysosomal storage disease. *Proc. Natl. Acad. Sci. USA* **96,** 2296–2300.

Daly, T. M., Ohlemiller, K. K., Roberts, M. S., Vogler, C. A., and Sands, M. S. (2001). Prevention of systemic clinical disease in MPS VII mice following AAV-mediated neonatal gene transfer. *Gene Ther.* **8,** 1291–1298.

Davidson, B. L., Stein, C. S., Heth, J. A., Martins, I., Kotin, R. M., Derksen, T. A., Zabner, J., Ghodsi, A., and Chiorini, J. A. (2000). Recombinant adeno-associated virus type 2, 4, and 5 vectors: Transduction of variant cell types and regions in the mammalian central nervous system. *Proc. Natl. Acad. Sci. USA* **97,** 3428–3432.

Dean, M. F., Muir, H., Benson, P. F., Button, L. R., Boylston, A., and Mowbray, J. (1976). Enzyme replacement therapy by fibroblast transplantation in a case of Hunter syndrome. *Nature* **261,** 323–325.

Déglon, N., Heyd, B., Tan, S. A., Joseph, J. M., Zurn, A. D., and Aebischer, P. (1996). Central nervous system delivery of recombinant ciliary neurotrophic factor by polymer encapsulated differentiated C2C12 myoblasts. *Hum. Gene Ther.* **7,** 2135–2146.

Desmaris, N., Bosch, A., Salaün, C., Petit, C., Prevost, M. C., Tordo, N., Perrin, P., Schwartz, O., de Rocquigny, H., and Heard, J. M. (2001). Production and neurotropism of lentivirus vectors pseudotyped with lyssavirus envelope glycoproteins. *Mol. Ther.* **4,** 149–156.

Downs-Kelly, E., Jones, M. Z., Alroy, J., Cavanagh, K. T., King, B., Lucas, R. E., Baker, J. C., Kraemer, S. A., and Hopwood, J. J. (2000). Caprine mucopolysaccharidosis IIID: A preliminary trial of enzyme replacement therapy. *J. Mol. Neurosci.* **15,** 251–262.

Elliger, S. S., Elliger, C. A., Aguilar, C. P., Raju, N. R., and Watson, G. L. (1999). Elimination of lysosomal storage in brains of MPS VII mice treated by intrathecal administration of an adeno-associated virus vector. *Gene Ther.* **6,** 1175–1178.

Eng, C. M., Guffon, N., Wilcox, W. R., Germain, D. P., Lee, P., Waldek, S., Caplan, L., Linthorst, G. E., and Desnick, R. J. (2001). Safety and efficacy of recombinant human alpha-galactosidase A—replacement therapy in Fabry's disease. *N. Engl. J. Med.* **345,** 9–16.

Evers, M., Saftig, P., Schmidt, P., Hafner, A., McLoghlin, D. B., Schmahl, W., Hess, B., von Figura, K., and Peters, C. (1996). Targeted disruption of the arylsulfatase B gene results in mice resembling the phenotype of mucopolysaccharidosis VI. *Proc. Natl. Acad. Sci. USA* **93,** 8214–8219.

Fairbairn, L. J., Lashford, L. S., Spooncer, E., McDermott, R. H., Lebens, G., Arrand, J. E., Arrand, J. R., Bellantuono, I., Holt, R., Hatton, C. E., Cooper, A., Besley, G. T., Wraith, J. E., Anson, D. S., Hopwood, J. J., and Dexter, T. M. (1996). Long-term in vitro correction of alpha-L-iduronidase deficiency (Hurler syndrome) in human bone marrow. *Proc. Natl. Acad. Sci. USA* **93,** 2025–2030.

Fischer, A., Carmichael, K. P., Munnell, J. F., Jhabvala, P., Thompson, J. N., Matalon, R., Jezyk, P. F., Wang, P., and Giger, U. (1998). Sulfamidase deficiency in a family of Dachshunds: A canine model of mucopolysaccharidosis IIIA (Sanfilippo A). *Pediatr. Res.* **44,** 74–82.

Fratantoni, J. C., Hall, C. W., and Neufeld, E. F. (1968). Hurler and Hunter syndromes: Mutual correction of the defect in cultured fibroblasts. *Science* **162,** 570–572.

Freeman, B. J., Roberts, M. S., Vogler, C. A., Nicholes, A., Hofling, A. A., and Sands, M. S. (1999). Behavior and therapeutic efficacy of beta-glucuronidase-positive mononuclear phagocytes in a murine model of mucopolysaccharidosis type VII. *Blood* **94,** 2142–2150.

Frisella, W. A., O'Connor, L. H., Vogler, C. A., Roberts, M., Walkley, S., Levy, B., Daly, T. M., and Sands, M. S. (2001). Intracranial injection of recombinant adeno-associated virus improves cognitive function in a murine model of mucopolysaccharidosis type VII. *Mol. Ther.* **3,** 351–358.

Fyfe, J. C., Kurzhals, R. L., Lassaline, M. E., Henthorn, P. S., Alur, P. R., Wang, P., Wolfe, J. H., Giger, U., Haskins, M. E., Patterson, D. F., Sun, H., Jain, S., and Yuhki, N. (1999). Molecular basis of feline beta-glucuronidase deficiency: An animal model of mucopolysaccharidosis VII. *Genomics* **58,** 121–128.

Ghodsi, A., Stein, C., Derksen, T., Yang, G., Anderson, R. D., and Davidson, B. L. (1998). Extensive beta-glucuronidase activity in murine central nervous system after adenovirus-mediated gene transfer to brain. *Hum. Gene Ther.* **9,** 2331–2340.

Ghodsi, A., Stein, C., Derksen, T., Martins, I., Anderson, R. D., and Davidson, B. L. (1999). Systemic hyperosmolality improves beta-glucuronidase distribution and pathology in murine MPS VII brain following intraventricular gene transfer. *Exp. Neurol.* **160,** 109–116.

Grabowski, G. A., Leslie, N., and Wenstrup, R. (1998). Enzyme therapy for Gaucher disease: The first 5 years. *Blood Rev.* **12,** 115–133.

Haskins, M. E., Aguirre, G. D., Jezyk, P. F., Desnick, R. J., and Patterson, D. F. (1983). The pathology of the feline model of mucopolysaccharidosis I. *Am. J. Pathol.* **112,** 27–36.

Haskins, M. E., Desnick, R. J., DiFerrante, N., Jezyk, P. F., and Patterson, D. F. (1984). Beta-glucuronidase deficiency in a dog: A model of human mucopolysaccharidosis VII. *Pediatr. Res.* **18,** 980–984.

Hoogerbrugge, P. M., Poorthuis, B. J., Mulder, A. H., Wagemaker, G., Dooren, L. J., Vossen, J. M., and van Bekkum, D. W. (1987). Correction of lysosomal enzyme deficiency in various organs of beta-glucuronidase-deficient mice by allogeneic bone marrow transplantation. *Transplantation* **43,** 609–614.

Hoogerbrugge, P. M., Brouwer, O. F., Bordigoni, P., Ringden, O., Kapaun, P., Ortega, J. J., O'Meara, A., Cornu, G., Souillet, G., Frappaz, D., *et al.* (1995). Allogeneic bone marrow transplantation for lysosomal storage diseases. The European Group for Bone Marrow Transplantation. *Lancet* **345,** 1398–1402.

Hughes, S. M., Moussavi-Harami, F., Sauter, S. L., and Davidson, B. L. (2002). Viral-mediated gene transfer to mouse primary neural progenitor cells. *Mol. Ther.* **5,** 16–24.

Jezyk, P. F., Haskins, M. E., Patterson, D. F., Mellman, W. J., and Greenstein, M. (1977). Mucopolysaccharidosis in a cat with arylsulfatase B deficiency: A model of Maroteaux-Lamy syndrome. *Science* **198,** 834–836.

Kakkis, E. D., McEntee, M. F., Schmidtchen, A., Neufeld, E. F., Ward, D. A., Gompf, R. E., Kania, S., Bedolla, C., Chien, S. L., and Shull, R. M. (1996). Long-term and high-dose trials of enzyme replacement therapy in the canine model of mucopolysaccharidosis I. *Biochem. Mol. Med.* **58,** 156–167.

Kakkis, E. D., Muenzer, J., Tiller, G. E., Waber, L., Belmont, J., Passage, M., Izykowski, B., Phillips, J., Doroshow, R., Walot, I., Hoft, R., and Neufeld, E. F. (2001a). Enzyme-replacement therapy in mucopolysaccharidosis I. *N. Engl. J. Med.* **344,** 182–188.

Kakkis, E. D., Schuchman, E., He, X., Wan, Q., Kania, S., Wiemelt, S., Hasson, C. W., O'Malley, T., Weil, M. A., Aguirre, G. A., Brown, D. E., and Haskins, M. E. (2001b). Enzyme replacement therapy in feline mucopolysaccharidosis I. *Mol. Genet. Metab.* **72,** 199–208.

Kamata, Y., Okuyama, T., Kosuga, M., O'Hira, A., Kanaji, A., Sasaki, K., Yamada, M., and Azuma, N. (2001). Adenovirus-mediated gene therapy for corneal clouding in mice with mucopolysaccharidosis type VII. *Mol. Ther.* **4,** 307–312.

Kaplan, A., Achord, D. T., and Sly, W. S. (1977). Phosphohexosyl components of a lysosomal enzyme are recognized by pinocytosis receptors on human fibroblasts. *Proc. Natl. Acad. Sci. USA* **74,** 2026–2030.

Kosuga, M., Takahashi, S., Sasaki, K., Enosawa, S., Li, X. K., Okuyama, S., Fujino, M., Suzuki, S., Yamada, M., Matsuo, N., Sakuragawa, N., and Okuyama, T. (2000). Phenotype correction in murine mucopolysaccharidosis type VII by transplantation of human amniotic epithelial cells after adenovirus-mediated gene transfer. *Cell Transplant* **9,** 687–692.

Kosuga, M., Sasaki, K., Tanabe, A., Li, X. K., Okawa, H., Ogino, I., Okuda, O., Arai, H., Sakuragawa, N., Kamata, Y., Azuma, N., Suzuki, S., Yamada, M., and Okuyama, T. (2001). Engraftment of genetically engineered amniotic epithelial cells corrects lysosomal storage in multiple areas of the brain in mucopolysaccharidosis type VII mice. *Mol. Ther.* **3,** 139–148.

Krause, D. S., Theise, N. D., Collector, M. I., Henegariu, O., Hwang, S., Gardner, R., Neutzel, S., and Sharkis, S. J. (2001). Multi-organ, multi-lineage engraftment by a single bone marrow-derived stem cell. *Cell* **105,** 369–377.

Krivit, W., Lockman, L. A., Watkins, P. A., Hirsch, J., and Shapiro, E. G. (1995). The future for treatment by bone marrow transplantation for adrenoleukodystrophy, metachromatic leukodystrophy, globoid cell leukodystrophy and Hurler syndrome. *J. Inherit. Metab. Dis.* **18,** 398–412.

Li, H. H., Yu, W. H., Rozengurt, N., Zhao, H. Z., Lyons, K. M., Anagnostaras, S., Fanselow, M. S., Suzuki, K., Vanier, M. T., and Neufeld, E. F. (1999). Mouse model of Sanfilippo syndrome type B produced by targeted disruption of the gene encoding alpha-N-acetylglucosaminidase. *Proc. Natl. Acad. Sci. USA* **96,** 14505–14510.

Li, T., and Davidson, B. L. (1995). Phenotype correction in retinal pigment epithelium in murine mucopolysaccharidosis VII by adenovirus-mediated gene transfer. *Proc. Natl. Acad. Sci. USA.* **92,** 7700–7704.

Maréchal, V., Naffakh, N., Danos, O., and Heard, J. M. (1993). Disappearance of lysosomal storage in spleen and liver of mucopolysaccharidosis VII mice after transplantation of genetically modified bone marrow cells. *Blood* **82,** 1358–1365.

Miana-Mena, F. J., Roux, S., Benichou, J. C., Osta, R., and Brulet, P. (2002). Neuronal activity-dependent membrane traffic at the neuromuscular junction. *Proc. Natl. Acad. Sci. USA* **99,** 3234–3239.

Moullier, P., Bohl, D., Heard, J. M., and Danos, O. (1993a). Correction of lysosomal storage in the liver and spleen of MPS VII mice by implantation of genetically modified skin fibroblasts. *Nat. Genet.* **4,** 154–159.

Moullier, P., Maréchal, V., Danos, O., and Heard, J. M. (1993b). Continuous systemic secretion of a lysosomal enzyme by genetically modified mouse skin fibroblasts. *Transplantation* **56,** 427–432.

Moullier, P., Bohl, D., Cardoso, J., Heard, J. M., and Danos, O. (1995). Long-term delivery of a lysosomal enzyme by genetically modified fibroblasts in dogs. *Nat. Med.* **1,** 353–357.

O'Connor, L. H., Erway, L. C., Vogler, C. A., Sly, W. S., Nicholes, A., Grubb, J., Holmberg, S. W., Levy, B., and Sands, M. S. (1998). Enzyme replacement therapy for murine mucopolysaccharidosis type VII leads to improvements in behavior and auditory function. *J. Clin. Invest.* **101,** 1394–1400.

Peltékian, E., Garcia, L., and Danos, O. (2002). Neurotropism and retrograde axonal transport of a canine adenoviral vector: A tool for targeting key structures undergoing neurodegenerative processes. *Mol. Ther.* **5,** 25–32.

Peters, C., Rommerskirch, W., Modaressi, S., and von Figura, K. (1991). Restoration of arylsulphatase B activity in human mucopolysaccharidosis-type-VI fibroblasts by retroviral-vector-mediated gene transfer. *Biochem. J.* **276,** 499–504.

Peters, C., Balthazor, M., Shapiro, E. G., King, R. J., Kollman, C., Hegland, J. D., Henslee-Downey, J., Trigg, M. E., Cowan, M. J., Sanders, J., Bunin, N., Weinstein, H., Lenarsky, C., Falk, P., Harris, R., Bowen, T., Williams, T. E., Grayson, G. H., Warkentin, P., Sender, L., Cool, V. A., Crittenden, M., Packman, S., Kaplan, P., Lockman, L. A., *et al.* (1996). Outcome of unrelated donor bone marrow transplantation in 40 children with Hurler syndrome. *Blood* **87,** 4894–4902.

Peters, C., Shapiro, E. G., Anderson, J., Henslee-Downey, P. J., Klemperer, M. R., Cowan, M. J., Saunders, E. F., deAlarcon, P. A., Twist, C., Nachman, J. B., Hale, G. A., Harris, R. E., Rozans, M. K., Kurtzberg, J., Grayson, G. H., Williams, T. E., Lenarsky, C., Wagner, J. E., and Krivit, W. (1998). Hurler syndrome: II. Outcome of HLA-genotypically identical sibling and HLA-haploidentical related donor bone marrow transplantation in fifty-four children. The Storage Disease Collaborative Study Group. *Blood* **91,** 2601–2608.

Pfeffer, S. R. (1991). Targeting of proteins to the lysosome. *Curr. Top. Microbiol. Immunol.* **70,** 43–65.

Poorthuis, B. J., Romme, A. E., Willemsen, R., and Wagemaker, G. (1994). Bone marrow transplantation has a significant effect on enzyme levels and storage of glycosaminoglycans in tissues and in isolated hepatocytes of mucopolysaccharidosis type VII mice. *Pediatr. Res.* **36,** 187–193.

Robbins, J. C., Lam, M. H., Tripp, C. S., Bugianesi, R. L., Ponpipom, M. M., and Shen, T. Y. (1981). Synthetic glycopeptide substrates for receptor-mediated endocytosis by macrophages. *Proc. Natl. Acad. Sci. USA* **78,** 7294–7298.

Ross, C. J., Ralph, M., and Chang, P. L. (1999). Delivery of recombinant gene products to the central nervous system with nonautologous cells in alginate microcapsules. *Hum. Gene Ther.* **10,** 49–59.

Ross, C. J., Bastedo, L., Maier, S. A., Sands, M. S., and Chang, P. L. (2000a). Treatment of a lysosomal storage disease, mucopolysaccharidosis VII, with microencapsulated recombinant cells. *Hum. Gene Ther.* **11,** 2117–2127.

Ross, C. J., Ralph, M., and Chang, P. L. (2000b). Somatic gene therapy for a neurodegenerative disease using microencapsulated recombinant cells. *Exp. Neurol.* **166,** 276–286.

Salvetti, A., Moullier, P., Cornet, V., Brooks, D., Hopwood, J. J., Danos, O., and Heard, J. M. (1995). In vivo delivery of human alpha-L-iduronidase in mice implanted with neo-organs. *Hum. Gene Ther.* **6,** 1153–1159.

Sando, G. N., and Neufeld, E. F. (1977). Recognition and receptor-mediated uptake of a lysosomal enzyme, alpha-1-iduronidase, by cultured human fibroblasts. *Cell* **12,** 619–627.

Sands, M. S., Barker, J. E., Vogler, C., Levy, B., Gwynn, B., Galvin, N., Sly, W. S., and Birkenmeier, E. (1993). Treatment of murine mucopolysaccharidosis type VII by syngeneic bone marrow transplantation in neonates. *Lab. Invest.* **68,** 676–686.

Sands, M. S., Vogler, C., Kyle, J. W., Grubb, J. H., Levy, B., Galvin, N., Sly, W. S., and Birkenmeier, E. H. (1994). Enzyme replacement therapy for murine mucopolysaccharidosis type VII. *J. Clin. Invest.* **93,** 2324–2331.

Sands, M. S., Erway, L. C., Vogler, C., Sly, W. S., and Birkenmeier, E. H. (1995). Syngeneic bone marrow transplantation reduces the hearing loss associated with murine mucopolysaccharidosis type VII. *Blood* **86,** 2033–2040.

Schiffmann, R., Murray, G. J., Treco, D., Daniel, P., Sellos-Moura, M., Myers, M., Quirk, J. M., Zirzow, G. C., Borowski, M., Loveday, K., Anderson, T., Gillespie, F., Oliver, K. L., Jeffries, N. O., Doo, E., Liang, T. J., Kreps, C., Gunter, K., Frei, K., Crutchfield, K., Selden, R. F., and Brady, R. O. (2000). Infusion of alpha-galactosidase A reduces tissue globotriaosylceramide storage in patients with Fabry disease. *Proc. Natl. Acad. Sci. USA* **97,** 365–370.

Schwarze, S. R., Ho, A., Vocero-Akbani, A., and Dowdy, S. F. (1999). In vivo protein transduction: Delivery of a biologically active protein into the mouse. *Science* **285,** 1569–1572.

Serguera, C., Sarkis, C., Ridet, J. L., Colin, P., Moullier, P., and Mallet, J. (2001). Primary adult human astrocytes as an ex vivo vehicle for beta-glucuronidase delivery in the brain. *Mol. Ther.* **3,** 875–881.

Sferra, T. J., Qu, G., McNeely, D., Rennard, R., Clark, K. R., Lo, W. D., and Johnson, P. R. (2000). Recombinant adeno-associated virus-mediated correction of lysosomal storage within the central nervous system of the adult mucopolysaccharidosis type VII mouse. *Hum. Gene Ther.* **11,** 507–519.

Shull, R. M., Kakkis, E. D., McEntee, M. F., Kania, S. A., Jonas, A. J., and Neufeld, E. F. (1994). Enzyme replacement in a canine model of Hurler syndrome. *Proc. Natl. Acad. Sci. USA* **91,** 12937–12941.

Skorupa, A. F., Fisher, K. J., Wilson, J. M., Parente, M. K., and Wolfe, J. H. (1999). Sustained production of beta-glucuronidase from localized sites after AAV vector gene transfer results in widespread distribution of enzyme and reversal of lysosomal storage lesions in a large volume of brain in mucopolysaccharidosis VII mice. *Exp. Neurol.* **160,** 17–27.

Slavin, S., and Yatziv, S. (1980). Correction of enzyme deficiency in mice by allogeneic bone marrow transplantation with total lymphoid irradiation. *Science* **210,** 1150–1152.

Sly, W. S., Vogler, C., Grubb, J. H., Zhou, M., Jiang, J., Zhou, X. Y., Tomatsu, S., Bi, Y., and Snella, E. M. (2001). Active site mutant transgene confers tolerance to human beta-glucuronidase without affecting the phenotype of MPS VII mice. *Proc. Natl. Acad. Sci. USA* **98,** 2205–2210.

Snyder, E. Y., Taylor, R. M., and Wolfe, J. H. (1995). Neural progenitor cell engraftment corrects lysosomal storage throughout the MPS VII mouse brain. *Nature* **374,** 367–370.

Soper, B. W., Lessard, M. D., Vogler, C. A., Levy, B., Beamer, W. G., Sly, W. S., and Barker, J. E. (2001). Nonablative neonatal marrow transplantation attenuates functional and physical defects of beta-glucuronidase deficiency. *Blood* **97,** 1498–1504.

Spellacy, E., Shull, R. M., Constantopoulos, G., and Neufeld, E. F. (1983). A canine model of human alpha-L-iduronidase deficiency. *Proc. Natl. Acad. Sci. USA* **80,** 6091–6095.

Stein, C. S., Ghodsi, A., Derksen, T., and Davidson, B. L. (1999). Systemic and central nervous system correction of lysosomal storage in mucopolysaccharidosis type VII mice. *J. Virol.* **73,** 3424–3429.

Stein, C. S., Kang, Y., Sauter, S. L., Townsend, K., Staber, P., Derksen, T. A., Martins, I., Qian, J., Davidson, B. L., and McCray, P. B., Jr. (2001). In vivo treatment of hemophilia A and mucopolysaccharidosis type VII using nonprimate lentiviral vectors. *Mol. Ther.* **3,** 850–856.

Taylor, R. M., and Wolfe, J. H. (1997). Decreased lysosomal storage in the adult MPS VII mouse brain in the vicinity of grafts of retroviral vector-corrected fibroblasts secreting high levels of beta-glucuronidase. *Nat. Med.* **3,** 771–774.

Thompson, J. N., Jones, M. Z., Dawson, G., and Huffman, P. S. (1992). N-Acetylglucosamine 6-sulphatase deficiency in a Nubian goat: A model of Sanfilippo syndrome type D (mucopolysaccharidosis IIID). *J. Inherit. Metab. Dis.* **15,** 760–768.

Van den Hout, H., Reuser, A. J., Vulto, A. G., Loonen, M. C., Cromme-Dijkhuis, A., and Van der Ploeg, A. T. (2000). Recombinant human alpha-glucosidase from rabbit milk in Pompe patients. *Lancet* **356,** 397–398.

Vogler, C., Sands, M. S., Levy, B., Galvin, N., Birkenmeier, E. H., and Sly, W. S. (1996). Enzyme replacement with recombinant beta-glucuronidase in murine mucopolysaccharidosis type VII: Impact of therapy during the first six weeks of life on subsequent lysosomal storage, growth, and survival. *Pediatr. Res.* **39,** 1050–1054.

Vogler, C., Levy, B., Galvin, N. J., Thorpe, C., Sands, M. S., Barker, J. E., Baty, J., Birkenmeier, E. H., and Sly, W. S. (1999). Enzyme replacement in murine mucopolysaccharidosis type VII: Neuronal and glial response to beta-glucuronidase requires early initiation of enzyme replacement therapy. *Pediatr. Res.* **45,** 838–844.

Watson, G. L., Sayles, J. N., Chen, C., Elliger, S. S., Elliger, C. A., Raju, N. R., Kurtzman, G. J., and Podsakoff, G. M. (1998). Treatment of lysosomal storage disease in MPS VII mice using a recombinant adeno-associated virus. *Gene Ther.* **5,** 1642–1649.

Whitley, C. B., Belani, K. G., Chang, P. N., Summers, C. G., Blazar, B. R., Tsai, M. Y., Latchaw, R. E., Ramsay, N. K., and Kersey, J. H. (1993). Long-term outcome of Hurler syndrome following bone marrow transplantation. *Am. J. Med. Genet.* **46,** 209–218.

Wilkerson, M. J., Lewis, D. C., Marks, S. L., and Prieur, D. J. (1998). Clinical and morphologic features of mucopolysaccharidosis type II in a dog: naturally occurring model of Hunter syndrome. *Vet. Pathol.* **35,** 230–233.

Wolfe, J. H., Schuchman, E. H., Stramm, L. E., Concaugh, E. A., Haskins, M. E., Aguirre, G. D., Patterson, D. F., Desnick, R. J., and Gilboa, E. (1990). Restoration of normal lysosomal function in mucopolysaccharidosis type VII cells by retroviral vector-mediated gene transfer. *Proc. Natl. Acad. Sci. USA* **87,** 2877–2881.

Wolfe, J. H., Sands, M. S., Barker, J. E., Gwynn, B., Rowe, L. B., Vogler, C. A., and Birkenmeier, E. H. (1992). Reversal of pathology in murine mucopolysaccharidosis type VII by somatic cell gene transfer. *Nature* **360,** 749–753.

Wolfe, J. H., Martin, C. E., Deshmane, S. L., Reilly, J. J., Kesari, S., and Fraser, N. W. (1996). Increased susceptibility to the pathogenic effects of wild-type and recombinant herpesviruses in MPS VII mice compared to normal siblings. *J. Neurovirol.* **2,** 417–422.

Wolfe, J. H., Sands, M. S., Harel, N., Weil, M. A., Parente, M. K., Polesky, A. C., Reilly, J. J., Hasson, C., Weimelt, S., and Haskins, M. E. (2000). Gene transfer of low levels of beta-glucuronidase corrects hepatic lysosomal storage in a large animal model of mucopolysaccharidosis VII. *Mol. Ther.* **2,** 552–561.

Xia, H., Mao, Q., and Davidson, B. L. (2001). The HIV Tat protein transduction domain improves the biodistribution of beta-glucuronidase expressed from recombinant viral vectors. *Nat. Biotechnol.* **19,** 640–644.

Xu, L, Haskins, M. E., Melniczek, J. R., Gao, C., Weil, M. A., O'Malley, T. M., O'Donnell, P. A., Mazrier, H., Ellinwood, N. M., Zweigle, J., Wolfe, J. H., and Ponder, K. P. (2002). Transduction of hepatocytes after neonatal delivery of a Moloney murine leukemia virus based retroviral vector results in long-term expression of beta-glucuronidase in mucopolysaccharidosis VII dogs. *Mol. Ther.* **5,** 141–153.

Yatziv, S., Weiss, L., Morecki, S., Fuks, Z., and Slavin, S. (1982). Long-term enzyme replacement therapy in beta-glucuronidase-deficient mice by allogeneic bone marrow transplantation. *J. Lab. Clin. Med.* **99,** 792–797.

Yoshida, M., Noguchi, J., Ikadai, H., Takahashi, M., and Nagase, S. (1993). Arylsulfatase B-deficient mucopolysaccharidosis in rats. *J. Clin. Invest.* **91,** 1099–1104.

Yu, W. H., Zhao, K. W., Ryazantsev, S., Rozengurt, N., and Neufeld, E. F. (2000). Short-term enzyme replacement in the murine model of Sanfilippo syndrome type B. *Mol. Genet. Metab.* **71,** 573–580.

Zhu, J., Kang, W., Wolfe, J. H., and Fraser, N. W. (2000). Significantly increased expression of beta-glucuronidase in the central nervous system of mucopolysaccharidosis type VII mice from the latency-associated transcript promoter in a nonpathogenic herpes simplex virus type 1 vector. *Mol. Ther.* **2,** 82–94.

# INDEX

## C

## H

**I**

## J



## K



## L

**M**

**N**

## O

**P**

## Q



## R

**W**



**X**



**Y**

# CONTENTS OF RECENT VOLUMES

## Volume 39



## Volume 40

## Volume 41

## Volume 42



## Volume 43



## Volume 44

## Volume 45

## Volume 46

## Volume 49

## Volume 50



## Volume 51

## Volume 52



## Volume 53

## Volume 54